Surface Mount and Related Technologies

ELECTRICAL ENGINEERING AND ELECTRONICS

A Series of Reference Books and Textbooks

1. Rational Fault Analysis, *edited by Richard Saeks and S. R. Liberty*
2. Nonparametric Methods in Communications, *edited by P. Papantoni-Kazakos and Dimitri Kazakos*
3. Interactive Pattern Recognition, *Yi-tzuu Chien*
4. Solid-State Electronics, *Lawrence E. Murr*
5. Electronic, Magnetic, and Thermal Properties of Solid Materials, *Klaus Schröder*
6. Magnetic-Bubble Memory Technology, *Hsu Chang*
7. Transformer and Inductor Design Handbook, *Colonel Wm. T. McLyman*
8. Electromagnetics: Classical and Modern Theory and Applications, *Samuel Seely and Alexander D. Poularikas*
9. One-Dimensional Digital Signal Processing, *Chi-Tsong Chen*
10. Interconnected Dynamical Systems, *Raymond A. DeCarlo and Richard Saeks*
11. Modern Digital Control Systems, *Raymond G. Jacquot*
12. Hybrid Circuit Design and Manufacture, *Roydn D. Jones*
13. Magnetic Core Selection for Transformers and Inductors: A User's Guide to Practice and Specification, *Colonel Wm. T. McLyman*
14. Static and Rotating Electromagnetic Devices, *Richard H. Engelmann*
15. Energy-Efficient Electric Motors: Selection and Application, *John C. Andreas*
16. Electromagnetic Compossibility, *Heinz M. Schlicke*
17. Electronics: Models, Analysis, and Systems, *James G. Gottling*
18. Digital Filter Design Handbook, *Fred J. Taylor*
19. Multivariable Control: An Introduction, *P. K. Sinha*
20. Flexible Circuits: Design and Applications, *Steve Gurley, with contributions by Carl A. Edstrom, Jr., Ray D. Greenway, and William P. Kelly*
21. Circuit Interruption: Theory and Techniques, *Thomas E. Browne, Jr.*
22. Switch Mode Power Conversion: Basic Theory and Design, *K. Kit Sum*
23. Pattern Recognition: Applications to Large Data-Set Problems, *Sing-Tze Bow*
24. Custom-Specific Integrated Circuits: Design and Fabrication, *Stanley L. Hurst*
25. Digital Circuits: Logic and Design, *Ronald C. Emery*
26. Large-Scale Control Systems: Theories and Techniques, *Magdi S. Mahmoud, Mohamed F. Hassan, and Mohamed G. Darwish*
27. Microprocessor Software Project Management, *Eli T. Fathi and Cedric V. W. Armstrong (Sponsored by Ontario Centre for Microelectronics)*
28. Low Frequency Electromagnetic Design, *Michael P. Perry*
29. Multidimensional Systems: Techniques and Applications, *edited by Spyros G. Tzafestas*
30. AC Motors for High-Performance Applications: Analysis and Control, *Sakae Yamamura*

31. Ceramic Materials for Electronics: Processing, Properties, and Applications, *edited by Relva C. Buchanan*
32. Microcomputer Bus Structures and Bus Interface Design, *Arthur L. Dexter*
33. End User's Guide to Innovative Flexible Circuit Packaging, *Jay J. Miniet*
34. Reliability Engineering for Electronic Design, *Norman B. Fuqua*
35. Design Fundamentals for Low-Voltage Distribution and Control, *Frank W. Kussy and Jack L. Warren*
36. Encapsulation of Electronic Devices and Components, *Edward R. Salmon*
37. Protective Relaying: Principles and Applications, *J. Lewis Blackburn*
38. Testing Active and Passive Electronic Components, *Richard F. Powell*
39. Adaptive Control Systems: Techniques and Applications, *V. V. Chalam*
40. Computer-Aided Analysis of Power Electronic Systems, *Venkatachari Rajagopalan*
41. Integrated Circuit Quality and Reliability, *Eugene R. Hnatek*
42. Systolic Signal Processing Systems, *edited by Earl E. Swartzlander, Jr.*
43. Adaptive Digital Filters and Signal Analysis, *Maurice G. Bellanger*
44. Electronic Ceramics: Properties, Configuration, and Applications, *edited by Lionel M. Levinson*
45. Computer Systems Engineering Management, *Robert S. Alford*
46. Systems Modeling and Computer Simulation, *edited by Naim A. Kheir*
47. Rigid-Flex Printed Wiring Design for Production Readiness, *Walter S. Rigling*
48. Analog Methods for Computer-Aided Circuit Analysis and Diagnosis, *edited by Takao Ozawa*
49. Transformer and Inductor Design Handbook, Second Edition, Revised and Expanded, *Colonel Wm. T. McLyman*
50. Power System Grounding and Transients: An Introduction, *A. P. Sakis Meliopoulos*
51. Signal Processing Handbook, *edited by C. H. Chen*
52. Electronic Product Design for Automated Manufacturing, *H. Richard Stillwell*
53. Dynamic Models and Discrete Event Simulation, *William Delaney and Erminia Vaccari*
54. FET Technology and Application: An Introduction, *Edwin S. Oxner*
55. Digital Speech Processing, Synthesis, and Recognition, *Sadaoki Furui*
56. VLSI RISC Architecture and Organization, *Stephen B. Furber*
57. Surface Mount and Related Technologies, *Gerald Ginsberg*
58. Uninterruptible Power Supplies: Power Conditioners For Critical Equipment, *David C. Griffith*
59. Polyphase Induction Motors: Analysis, Design, and Application, *Paul L. Cochran*

Additional Volumes in Preparation

Linear Circuits and Systems, *Nubuo Nagai*

Electrical Engineering-Electronics Software

1. Transformer and Inductor Design Software for the IBM PC, *Colonel Wm. T. McLyman*
2. Transformer and Inductor Design Software for the Macintosh, *Colonel Wm. T. McLyman*
3. Digital Filter Design Software for the IBM PC, *Fred J. Taylor and Thanos Stouraitis*

Surface Mount and Related Technologies

Gerald Ginsberg
Component Data Associates, Inc.
Lafayette Hill, Pennsylvania

Marcel Dekker, Inc. • New York and Basel

Library of Congress Cataloging-in-Publication Data

Ginsberg, Gerald
Surface mount and related technologies.
p. cm. -- (Electrical engineering and electronics ; 57)
Bibliography: p.
Includes index.
1. Printed circuits--Design and construction. 2. Surface mount technology. 3. Electronic packaging. I. Title. II. Series.
TK7868.P7G56 1989
621.381'74--dc19 89-1157
ISBN 0-8247-8073-6 (alk. paper)

This book is printed on acid-free paper.

MARCEL DEKKER, INC.
270 Madison Avenue, New York, New York 10016

Current printing (last digit):
10 9 8 7 6 5 4 3 2 1

PRINTED IN THE UNITED STATES OF AMERICA

Preface

Will present technologies be adequate to accommodate the demands of the 1990s or will some new techniques supplant the established ones? This question is timely because the past several years have seen considerable activity in the development of new integrated circuits, component packages, and improved interconnection techniques.

This intensity is prompted by four factors. First is the rapidly growing capabilities of state-of-the-art integrated circuits, achieved primarily by the decrease in feature size through the use of more sophisticated photolithography techniques and the increase in chip size. This trend will continue for the foreseeable future.

Second is the emergence of new integrated circuit component packages for these complex VLSI (Very Large Scale Integration) devices. As VLSI circuits grow in complexity, more input/output (I/O) capabilities for the interconnection among these circuits and with the outside world are required.

The relatively static state of component interconnection technology during the past decade is the third factor. There is no question that considerable progress has been achieved in extending the capabilities, conductor densities, and yields of printed wiring technology. Presently, however, further improvements in the state-of-the-art are causing diminishing returns and cost penalties. Whether more stringent future demands can be met by the direct extension of the methods now employed is disputable.

Surface mount technology is the final factor affecting changes in circuit technology. This is because VLSI devices are planar, as are the structures that interconnect them. However, the form factor of the end product equipment is essentially cubic, a basic incongruity requiring enormous ingenuity to provide a single reliable transition.

This book is directed toward the population of the electronic packaging industry that is devoted to solving these problems. In order to provide guidance, the book first describes recent developments in integrated circuit technology and the surface mount components that have been developed to optimize their packaging and interconnection. Next follows a thorough discussion of all major considerations associated with the implementation of surface mount technology with respect to design, fabrication, assembly, and testing of both packaging and interconnection structures and complete assemblies. Last, the book provides insight into the most advanced form of surface mounting—chip-on-board technology, whereby performance is optimized through the elimination of the component package and the mounting of bare integrated circuit chips on the packaging and interconnecting structure.

Gerald Ginsberg

Contents

Surface Mount and Related Technologies

1

Introduction

1.1 INTERCONNECTION SYSTEMS [1, 2]

Advances in interconnection systems have occurred in response to the evolution of electronic technology, particularly the growth in component sophistication, complexity, and number of electronic circuit component terminals to be interconnected. But a basic concern has developed with the progress of interconnection technology. A variety of factors have contributed to this growing concern, among which are the following:

1. *Interconnection costs* [3]: These have not followed the decline in device costs, Figure 1.1. Today a complete computer-on-a-chip costs less than the surrounding components and interconnection system.
2. *Operating speeds*: There is a strong trend toward increased use of the faster-operating integrated circuits (ICs). The interconnection system serving them should have configurations that make it possible to locate all such devices within a very limited area in order to avoid or minimize excessive propagation delays in the wiring.
3. *Integration of functions*: Further integration of circuit functions within dual-inline packages (DIPs) is becoming more limited, in

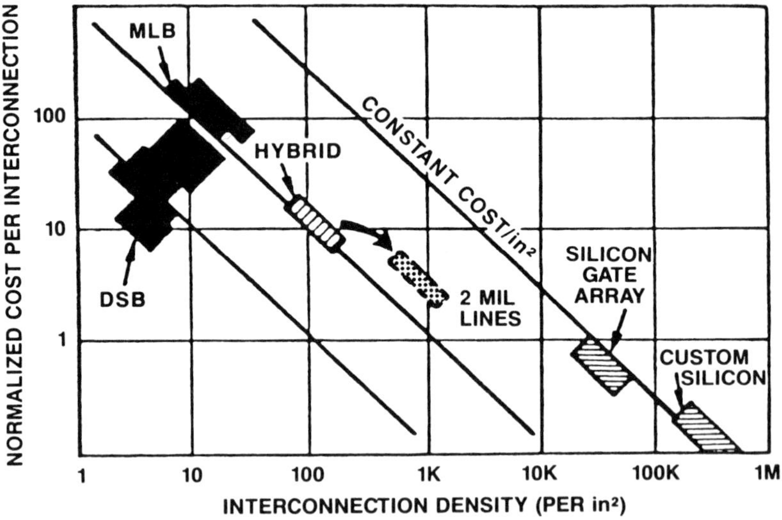

FIGURE 1.1 Interconnection cost versus density. DSB = double-sided board; MLB = multilayer board. (Courtesy of Bell Laboratories.)

part by the practical considerations of the economical size of the ICs and DIPs suitable for easy design change and field repair.

4. *Lead spacing*: Any increase in the number of leads per IC package calls for more interconnections between them (Table 1.1). Large-scale integrated circuits (LSIs) having from 48 up to 150 leads cannot be efficiently packaged in the familiar DIP configuration with two rows of leads on 0.1-in. centers. These devices will require the new packages that are coming onto the market.

1.2 SEMICONDUCTOR DEVICE TECHNOLOGY [4, 5]

The application of the emerging higher-speed, higher-density ICs offers great potential for increasing the performance of complex electronic systems. This is especially true in applications where size, weight, and power requirements are critical. In order to fully utilize the potential of these new devices, compatible packaging technology is required. This demands a new definition of the interconnection system. The significance and magnitude of these new interconnection system requirements can be seen by appreciating the development of semiconductor technology (Figure 1.2).

TABLE 1.1 Connectivity Capacity Requirements for Dual-Inline Packages (DIPs) and Small Chip Carriers (Dimensions Only in Inches)

Case	Typical packaging situation	Approximate number of pins per in.^{2}	Required connective capacity, in./in.^{2}	Wiring level		
				50-mil grid	25-mil grid	20-mil grid
1	1-16 pin DIP per in.^{2}	15	34	2 (1.7)	1 (0.8)	1 (0.7)
2	22-pin DIP on 1-in. centers	20	45	2.3 (2.3)	1+ (1.13)	1 (0.9)
3	2-16 pin DIPs per in.^{2}	30	67	3-4 (3.4)	2 (1.68)	2 (1.34)
4	40-pin DIP on 1-in. centers	20	45	2-3 (2.3)	1+ (1.13)	1 (0.9)
5	24-pin chip carriers on 0.7-in. centers	50	112	∿6 (5.6)	∿3 (2.8)	∿3 (2.2)
6	24-pin chip carriers on 0.5-in. centers	96	216	∿11 (10.8)	∿6 (5.4)	∿5 (4.3)
	Connective capacity per level, in./in.^{2}			20	40	50

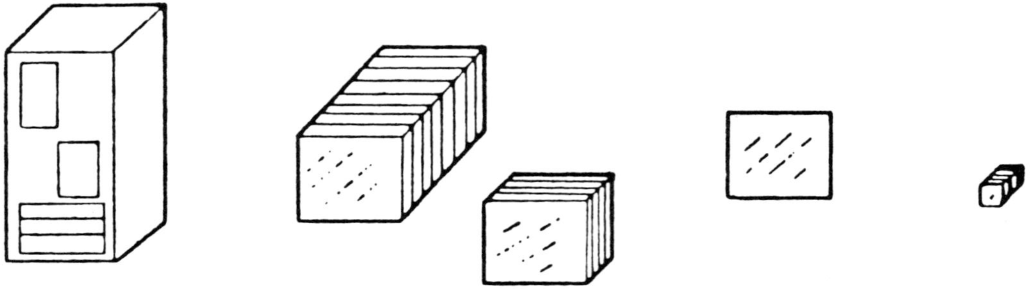

Phases	1978	1981	1983	1985	1988
Semiconductor Technology	ECL SSI/MSI	ECL LSI	2 μm CMOS	1¼ μm VHSIC	0.5 μm VHSIC
Gates/Chip	10–50	400	6K	30K	100K
Clock Rate	10 MHz	20–40 MHz	30–40 MHz	40 MHz	100 MHz
No. of Bds	60 Bds	24 Bds	6 Bds	1 Bd	4–6 Chips
Size	8 ft^3	0.86 ft^3	0.2 ft^3	0.1 ft^3	0.01 ft^3
Power	3 kW	1.6 kW	300 W	50 W	3W

FIGURE 1.2 Semiconductor technology development (from ref. 3).

As the semiconductors in electronic equipment go to the use of VLSI (very large scale integration), there will be a greater percentage of the total system cost, of the total design and manufacturing time, and of the testing costs affected by the packaging/interconnection system technology. Also, chip packaging levels will increase to over 10,000 gates/chip and board pinouts will approach 300, with proportional increases in power to be dissipated per unit volume. The use of peripheral devices such as DIPs and chip carriers, or matrix devices, such as pin-grid arrays, will significantly affect packaging density and, therefore, the interconnection system.

1.3 SURFACE MOUNTING [1]

Present component interconnection technology is based almost exclusively on the use of plated-through holes, and therefore all devices or their packages must have leads that will be eventually soldered into these holes. These component leads add to material costs, assembly labor, and manufacturing problems, but they are conceptually totally avoidable.

The newly emerging packages are radically resolving this problem, since they are designed for the surface mounting of planar components on a planar surface and do not require discrete component leads and plated-through holes for their termination. This development will have a profound impact on future interconnection methods, because the surface-attachment approach achieves the following:

1. Eliminates the cost and labor for component leads and their forming.
2. Simplifies the problems of component shipping and handling by the elimination of their delicate leads.
3. Simplifies assembly, since there are no leads to preform. Also, the self-centering capability of many of these new devices during the soldering operation makes them tolerant to some component placement misalignment.
4. Increases the component placement rate from 300 units/hour for manual assembly of 3,000 or more units/hour for automated methods, by one more order of magnitude, e.g., 200 units in 7 sec on an 8- by 10-in. board is achievable.
5. Simplifies the soldering process by assembling with solder-reflow systems that use vapor-condensation methods without the need to worry about leads protruding on the underside of the substrate.
6. Improves connectivity/unit area on the underlying substrate by reducing hole diameters, or by totally eliminating the need for holes.

REFERENCES

1. G. Messner and C. Lassen, "Substrates for Surface Mounted Components," *Circuit World*, August 1981, pp. 10-14.
2. B. T. Patterson, "The Challenge of Interconnection, Part One: The Interconnection Crisis," *Circuits Manufacturing*, February 1981, pp. 22-24.
3. Guidelines for Component Packaging and Interconnection with Emphasis on Surface Mounting, IPC-SM-780, Institute for Interconnecting and Packaging Electronic Circuits, 1988.
4. C. W. Brooks and M. R. Lucas, "The Packaging Challenge of High Speed VLSI Circuitry," Proceedings, Nepcon West, February 1981, pp. 150-154.
5. J. W. Balde, "Technical and Economic Tradeoffs in Packaging with VLSI," *Electronic Packaging and Production*, September 1980, pp. 107-110.

2

Semiconductor Developments

Much has been written about Surface Mount Technology (SMT) and how, in general terms, it better satisfies the growing demand for packaging and interconnecting high-technology integrated circuits, than does conventional through-hole technology [1]. However, now that an SMT state-of-art is beginning to emerge, this is an appropriate time to become more specific about the new developments in semiconductor technology that are the driving force behind SMT and, thereby, behind new printed board technology.

During the last few years, the number of approaches available to circuit designers to use (or create) integrated circuits has blossomed into a wide variety (Figure 2.1). Therefore, this chapter first describes the standard semiconductor logic families and the sophisticated custom/semicustom ways in which they are being combined by integrated circuit (IC) manufacturers. This is followed by a discussion of how this impacts end-product applications. Finally, their ultimate impact on surface mount-related issues is covered.

2.1 INTEGRATED CIRCUIT LOGIC FAMILIES [2-4]

A circuit designer can work with one of several integrated circuit transistor structures. The two most popular are bipolar and metal oxide (MOS) silicon chip technologies. These, in turn, are beginning to receive competition from the newly emerging gallium arsenide (GaAs) chip structures.

The demise of the standard silicon-based logic families has been forecast for almost as long as standard logic has been around. And, true, the standard logic families that exist today are a far cry from those of the 1960s. Yet their use persists. Simple one- and two-gate chips are gradually being made obsolete by standard medium-scale integration (MSI) and, in some cases, large-scale integration (LSI) chips with larger functions. However, the most basic building block functions will remain as extra "glue" circuits for many years to come.

Gate-switching speed, power dissipation, and functional density are the basic differences separating the integrated circuit technologies at the low-integration levels (Figure 2.2). However, these are no longer valid in the very-large-scale integration (VLSI) era. Thus, the answers to selection questions lie in understanding the available technologies from the new perspective of identifying the applications areas most generally suited for each technology and the key system design trade-offs involved in the use of each technology.

2.1.1 Bipolar Circuit Technology

Bipolar transistors can be formed by either alloy or double-diffused techniques. However, all major bipolar "integrated" circuits use double-diffused transistors formed in an epitaxial (high-resistivity film) layer deposited on the silicon wafer. The epitaxial layer serves as the "collector" region of the transistor and allows for easy isolation of the separate circuit components with the addition of a deep diffused "fence."

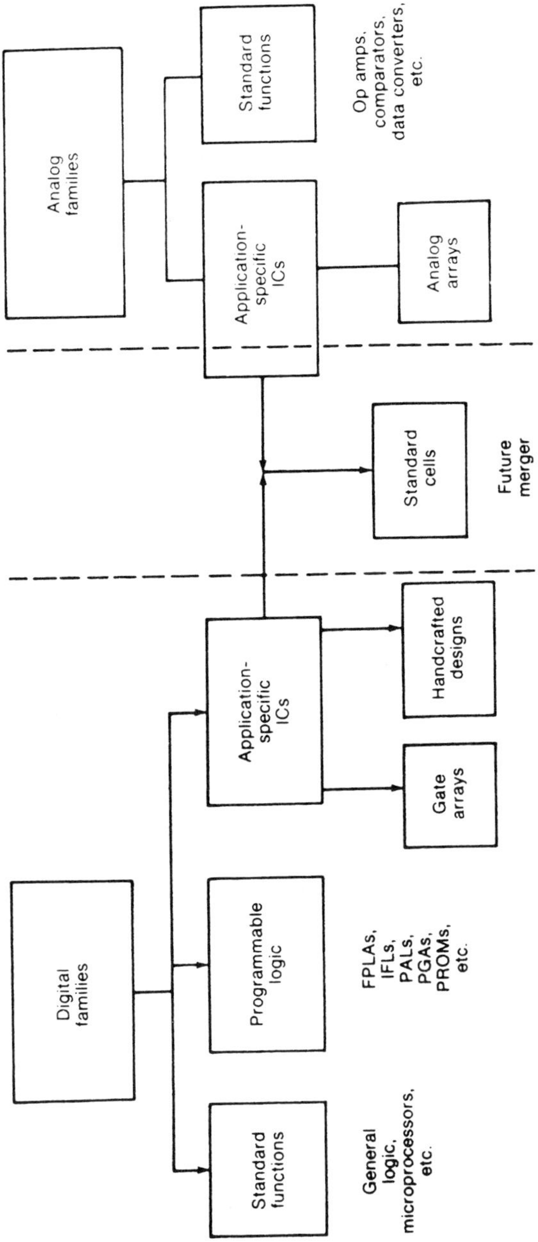

FIGURE 2.1 The wide variety of integrated circuit technologies.

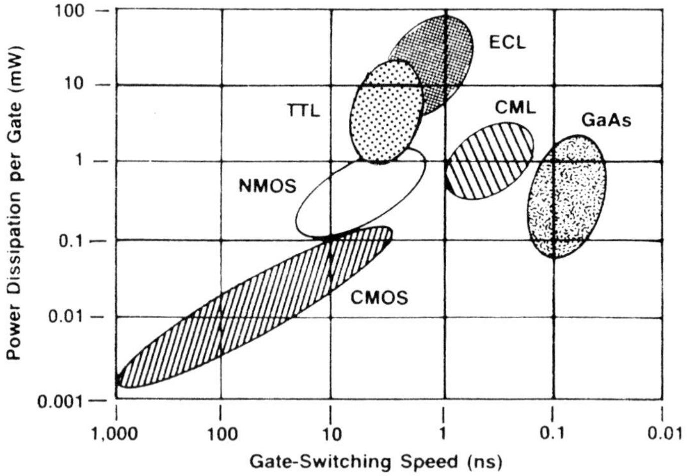

FIGURE 2.2 Speed versus power comparisons for various technologies.

Because they are formed below the wafer surface, bipolar transistors are not prone to contamination. Unsensitivity to contamination is not the only factor, however: They can operate at voltage and current values that are compatible with other electrical devices. The switching speeds of bipolar transistors are also faster than those of the simpler MOS discrete transistors.

Using conventional bipolar structures, circuit speed can be improved by "wiring" together the various elements into logic cells. Transistor-transistor logic (TTL) and emitter-coupled logic (ECL) are the two most popular cell types.

TTL circuits have been popular with most system builders because of a combination of acceptable performance and ease of design. Alternately, ECL design has been a standard process only for computer builders who need high performance badly enough to put up with the special headaches created by the use of the higher-speed ECL circuits. [For example, printed board conductors must be designed with waveguide (controlled impedance) techniques because the rise time of ECL signals puts them in the microwave frequency range. More will be said about this later.]

2.1.2 MOS Circuit Technology

The formation of MOS integrated circuit devices is different from bipolar technology. The basic MOS fabrication process starts with an incoming wafer that is processed through oxidation and goes directly to masking. Thus, it can eliminate the epitaxial layer and isolation diffusion steps required in bipolar technology. The absence of the isolation structure allows a higher MOS component density than bipolar technology.

In both bipolar and MOS technology, the area of the circuit that does not contain active devices is called the "field." A particular problem arises over the field in MOS circuits. Metal conductors running on top of the field oxide form a capacitor with the silicon below. If the voltage on the conductors becomes high enough, the "field capacitor" will create charge in the underlying silicon and cause shorted devices. Additional wafer processing is required to overcome this problem. (More careful handling during the printed board assembly process is also required.)

Unlike the simpler metal gate processing, complementary MOS (CMOS) requires more processing steps than the bipolar process. However, this additional processing results in CMOS devices that have lower power consumption and increased speed, making them attractive for many applications.

The U.S. Department of Defense and the aerospace industry have stepped up their requirements for increasingly higher tolerances against radiation effects for integrated circuits. These stricter mandates are expected to yield greater system survivability in hostile (nuclear) environments. Bipolar technology is inherently more radiation-resistant than most CMOS processes. Thus, the advantages of designing with CMOS have led to additional processing steps to enhance CMOS survivability.

Other CMOS processing variations prevail to customize particular features of the end product. This, in many cases, makes CMOS the most complicated and most involved technology in the industry.

2.1.3 Gallium Arsenide Technology [5-8]

Gallium arsenide (GaAs) is an ideal semiconductor medium for achieving very high speed in electronic devices and integrated circuits. This is because its energy band structure is such that electrons in GaAs are exceedingly "light" and highly mobile. Thus, electron velocities measured in GaAs transistor structures range up to about five times those achieved in silicon-based devices.

Furthermore, GaAs is readily available in a semi-insulating substrate form that substantially reduces parasitic capacitances, so that its outstanding device speeds can be fully realized in integrated circuits (Table 2.1). This high speed, plus power dissipation that often tends to be substantially lower than that of high-speed silicon devices, accounts for the growing interest in gallium arsenide.

TABLE 2.1 Examples of High-Speed Integrated Circuits [7]

	GaAs	ECL	CMOS
Critical dimensions (μm)	1.0	2.0	2.0
Chip size (mm^2)	4.6 × 4.3	6.6 × 6.8	10 × 10
No. of equivalent gates	1200	2500	20K
Time delay (psec/gate)	375	510	1500
Power dissipation (μW/gate)	190	1200	310
(50-MHz clock)			
Speed-power (fJ)	71	610	465

Although a number of different GaAs integrated circuit families exist, it is possible to draw some general conclusions about the advantages of GaAs technology over silicon technology.

Currently, for the same power consumption, GaAs is about half an order of magnitude faster than emitter-coupled logic (ECL), the fastest silicon-based family.

GaAs is more radiation-hard than silicon. At this time, however, the difference is difficult to quantify.

GaAs is better suited to the efficient integration of electronic and optic components. The usefulness of GaAs for this type of integration is still under investigation. However, if developed to the appropriate level, this may have a major impact on system design in general.

Basic disadvantages of GaAs technology, as compared to silicon technology, are as follows:

GaAs wafers presently exhibit a large "density of dislocations." That is to say, there are a large number of irregularities per unit area. Consequently, GaAs chips (1) have to be smaller in area, (2) have a smaller transistor count, and (3) have a poorer yield. Of course, chip size can be traded off against production yield to some extent.

GaAs substrates are about two orders of magnitude more expensive than silicon substrates. Moreover, GaAs is brittle: Wafers can be damaged easily during the fabrication of integrated circuits.

The noise margin of GaAs at present is not as good as that of silicon. Thus, it is often necessary to trade off chip area for higher reliability.

Lastly, some companies are currently reporting problems with the testing of designs for high-speed GaAs integrated circuits. Testing facilities that are fast enough are simply not yet available. But this situation is expected to improve.

2.2 APPLICATION-SPECIFIC INTEGRATED CIRCUITS [1, 9, 10]

Electronic system design is becoming more complex, with architectures reaching new heights of sophistication and expanding integrated circuit choices. Yesterday's systems comprised off-the-shelf, small-scale (SSI) and medium-scale (MSI) levels of circuit integration. However, many of today's designs are implemented with customized large-scale (LSI) and very-large-scale (VLSI) integration. Leading the migration toward LSI/VLSI are application-specific integrated circuits (ASICs) with functions tailored to the user's requirements.

Therefore, in order to keep their competitive edge, system manufacturers must make informed decisions when selecting a particular type of customized integrated circuit technology. This often requires an in-depth trade-off analysis.

There are four general ASIC categories: (1) programmable logic devices (PLDs); (2) gate/macrocell arrays; (3) cell-based devices (both standard/macrosystem cells and silicon-compiled cells); and (4) handcrafted, fully custom integrated circuits (Figure 2.3). (The gate arrays and cell-based devices are often called semicustom integrated circuits.) Also, recent advances in integrated circuit fabrication are spawning a new kind of ASIC between a PLD and a gate array. It is important to determine which type is best suited to a designer's requirements (Table 2.2). The first consideration is the density of the device needed. For example, PLDs have relatively low densities of from 50 to 5000 equivalent gates per chip.

Macrocell and gate arrays are next, with densities ranging from 500 to 10,000 gates per chip, sometimes more for specific purposes. Standard cells have still higher densities, many thousands of equivalent gates per chip. Hand-packed custom circuits are the densest, with 70,000 switching transistors per chip being not uncommon.

The time to prototype and develop an ASIC is a major concern when making the choice as to which ASIC is best suited for an application. Because PLDs are readily available off the shelf, prototypes can be obtained almost immediately at a relatively negligible development cost. Gate array and macrocell array prototypes development

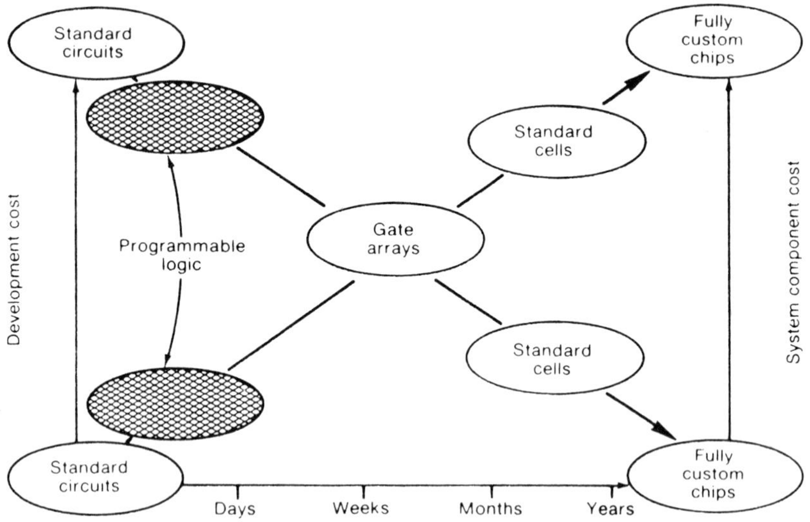

FIGURE 2.3 Integrated circuit development time versus cost. (Source: Advanced Micro Devices, Inc.)

time is typically from 7 to 13 weeks, with prototype costs from $10,000 to $40,000. Standard cells take even longer, typically from 13 to 26 weeks, because their wafers are made from scratch instead of being preprocessed like macrocell and gate array chips.

Additionally, standard-cell prototyping costs are usually between $40,000 and $100,000. Handpacked, fully custom chips have the longest prototyping times, taking from 39 weeks to 2 years, with costs ranging from $100,000 to more than $500,000.

PLDs are often the logical choice when the product is required immediately or when frequent design changes are to occur. They are limited, though, to relatively simple requirements when compared to the functionality and densities of the other ASIC technologies. If functional and volume requirements are high, gate array and macrocell solutions are usually the most feasible, with standard cell circuits the choice for the highest-volume demands. However, the better silicon usage characteristics of standard cells may offset their higher development costs if the volumes are suitable.

Silicon-compiled ASICs, though still in their infancy, also deserve consideration. Handpacked, custom circuits are often acceptable when the largest production quantities are needed and if the long turnaround time can be tolerated by the user.

TABLE 2.2 ASIC Technology Selection Factors

	ASIC types			
	Programmable logic	Gate/macrocell arrays	Standard cell/macrosystems	Handpacked custom
Chip density (number of gates)	Few	16–100,000	Hundreds to tens of thousands	From hundreds to tens of thousands
Percent of wafer preprocessed	100	80–90	0	0
Time to prototypes (weeks, in 1983)	Off-the-shelf	7–13	13–26	39–104
Option development cost per chip (depends on complexity)	None	$10,000–$40,000	$40,000–$100,000	$100,000–$500,000
Ability to make design changes or correction	Minimal cost	Easy, fast, and inexpensive	Easy, but somewhat more expensive and slower than arrays	Hard, slower, and more expensive than standard cell/macrosystems
Unit production cost of chip	Low	High	Medium	Lowest

2.2.1 Programmable Logic Devices [11, 12]

Until recently, programmable logic devices (PLDs) were considered for use primarily as TTL "glue" replacements in order to compact several gates into a single package when there were shortages in printed board space. They were not really accepted as a valid ASIC choice. But that has changed, due to four basic reasons:

1. The first is the trend toward acceptance of PLDs as a mainstream system building block, such that the designer begins a design with PLDs in mind, as opposed to their being an afterthought.
2. The second reason is the sudden proliferation of PLD manufacturing technologies. Just as digital logic gates began with bipolar chips, so did PLDs. Now, however, PLDs are available in a selection that also includes CMOS and ECL technologies.
3. Third is the growth in the complexity and functionality of PLDs. Early PLDs (in 20-pin, dual-inline packages, DIPs) had usable gate counts in the 100- to 300-gate range. Now, gate densities have climbed to as high as a few thousand. (Pin counts have also increased to as high as 84.) The result is that such devices now offer logic densities that once were the exclusive domain of gate arrays and cell-based parts.
4. The fourth reason is the increase of PLD manufacturers, from three in 1979 to more than 20 today. This has served both to increase the selection of functions available to the designer and to make the cost of the devices more competitive.

Applying CMOS and erasable-cell technologies to the basic PLD configuration has created erasable programmable logic devices (EPLDs) with a new level of versatility in logic design. In certain applications they match the density of gate arrays without the associated tooling costs, extended development schedules, and loss of in-house design control. Thus, EPLDs fit between PLDs and gate arrays in terms of complexity, versatility, and cost per function (Figure 2.4).

2.2.2 Gate Arrays [4, 13-15]

Gate arrays are distinguished by their use of prefabricated chips containing transistors and "feedthroughs," none of which are interconnected by signal wiring (Figure 2.5). As a result, the chip vendor can make large quantities of identical unwired gate array chips, fabricating the wiring only when a customer supplies the circuit schematics or interconnect lists.

Since the fabrication of wiring layers is a relatively low-risk, inexpensive operation, manufacturing yields tend to be high and processing times short. More important, design costs are low, since a manual layout phase is not needed.

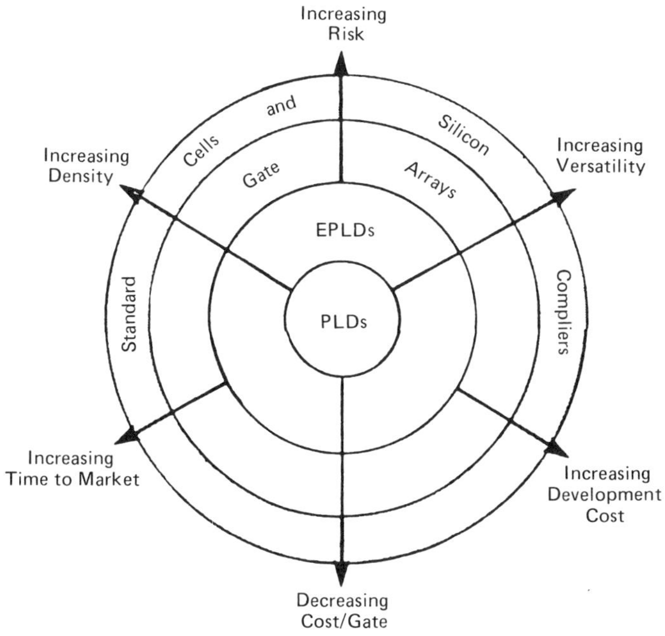

FIGURE 2.4 Trade-offs when specifying an ASIC technology.

Instead, design automation tools, which assign transistors and feedthroughs to circuit functions and determine the wiring paths, perform the layout task. The use of these tools is straightforward, consumes a relatively small amount of computational resources, and tends to produce final chips that are efficient in space and speed.

Arrays are available in all of, and combinations of, the basic integrated circuit fabrication technologies. The mainstay of the fastest digital systems, silicon ECL gate arrays, faces competition from the other technologies as new processes and materials become available. However, the same technological advances that give MOS its advantages will help bipolar arrays remain ahead of MOS and mixed bipolar-MOS and in stride with the subnanosecond speeds of gallium arsenide. In fact, ECL gate arrays will soon be available with as many as 10,000 gates along with gate delays of only 200 to 300 psec.

Nonetheless, ECL does stand to lose its edge in certain fields. For instance, printed boards built previously with SSI and MSI ECL chips can now be replaced by high-density CMOS gate arrays and mixed bipolar-CMOS chips with delays of several hundred picoseconds and integration levels of 20,000 gates and higher. At the same time, GaAs

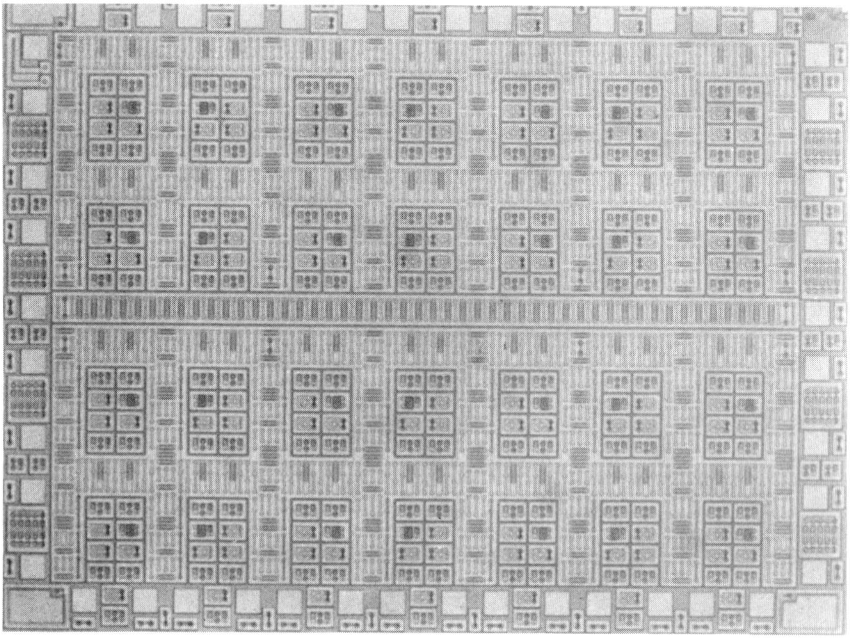

FIGURE 2.5 Bipolar 28-array integrated circuit.

may outrun ECL in some designs calling for extremely high speed. GaAs logic arrays and standard cells now have delays of only 200 psec or less with complexities up to a few thousand gates (Table 2.3).

2.2.3 Cell-Based Technology [16]

Cell-based ASICs can be implemented either as standard cells or as compiled cells. Both technologies provide high levels of integration, flexibility, and performance. The trade-offs are in degree of user involvement and functional capability.

Turnaround time and nonrecurring engineering costs are similar for the two approaches. However, designing with standard cells relinquishes more of the design effort to the device vendor than does the compiled-cell approach, which allows the customer to be more involved. Also, although the performance of standard and compiled cells may be the same at the transistor level, performance and density differences emerge at the chip level (Table 2.4).

It should be noted, however, that neither approach is inherently superior or preferable to the other. Rather, the decision to use one or the other depends on the application or parts of the application.

TABLE 2.3 Typical State-of-the-Art Gate Arrays

	Number of gates	Gate delays	Package pins	Number I/Os
Bipolar gate array	8,000	210 psec	235 PGA	188
CMOS gate array	20,000	400 psec	284 TAB or PGA	238
GaAs gate array	2,000	230 psec	88 LCC	80

For example, some on-chip memory is more easily implemented using compilers, while many logic functions are better implemented in standard cells. A few vendors even provide the ability to implement both approaches on the same chip for optimum flexibility.

A standard cell uses a library of simple, fixed cells and soft macrocells (predefined combinations of cells). User involvement typically includes logic partitioning, schematic entry, simulation, and net list verification. Fixed-height cells are then arranged by automatic placement and routing programs in fixed-height rows separated by variable-width routing channels.

Standard cell placement and routing is usually handled by the device vendor rather than by the customer. After the place-and-route operation, the design is resimulated using actual wiring delay figures to help ensure accurate performance estimation.

TABLE 2.4 Three Implementations of the Same Circuit

	Cell compiler	Standard cell	Gate array
Core size	1085 × 1112	1872 × 1740	2160 × 2112
Number of transistors in core	980	1286	1286
Transistor density (sq. mils per transistor)	1.91	3.93	5.49
Design time	4 days	7 days	7 days
Performance (critical path delay per gate)	45 nsec	68 nsec	49 nsec[a]

[a]The gate array design is implemented in a faster CMOS process technology.

A cell compiler is a design tool for automatically generating the layout, simulation models, and schematic symbols targeted for a specific function. However, since one compiler cannot generate layouts for all functions, a design environment usually includes a library of cell compilers or parameter-based software modules.

In concept, a system engineer can design a chip by identifying compiler functions that correspond to the block diagram description of a circuit. These functions are compiled by the software. The design process then proceeds in a manner that is similar to that for standard cells.

2.2.4 Fully Custom (Handcrafted) Devices [13]

A fully custom device implies that little or no formality is exploited in easing the task of chip design. For VLSI circuits, custom design uses hierarchical design methods whereby the bottom-most design element or cell contains the basic circuits, much like cells in standard cell devices. The primary difference is that there is little or no predefined structure to a custom-designed cell.

These cells are then interconnected to form larger cells, which are interconnected to form still larger cells, and so on until the entire chip is completed. The cells are not constrained into predetermined rows but can be located anywhere on the chip to minimize wasted space. A typical custom-designed chip is shown in Figure 2.6.

Cell placement and net routing in a custom layout are complex tasks, especially if maximum cell packing is desired. Because regular channels for wiring are usually not created, the task of routing is less constrained than in channel routing.

Even in a custom layout, however, routing can be done in two steps. The first step, as in gate array and standard cell designs, is to assign nets to particular regions of the chip or channels, if all cells are rectilinear. Then each region is routed by assigned routes to tracks.

Cell placement on a custom chip is similar to the procedure for standard cell chips. Again, as in the standard cell approach, the object is to minimize wiring length and congestion, assuring 100% wireability, and to minimize the area used by cells on the chip.

Custom design and layout, the most time-consuming of the ASIC styles discussed in this chapter, produce the most area-efficient chips and usually the fastest circuits. The trade-offs are with respect to design time, performance, chip size, production yield, and costs (both component and system).

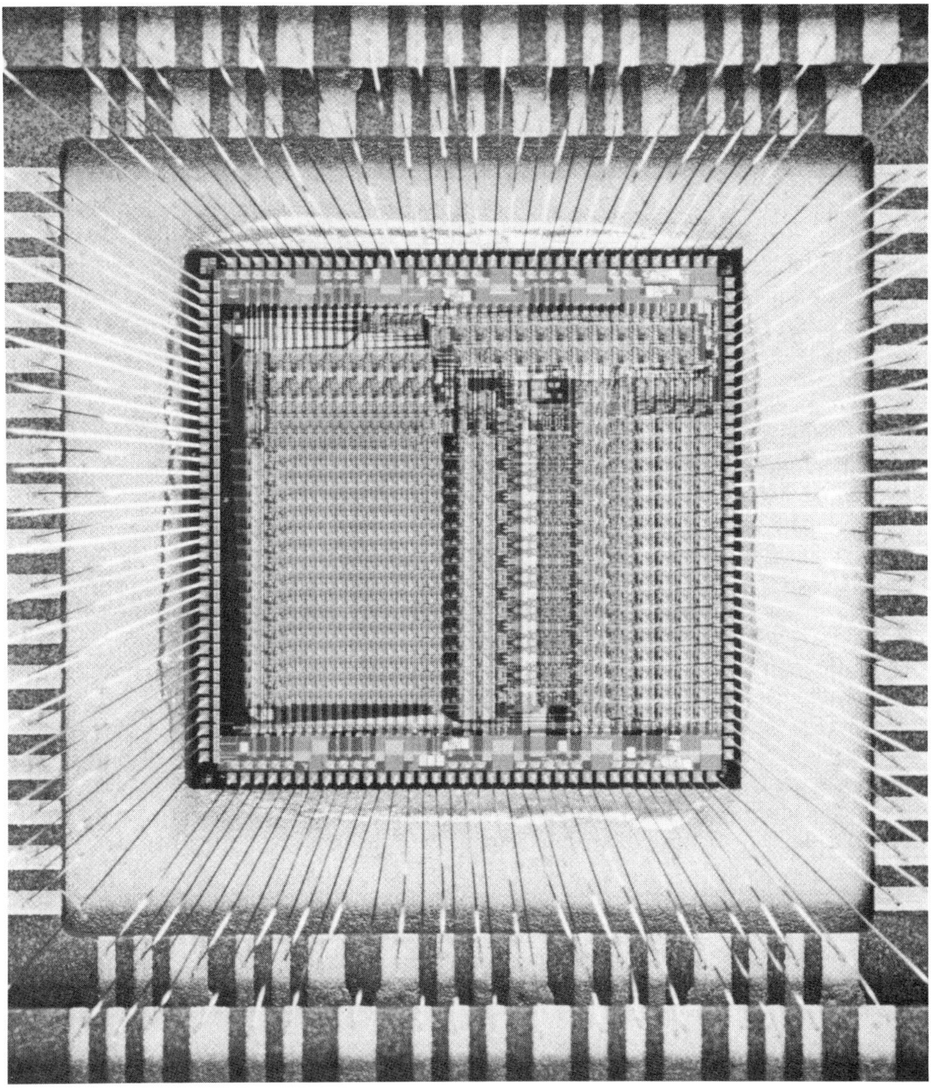

FIGURE 2.6 Custom VLSI device with 5200 gates, 30,000 transistors, and a 144-I/O pin-grid array package. (Source: Honeywell, Inc.)

2.3 SPECIALIZED TECHNOLOGY

In addition to the state-of-the-art integrated circuit technologies and architectures just described, other, more specialized approaches are being undertaken to obtain more cost-effective performance than is presently possible. Efforts in these areas are either vendor-driven (wafer-scale integration, silicon-on-sapphire, bubble memories, etc.) or user-driven (e.g., very-high-speed integrated circuits). Although in their relative infancy, developments in these technologies have a potentially great impact on printed board technology.

2.3.1 Wafer-Scale Integration [17, 18]

Wafer-scale integration (WSI) is conceptually the ability to interconnect electrically sound circuits on a tested and mapped semiconductor wafer using a wafer-level interconnect scheme. If achieved, WSI could open a market with almost limitless applications. After nearly two decades of effort by several companies, however, a number of problems still exist (particularly in the wiring approach) that make these devices irreparable, expensive, and, in general, unsuited to volume production. But because the benefits of WSI are so alluring, new companies (and approaches) have emerged to address the issues associated with this challenging integration concept.

Attempts so far to implement wafer-scale parts have concentrated on integrating highly complex devices with a random point-to-point wafer-level interconnect. Although integrating such devices will eventually result in wafer-level products with performance characteristics equivalent to a mainframe computer, testing the viability of such complex devices may be overly optimistic for such an immature technology. Integrating less complex, higher-yield circuits in a more structured or bus-oriented architecture might be a more realistic endeavor.

Memory devices are the most logical and likely candidates to usher in this new generation of WSI system designs. For one thing, memory devices, in general are bus-oriented. This makes routing less complex and less prone to error. Also, memory circuits are often constructed with a single layer of metallization, allowing the multichip macrocircuits to be interconnected with a second layer. (This type of double-layer metallization is a commonly used fabrication technology.) One of the most likely candidates for a WSI memory device is in airborne- or satellite-based electronics that would benefit from the decreased size and weight that WSI memory devices could provide along with reduced power supply costs, lower cooling costs, and improved space optimization.

Conversely, there are studies that indicate that WSI is not practical for most applications. For example, there are those who feel

TABLE 2.5 Figures-of-Merit for WSI Versus Typical Packaging Technologies (100-MHz Clock Rate)

Package	Power versus delay (normalized)	Size or weight (normalized)	Cost (normalized)	Overall figure-of-merit
Printed circuit board	1.00	1.00	1.00	1.00
Thick-film multilayer on ceramic	1.08	0.42	1.02	0.46
Ceramic multilayer hybrid	0.34	0.20	0.65	0.044
Thin-film multilayer hybrid on various substrates, populated on one side	0.19	0.14	0.60	0.016
Wafer-scale integration	0.01	0.09	0.46	0.0041
Thin-film multilayer hybrid on various substrates, populated on both sides	0.08	0.07	0.44	0.0025

Source: General Electric Corp.

that existing advanced VLSI-hybrid packaging approaches can achieve circuit density, performance, and cost-effectiveness levels comparable to those that WSI potentially possesses (Table 2.5). It is also claimed that, in terms of packaging density, thin-film multilayer hybrid circuits on various substrates, when populated on both sides, can package almost 25% to 30% more components in about the same area as a wafer-scale device. The two technologies are almost equal as far as package delay is concerned. (Double-sided thin-film hybrids rate 10.5-nsec versus 11.8-nsec for WSI when both have 450 output buffers per package.)

In terms of power per chip, both single- and double-sided multilayer hybrid approaches are competitive with WSI. When the product of power times delay (an accepted figure-of-merit used to compare such technologies) is considered, double-sided thin-film hybrid circuits rate better than wafer-scale integration.

This is not to say that wafer-scale logic devices are not worth developing. On the contrary, if and when wafer-scale technologies have been fine-tuned, logic and system-level WSI might find a niche for itself. When that time comes, however, comparable advances may have been made to improve the capabilities of the competing technologies. With respect to printed board technology, it does not appear that WSI will have any serious impact within the near future. As Table 2.5 shows, hybrid technologies are, and will continue to be, of more concern.

2.3.2 Silicon-on-Sapphire Devices [19, 20]

In a silicon-on-sapphire (SOS) circuit, an isolating monolithic sapphire substrate is used as the circuit carrier. A thin silicon epitaxial film is grown on the sapphire surface, and transistors are then fabricated in the film by means of conventional CMOS technology.

Complete isolation is achieved between each device on the chip as a result of the use of the sapphire substrate. By virtue of this dielectric isolation, SOS is inherently immune to the destructive, high-current condition known as "latch-up." (Conventional CMOS must be processed with a special epitaxial layer to prevent latch-up.) This also helps to account for its exceptional resistance to transient radiation and "single-event upset" faults, which have been known to cause faults in integrated circuit memory cells.

The sapphire substrate also results in very low parasitic capacitances. This, in turn, results in higher circuit speed and lower power dissipation than is possible with bulk silicon MOS devices. Other comparisons between SOS and bulk silicon are shown in Table 2.6.

Unfortunately, the use of SOS as not been price-competitive in the marketplace because of higher manufacturing costs (wafer material, photomask replacement and dicing), lower demand, and fewer companies producing sapphire wafers. This will perhaps change when more semiconductor manufacturers adopt the technical advantages of SOS.

Fortunately, an SOS process requires fewer masking steps than a silicon-gate bulk process, which of course gives SOS a yield advantage. This advantage increases with reductions in feature size. This will probably be the determining factor in wider acceptance of CMOS/SOS circuits.

2.3.3 Magnetic Bubble Memories [21, 22]

Magnetic bubble memory devices are manufactured using techniques that are typical of the semiconductor industry at large. The heart of the device is a chip whose top surface consists of a thin film of a magnetically uniaxial material on which is deposited an array of control

TABLE 2.6 Silicon-on-Sapphire Versus Bulk Silicon CMOS

Parameter	SOS	Bulk
Design flexibility	Higher	Lower
Speed	Higher	Lower
Power dissipation	Lower	Higher
Flicker noise	Higher	Lower
Leakage current	Higher	Lower
High-temperature performance	Better	Worse
Latch-up	No	Yes
Radiation tolerance	Higher	Lower
Packing density	Higher	Lower
Wafer material cost	Higher	Lower
Yield (4-μm)	Equal	Equal
Yield (2-μm)	Higher	Lower

elements. The magnetic material has been specifically created so that its "easy" axis of magnetization is normal to the plane of the thin film.

In the absence of any external magnetic field, the domain structure of the thin film has an antiparallel "stripe" pattern arranged, for minimum energy, to give zero net magnetism (Figure 2.7). If a bias field is applied perpendicular to the plane of the film, it becomes energetically favorable for the antiparallel domains to shrink.

When the field is increased still further, these islands contract under pressure from the applied field to form cylindrical domains. These cylindrical domains are the "bubbles." If the field is increased still further, the bubbles will totally collapse, leaving a magnetically saturated sample. Thus, binary data is stored as the presence of absence of bubbles at addressable locations.

As sophisticated (and complicated) as all of this may seem, magnetic bubble memories offer an attractive alternative to other media, such as tape and disk, for reliable data storage in harsh environments. Their high packing density, ruggedness, reliability, nonvolatility, and nuclear hardness also make them well suited for military applications.

Although not a technical issue, cost is a key consideration when selecting a memory device. For severe environment applications, for

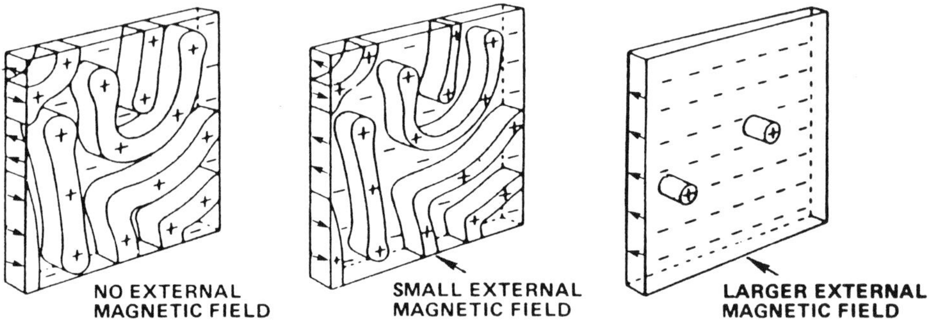

FIGURE 2.7 Magnetic bubble memory formation.

which magnetic bubble memories are best suited, ruggedized disk, and tape drives are slower but offer a lower cost per bit (Figure 2.8). At the other end of the spectrum, the access times of magnetic bubble memories cannot compete with random-access semiconductor memories (RAMs).

Therefore, if magnetic bubble memories are to find a place for themselves in the marketplace, it will have to be in applications that can accept their slower access times or higher costs per bit in exchange for improved data security, environmental stability, radiation tolerance, and potential savings in reduced maintenance costs (Unfortunately, although several device manufacturers have entered the field, most of them have retracted as it became evident that the technology has a smaller market than was originally predicted.)

2.3.4 Very-High-Speed Integrated Circuits [23, 24]

The most significant "user-driven" attempt to provide a technological edge is the very-high-speed integrated circuit (VHSIC) program formulated in 1978 by the U.S. Department of Defense and initiated in 1980 to design and insert advanced integrated circuitry into military systems. The VHSIC program was based on the premise that the VLSI devices available at the time could not meet military demands for increased speed and operation in harsh environments. Therefore, VHSIC was undertaken to achieve reduced integrated circuit line widths (thereby reducing resistance and capacitance timing delays), low power consumption, radiation hardening, built-in error checking/testability, and maintainability.

Phase 0, begun in March 1980, was a 9-month study that identified the needed technology for long-range VHSIC applications using

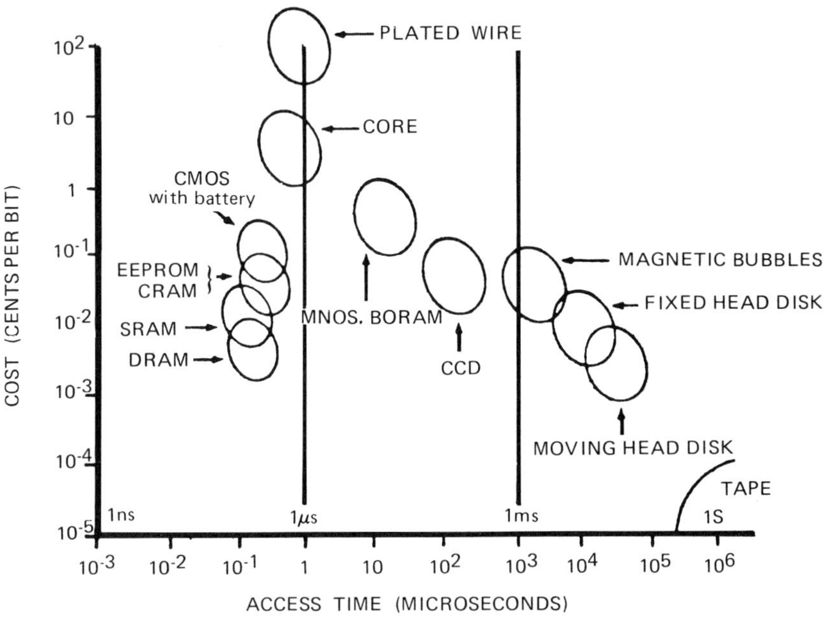

FIGURE 2.8 Memory device cost-per-bit versus access time.

1.25- and 0.5-μm (millionths of a meter) circuitry. (Nine awards were made totaling $10.5 million.)

Phase 1 was a 3-year effort that began in 1981. Participating companies in this part of the program were given total freedom in selecting the types of chips that would be developed (Table 2.7). This $167.8 million phase of the program consisted of three parts:

1. Developing integrated circuits with minimum feature size in the 1.25-μm range combined with hardware and software capabilities for computer-aided modeling and design of these circuits.
2. Developing pilot lines that extended this technology to the limited production of these devices.
3. Completing useful military breadboards that ensured the usefulness of these devices.

After the satisfactory completion of Phase 1, the 5-year Phase 2 of the program began in October 1984. This effort (funded for at least $336.6 million) is segmented into four parts:

1. Developing subsystem demonstrations of chips with 1.25-μm feature sizes based on Phase 1 prototype brassboards.
2. Speeding the insertion of VHSIC technology into military systems.
3. Standardizing the computer languages used in these systems.
4. Developing pilot-line production of integrated circuit chips with 0.5-μm design rules and developing subsystem demonstrations based on these devices.

The technological achievements of VHSIC Phase 1, and potentially from Phase 2, represent significant advancements in the integrated circuit state-of-the-art. VHSIC has also resulted in significant advances

TABLE 2.7 VHSIC Phase 1 Packaging

Contractor	DOD application brassboard	Chip technology	Line width (μm)	Power dissipation (W)	Chip I/O count
Honeywell	Electro-optical signal processor	Bipolar	1.25	2-2.2	
Hughes	Battlefield information distribution system	CMOS/SOS	1.25	0.0-0.7	25-104
IBM	Acoustic signal processor (sonar)	NMOS	1.25	3	240
TI	Multimode fire and forget missile	Bipolar and CMOS	1.25	0.47-1.6	32-84
TRW	Electronic warfare signal processor (airborne)	Bipolar and CMOS	1.25	0.250-3	100-131
Westinghouse	Airborne radar	CMOS	1.25	0.5-1	44-224

in the packaging associated with such devices, as indicated in Table 2.7. At the packaging and interconnecting structure level, some major advances will have to be made in order to facilitate the "insertion" of these devices into end-product assemblies.

2.4 SUMMARY AND CONCLUSIONS [25-27]

There appears to be no near-term letup in the increase in capability and complexity that will be achieved by semiconductor technology. In fact, if anything, the pace of new developments seems to be quickening.

Max. clock rate (MHz)	No. of gates (× 1000)	Chip size (mils)	No. of I/O pins	Packaging and mounting technology
33	13-19.8	310^2 typ.	180	Pin-grid array on multilayer PC board; area tab
25	N.a.	370^2 typ.	148	Leaded ceramic chip carrier with 25-mil centers; wire bonding on multilayer PC board
N.a.	30-38	315^2	244	Pin-grid array on multilayer PC board; flip-chip with area decal
25	3.2-44	350^2	160	Leadless chip carrier with 25-mil centers; wire bonding
25	N.a.	310^2 typ.	132	Custom leaded ceramic chip carrier on multilayer PC board; wire bonding 25-mil centers
25-40	5-30	340^2	44-224	Leaded ceramic chip carrier with 20-mil centers; wire bonding

Added to this is the general trend toward having more custom and semicustom integrated circuits available to the circuit designer. These new devices offer many new capabilities to the designer, along with a more complex selection process.

In the past this selection process by the circuit designer was done, to a great extent, without giving major consideration to its impact on printed board technology. It was generally felt that board technology would continue to evolve to more sophistication along with advances in chip-level technology.

However, the new options available to the circuit designer are becoming more dependent on the cost and performance of the next-level packaging/interconnecting structure. For example, consider a digital system requiring a subsystem that is to be implemented with 2300 gates and 125 input/output devices. Available are arrays with 1200, 1900, and 3200 gates each. If the circuit designer partitions the system into one, two, or three chips, the implementation listed in Table 2.8 results. If the designer were concerned only with integrated circuit costs, the circuit would, theoretically, be implemented with two arrays, one with 1200 gates and one with 1900 gates. The three-chip alternative would cost 14% more; the one-chip alternative—with a larger, more expensive chip and package—would cost more than twice as much.

These calculations, however, consider only the procurement costs for packaged chips and do not take into account system implementation parameters that are affected by the different subsystem versions. Such parameters include size, the cost of mounting and interconnecting the chips on a printed board, power, reliability, and sometimes supplementary chip design costs. If it is assumed that the cost of mounting the chips is proportional to the total number of package

TABLE 2.8 Alternative Partitioning Costs

					Relative cost		
Chip	Gates	I/Os	Array size	Package pins	Chip	Pkg.	Total
1	2300	125	3200	132	23.7	25.5	49.2
2	940	80	1200	84	5.1	4.4	21.6
	1500	75	1900	84	7.7	4.4	
3	900	75	1200	84	5.1	4.4	24.7
	900	39	1200	40	5.1	1.7	
	900	65	1200	68	5.1	3.3	

pins, then the more chips it takes to implement the subsystem, the more cost has to be added to the total implementation cost.

Thus, as the cost of system integration, such as printed board costs, goes up, the economics of reducing total chip count tends to push toward systems with fewer, more complex chips. There is also a general improvement in circuit performance with the use of fewer devices. Therefore, a final design decision as to which implementation of the subsystem to develop must take into account factors such as these and, hence, is not a trivial task. Therefore, the circuit designer is becoming more of a system designer, and vice versa, than he has been in the past (and probably more than he would like to be). Concurrently, it is important that the printed board manufacturer become more of a system fabricator, rather than being just an interconnecting component supplier.

In addition to the obvious cost considerations, printed boards (and competing technologies) are elements of the overall performance of the end-product system. As such, their "component" cost/performance parameters must be entered into the overall system design trade-off analysis and decision-making procedure.

In addition to these generalities, the following elements will affect the success or failure of a printed board manufacturer, material vendor, or equipment supplier within the next 5 to 10 years:

"Mils" (thousandths of an inch) terminology will have to be replaced with "micron" (0.04 mil) terminology. This will be necessary in order to get the printed board industry to think more in terms of semiconductor magnitudes and in terms of higher levels of precision and accuracy. As an example, drilling equipment spindles must become capable of runout tolerances at the drill bit tip below 1 mil, perhaps as small as several microns or less. This will undoubtably require the increased use of optical alignment accessories, as it will in other areas of the industry.

Imaging and registration systems will need to be developed to allow for the producing of still finer line circuitry than is presently realized. Some leading-edge OEMs are seriously considering the need for line widths at 0.002 in. and less in order to achieve even higher conductor routing densities and improved circuit performance. The capability of imaging films and image-transfer processes might also come into question, especially if further advances are made in "direct imaging" technology. This will have to be done along with a reconsideration of the suitability of the metallization technique being used. Is thin-film "sputtering" viable at the printed board level?

Material suppliers must develop, or further refine, their laminates to have close to zero movement, low dielectric constants, and good machinability. It is also quite possible that heterogeneous (mixed-laminate material) printed boards will evolve. This will further complicate laminate usage.

In addition to thinking in terms of "microns," the printed board industry must also adopt more of the manufacturing disciplines associated with integrated circuit fabrication. Most important, this means greater strides toward "clean room" environments and procedures in the work place. Also, employees will have to be apprised of the need for greater skill (or care) levels or, at least, they should have a better understanding of their role in the overall production of "high-tech" products.

In order to achieve these specific and other implied goals, the printed board industry must have thorough knowledge of the overall goals of the electronics industry and increase its expertise so that it is better able to understand and appreciate the objectives of its customers, and vice versa. This will probably lead to a point where ASICs are packaged and interconnected on ASPCs (application-specific printed circuits). Therefore, the overall printed board industry will have to think more along custom and semicustom lines, not only for its end products, but also with respect to its usage of materials, processes, equipment, and personnel.

REFERENCES

1. D. Bursky, "Semicustom ICs—1984 Technology Forecast," *Electronic Design*, January 12, 1984, pp. 206-228.
2. P. VanZant, "Integrated Circuit Fabrication," *Microelectronic Manufacturing and Testing*, October 1985, pp. 16-18.
3. W. Andrews, "Options Grow as Programmable Devices Join Standard Parts & Semicustom Circuits," *Electronic Engineering Times*, September 29, 1986, pp. T4, T43, T46.
4. L. Jack et al., "GaAs or Silicon: Which Should You Use?", *Electronic Engineering Times*, July 14, 1986, pp. T54-T56.
5. C. Brown, "ECL and GaAs Move into the Commercial Mainstream," *Electronic Engineering Times*, July 21, 1986, pp. 37, 44.
6. R. C. Eden et al., "Integrated Circuits: The Case for Gallium Arsenide," *IEEE Spectrum*, December 1983, pp. 30-37.
7. T. M. Reeder and A. G. Rode, "High Speed Systems Look to GaAs for Low Power LSI," *Computer Design*, September 1984.
8. V. Miltinovic, "GaAs Microprocessor Technology," *Computer*, October 1986, pp. 10-13.

9. R. Collett, "ASICs: Take Your Pick," *Digital Design*, June 1986, pp. 29-36.

10. D. Chakravarty, "Semicustom ASICs Solve Costly Problem of Lengthy Design Times," *Electronic Products*, March 1, 1985.

11. J. Vithayathil, "PLDs Are Now a Real Option for ASIC Users," *Electronic Engineering Times*, September 29, 1986, pp. T14-T18.

12. D. A. Laws, "EPLDs as an Alternative to Gate Arrays for Custom Logic Design," *Electronic Engineering Times*, May 12, 1986, pp. T28, T73.

13. H. W. Carter, "Computer-Aided Design of Integrated Circuits," *Computer*, April 1986.

14. D. Bray and P. Irissou, "A New Gridded Bipolar Linear Semi-Custom Array Family with CAD Support," Proceedings of 5th International Conference on Custom & Semi-Custom ICs, November 5-7, 1985.

15. D. Bursky, "Subnanosecond Silicon ECL Gate Arrays Face Challenge from GaAs and CMOS," *Electronic Design*, June 12, 1986, pp. 74-83.

16. S. Dholakia et al., "The Relative Merits of Standard and Compiled Cells," *Electronic Engineering Times*, May 12, 1986, pp. T18, T24.

17. R. Varshney, "Wafer-Scale Integration Graduates to Rank as Viable Technology," *Electronic Engineering Times*, March 3, 1986.

18. C. A. Neugebauer, P. O. Carlson, et al., "Multilevel Interconnections for Wafer Scale Integration," *Journal of Vacuum Science Technology*, November/December 1986.

19. I. Hoglund, "Silicon-on-Sapphire Reconsidered," *Integrated Circuits Magazine*, April 1985, pp. 12-17.

20. J. Handen and E. J. Schmitt, "CMOS/SOS Approach for Rad-Hard Needs," *Electronic Engineering Times*, July 14, 1986, pp. T40-T42.

21. P. V. Cooper and P. M. MacIntosh, "Bubbles on the Rise," *Defense Electronics*, June 1986, pp. 139-149.

22. C. Weisenstein, "Military Floats Use of Bubble Memories," *Electronic Engineering Times*, July 14, 1986, p. T20.

23. R. N. Castellano, "VHSIC Program Spurs U.S. IC Technology, *Defense Electronics*, July 1986, pp. 114-127.

24. O. P. Layden, "VHSIC Chip and Module Packaging Stystus," Proceedings, National Electronic Packaging & Production Conference, February 26-28, 1986.

25. D. P. Johnson and J. Lipman, "IC Packaging: An Introduction for the VLSI Designer," *VLSI Systems Design*, June 1986, pp. 108-115.

26. P. Schroeder, Cray Research, Inc., September 3, 1986, personal communication.

27. J. L. Turino, Logical Solutions Technology, Inc., July 30, 1986, personal communication.

3

Surface Mount Electrical Components

Electronic component packages serve to protect the devices within them from the environment, provide communication links, remove heat, and provide a means for handling and testing [1-3]. With larger, higher-density integrated circuits (ICs), these functions will not only continue to be important, they will be more challenging to meet. At the same time, there will be an emphasis on maintaining or decreasing the costs of the more complex packages. This will require increases in functional density and higher-performance designs as device density continues to improve dramatically (Figure 3.1).

The material composition, finish, and configuration of both the body and the terminations of a component package must be considered in the choice of assembly methods. Table 3.1 contains a summary of the body and termination characteristics of the major classes of components.

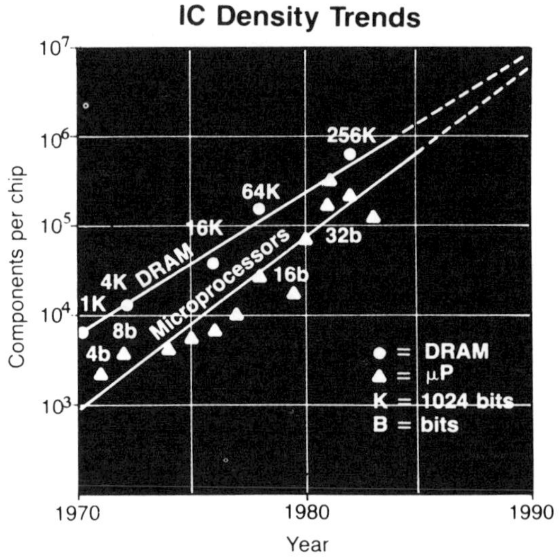

FIGURE 3.1 Active elements per integrated circuit chip.

3.1 PACKAGE DESIGN FACTORS

There are several interrelated package design factors for advanced IC components (Figure 3.2). Some of the interrelated factors impact others very strongly. In the package design stage, one such factor may be emphasized for a given application. Examples are package electrical performance for a high-speed chip, or cost for a lower-performance device. Such trade-offs are common depending on the application. Therefore, the characteristics of each package must be clearly understood in order to make the optimum selection (Table 3.2).

3.1.1 Cost

One must consider, in addition to the package cost itself, the sum of the cost increments of the interfacing factors of handling, testing, insertion equipment repair procedures, inventory, multiple sourcing, and reliability criteria. In other words, the packaging cost factor must be viewed not only in terms of the related design trade-offs, but also in terms of the change in overall system cost.

Other questions may also affect cost. Can packaging influence sales? Are compatible packaging families to be made available for discrete, bipolar, and MOS technologies for uniform board usage?

TABLE 3.1 Component Physical Characteristics

			Termination		
Component type	No. of leads	Body material	Organization	Composition	Finish
Axial leaded	2	Plastic/composition	Linear	Metal lead	Solder
Radial leaded	2	Plastic/composition	Linear	Metal lead	Solder
Chip component	2-4	Plastic/ceramic	Linear	Metallized end cap	Solder/gold/Ag-Pd
				Ribbon lead	Solder/gold
Multileaded radial	3-12	Plastic/metal can	Perimeter	Metal lead	Solder
SOT	3-4	Postmolded plastic	Linear	Formed metal lead	Solder/silver
SOIC	4-28 (40)	Postmolded plastic	Linear	Formed metal lead	Solder/silver
SIP	4-34	Plastic/metal/ceramic	Linear	Metal lead	Solder/gold
DIP	4-68	Postmolded plastic/ceramic	Linear	Formed metal lead	Solder/gold
Ribbon leaded	2-50	Plastic	Perimeter	Metal lead	Solder/gold
Leaded chip carrier	18-164	Plastic/ceramic	Perimeter	Formed metal lead	Solder/gold
Leadless chip carrier	18-164	Plastic/ceramic	Perimeter	Metallized pad	Solder/gold
Pin-grid array	100+	Plastic/ceramic	Area	Metal lead	Solder
Pad-grid array	100+	Plastic/ceramic	Area	Metallized pad	Solder/gold
Connector	100+	Plastic	Linear	Metal lead	Solder
Socket	100+	Plastic	All	Metal lead	Solder

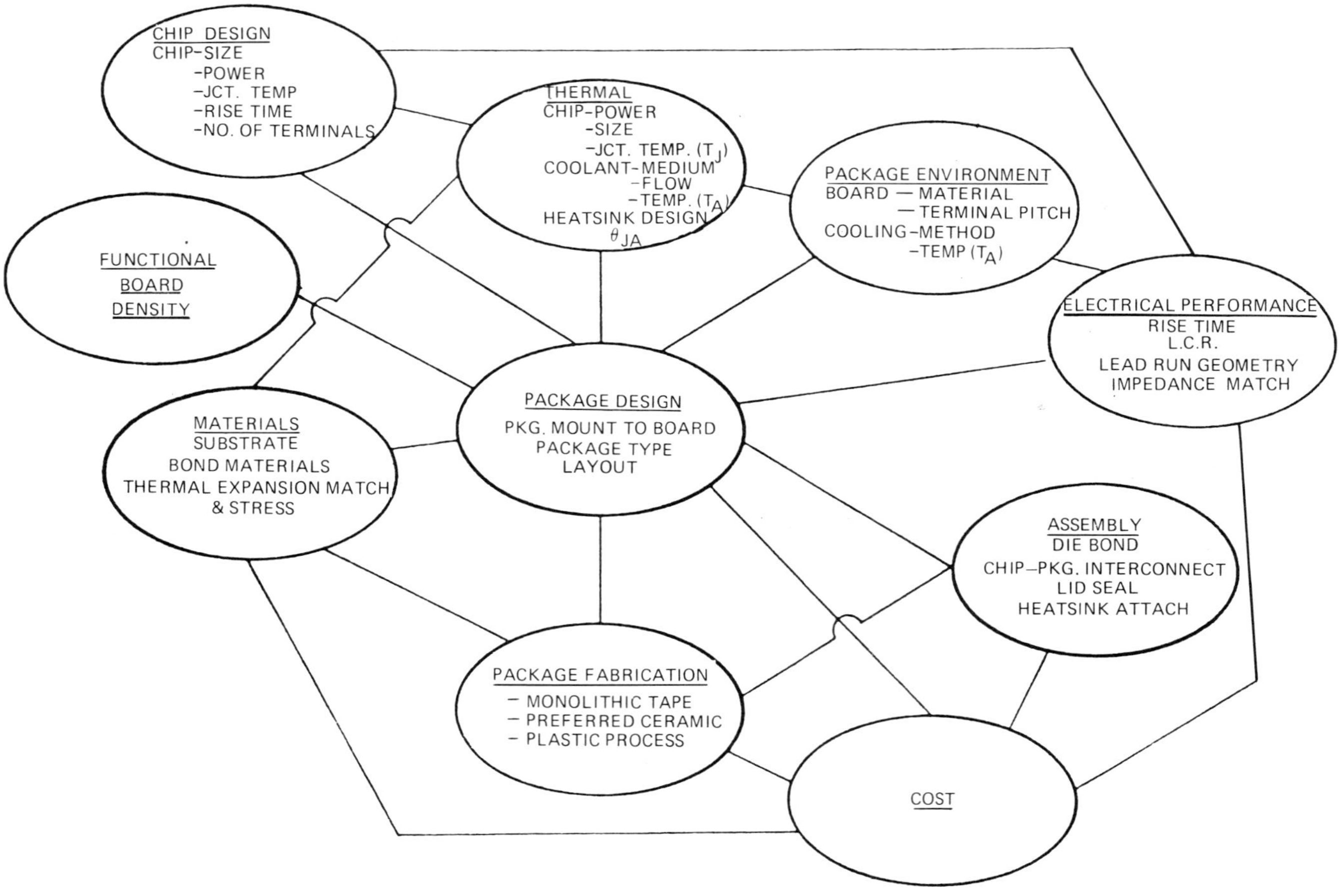

FIGURE 3.2 Interrelated package design factors.

TABLE 3.2 Summary of Package Characteristics (68 Pins)

Package type	Ceramic DIP	Plastic DIP	QUIP	Pin grid	Ceramic chip carrier, leadless (type A)	Plastic chip carrier, leaded (type A)
Relative cost,[a] package only	2	5[b]	3	1	4	6[b]
Relative cost,[a] total, including package, connector, etc.	3	6	1	2	4	5
Lead resistance (R max. in. Ω)	1.1	0.1	0.3	0.1	0.1	0.1
Lead capacitance (C max. in. pF)	7	4	3	2	2	2
Thermal resistance junction to ambient (Θ_{J-A} max. in. $^0C/W$)	35	35	35	35	30	30
PC board area consumption in.^{2} (cm^2)	2.9 (18.7)	2.9 (18.7)	2.1 (13.5)	1.0 (6.45)	1.4 (9.0)	1.0 (6.45)

[a] 1 = most, 6 = least.
[b] Very close.
Courtesy of Motorola, Inc.

Are newer, smaller packages more amenable to automation in handling, testing, and attachment to the substrate?

3.1.2 Electrical Performance

Another driving force in package improvement is system electrical performance. Package size reduction may go hand in hand with lower cost and higher speed due to the use of chip carriers. Also, lead lengths are more uniform in the chip carrier designs than in the small-outline IC (SOIC) configurations.

The cost savings for plastic packages can be substantial. However, for high-power, hermeticity, and maximum reliability, ceramic packages might be used.

3.1.3 Functional Density

Functional density increases have occurred as IC chip complexity and lead count have increased, through reduced lead-to-lead spacing on the device package. As this trend continues, one of the consequences is higher tolerances on the package features and substrate terminal areas.

While not all high-density packages are surface-mounted, many are for good reasons. In addition to those reasons previously mentioned, there is greater conductor routability on the substrate. This increased routability in turn is a necessity when using the higher-density surface-mount packages in order to avoid or minimize fan-out. If the fanout is not controlled, the "footprint" of the package will be large and the desired functional density will not be realized.

Improved functional density can also be achieved by attaching the surface mount components to both sides of the packaging/interconnection (P&I) structure. This obviously will put increasing demands on the fabrication, assembly, and testing processes being used.

3.2 SEMICONDUCTOR PACKAGES [4]

A very-large through-hole-mount DIP can supply the more than 40 pins that most complex IC chips demand. But since 3-in.-long components take up a lot of space, designers are turning to more exotic but more appropriate packages (Table 3.3).

To pass on all of the high performance of large-scale integration (LSI) to the overall system, the package must be capable of holding a 0.350-in. (350 mil) square chip, with lead counts of up to and over 80, while dissipating as much as 5 W of power. Until recently, DIPs

TABLE 3.3 Typical Integrated Circuit Packages

LSI package type	Maximum number of leads	Method of attachment	Removal from board	Board area, including leads (in.)	Hermetic seal
Cofired ceramic DIP	64	Wave solder or socket	Difficult	3.2 × 0.900	Yes
Cerdip	40	Wave solder or socket	Difficult	2 × 0.600	Yes
Plastic DIP	40	Wave solder or socket	Difficult	2 × 0.600	No
Cofired ceramic chip carrier	156 @ 0.050 in.	Socket	Simple	0.460 × 0.430 (40 lead)	No
Leaded chip carrier	156 @ 0.050 in.	Reflow solder	Simple	0.770 × 0.770 (36 lead)	Yes
Minipak	28	Reflow solder	Simple	0.500 × 0.500	No
Flatpack with leads out all four sides	64	Reflow solder	Simple	Approx. the same as leaded chip carrier	Yes
Ceramic substrate with clips on four sides	Up to 156	Reflow solder or socket	Simple	0.650 × 0.650 (44 lead)	Yes
Plastic premolded chip carrier	Up to 156	Reflow solder or socket	Simple	0.650 × 0.650 (44 lead)	No
Leadless inverted device (LID)	40	Reflow solder	Simple	0.450 × 0.450	No
Film carrier	40/64	Reflow solder or wire-bond	Simple	About 0.312 × 0.132 (40 lead)	No

have housed about 95% of all LSIs. However, molded-plastic DIPs are comfortable with no more than 40 leads and only at the lower, metal oxide semiconductor (MOS) levels of power dissipation.

Cofired, heat-sunk ceramic DIPs with up to 64 leads do better. But the rival ceramic chip carrier occupies about one-third the space and also degrades chip performance less because its lead resistance is lower.

Ease of manufacture and handling is the rationale for using the tape-automated bonding (TAB) film carrier. Strictly speaking, however, the TAB film carrier is not a package at all. It is a series of micro-interconnections, usually etched from the copper surface on a tape, to which the IC chips are automatically bonded. Then reels of the tape with chips attached can be automatically bonded to the substrate or directly enclosed in a package.

3.2.1 Dual-Inline Packages (DIPs) [5]

Four significant factors have, in large part, caused the popularity of the DIP (Figure 3.3):

1. The DIP has served both electrical and mechanical constraints of the industry well throughout its 20 years of use.
2. Pin counts (I/Os) have ranged from 8 through 40 leads. These are far and away the most popular pin counts, with over 75% of

FIGURE 3.3 Forty-lead dual-inline package. (Source: Motorola, Inc.)

the devices built being housed in these types with 14- and 16-lead pin configurations.

3. Intensive capital investments have been made in all types of handling, assembly, and test equipment based on the DIP package.
4. DIPs are readily pluggable, with easy socketing for testing and maintenance.

Still, what of the future? Can DIPs remain the standardbearer and still meet tomorrow's needs? Examining the new generations of ICs, it becomes readily apparent that the DIP cannot function as well for input-output (I/O) counts above 48 (Figure 3.4). As semiconductor manufacturers achieve submicron feature size in chip geometries and package hundreds of thousands of active devices on a chip with over 60 I/Os, the DIP will not make it.

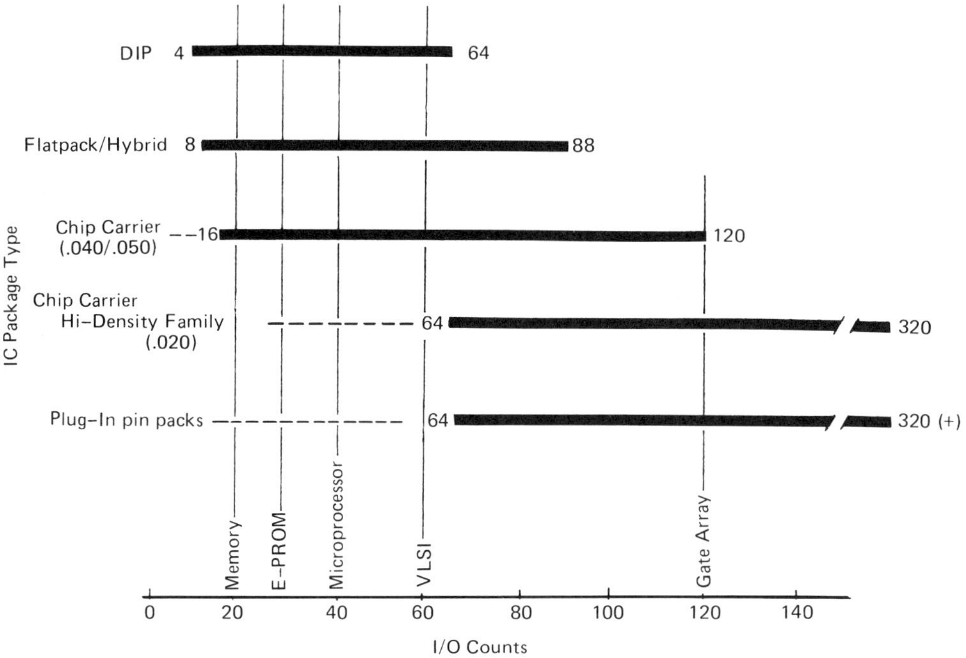

FIGURE 3.4 I/O interconnection range for various integrated circuit package types.

3.2.2 Chip Carriers [6]

Chip carriers can be generally described as being low-profile, rectangular (usually square) packages with I/O connections on all four sides. The I/O connections consist of metallized terminations on "leadless" versions and leads formed around or attached to the side of the package in the "leaded" versions. General selection considerations for these packages are listed in Table 3.4.

Recognizing that no one outline can satisfy all semiconductor packaging requirements, the Joint Electronic Device Engineering Council (JEDEC) has established a standard for chip carriers (CCs) that allows for multiple design approaches, manufacturing techniques, and attachment means, thus enabling designers to choose and tailor packages to their applications. The JEDEC has standardized two basic package styles (one with 0.050-in. center terminal spacing and another with 0.040-in. spacing) and provided interchangeability within

TABLE 3.4 Chip Carrier Application Considerations

Consideration	Leadless CCs	Leaded CCs
Thermal expansion to match	Critical	Less critical
Removal and replacement	Comparatively easy with special tools	Less risk of damaging
Solder joint inspection	Difficult	Less difficult
Flux removal after soldering	Difficult	Less difficult
Socket compatible	Yes (except type C)	Yes (except type B)
Lead length	Minimal	Moderate (inductance greater)
Condutive cooling	Good, with direct lead conduction path (lower profile height)	Poor (higher profile height)
Preparation for soldering	Solder coating of terminals required	None except for solder coating as required for solderability
Self-centering	Usually	Rarely
Flexure of substrate	Critical	Less critical

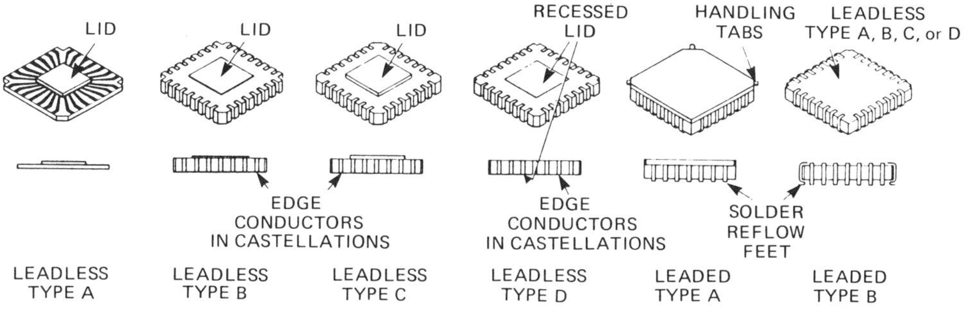

FIGURE 3.5 The 50-mil center JEDEC packages.

similar terminal spacing outlines, so that designers can interchange types to suit changing requirements without redesigning the interconnecting assembly.

The original 50-mil (0.050-in) center family contained six variations (Figure 3.5). Four leadless types, A (Figure 3.6), B, C (Figure 3.7), and D, mount in different orientations depending on the type, mounting substrate, and preferred thermal orientation. These packages are ceramic, with hermetically sealed metal or ceramic lids.

FIGURE 3.6 Leadless ceramic chip carriers: (left) type A, 50-mil centers/69-I/O; (center) 25-mil centers/132-I/O; (right) 20-mil centers/156-I/O.

FIGURE 3.7 Leadless type C chip carrier.

The leaded type A category comprises two standards. The first is a premolded plastic version (Figure 3.8) that is designated MS006. The second leaded type A package, MS007, has a postmolded plastic body (Figure 3.9). When the leadless types mount with substrate clips (Figure 3.10), they become the leaded type B package.

The differences among leadless-type chip carriers center on internal features and mounting orientations with respect to substrate and thermal performance. The latter primarily concerns designers who can choose a chip carrier based on its cavity-up or cavity-down thermal orientation. These terms define the cavity position and the back side of the die, which serves as the package's primary heat transfer or heat removal surface.

The cavity-up configuration places the heat-radiating surface down, next to the mounting surface. This arrangement limits the heat transfer effectiveness of the forced-air cooled systems and the attachment of heat sinks, but it does allow for a larger cavity size and thus a larger die.

The cavity-down orientation allows heat sinks to be mounted on the primary heat-radiating surface but, as noted, results in a smaller cavity size. Designers must make the trade-off—thermal performance/die orientation versus cavity size--for each application.

Although most of the industry is concentrating heavily on the 50-mil center chip carriers, the 40-mil center devices were the family's forerunners, and today they are part of the JEDEC standards. The

FIGURE 3.8 Premolded plastic-leaded chip carrier.

FIGURE 3.9 Postmolded plastic-leaded chip carrier.

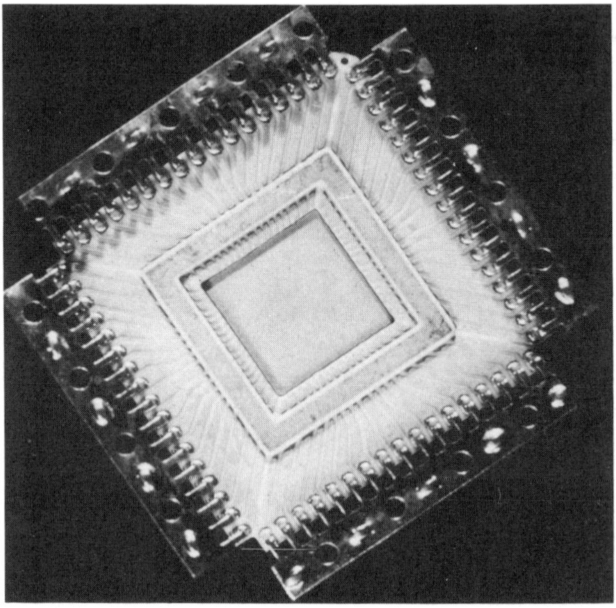

FIGURE 3.10 Leaded type A chip carrier with clip leads.

40-mil center chip carriers have been and continue to be applied mainly in military programs, where the need for the highest possible packaging density rates high on the requirements list. Thus, specifically for military-oriented users, the 40-mil center spacing device was defined and registered and carries JEDEC's standard number MS009.

Turning again to the 50-mil center family, note that all of these devices mount on a common pattern. It is therefore possible to attach them directly to the mounting substrate or put them in a socket.

The 50-mil leadless type A package, when used in the cavity-down configuration, must be socketed on both ceramic and printed wiring substrates because of its package lid. The other leadless types, however, offer the choice of socketing or soldering on ceramic substrates. The leaded type A, on the other hand, is unusual in that it can be soldered or socketed directly onto either mounting substrate, as opposed to the leaded type B, which requires soldering because its lead configuration does not permit socketing.

Although the question of soldering or socketing might present some trade-off considerations, there is little need for concern about the number of pinouts available. There is a variety of sizes in the 50-mil center family, including 28-, 44-, 52-, 68-, 84-, 100-, 124-, and 156-input/output terminal sizes (Table 3.5). The family also includes smaller terminal counts, and the 40-mil center family provides smaller sizes and additional terminal counts.

Note that the chip-carrier concept can apply to multichip hybrid circuitry as well as monolithic dice. The chip carrier provides an alternative approach to high-density designs that hybrids have traditionally serviced. A chip carrier/hybrid can be assembled using conventional reflow soldering techniques and, because the chip carriers attach to the substrate by this method, the rework operation basically consists of a heat, remove, replace, and reflow operation. Thus, the temperature excursion experienced by the chip carrier/hybrid is limited to the reflow temperature of the solder, which minimizes temperature-induced damage to the circuit.

Several semiconductor companies have outlines for a package that is identified as leadless type E. This rectangular configuration is specifically targeted for memory applications. The pinout versions

TABLE 3.5 JEDEC Package Terminal Sizes

No. of terminals	Leadless type A	Leadless type B	Leadless type C	Leadless type D	Leaded type A	Leaded type B	0.040-in. family
16	-	-	X	-	-	-	X
20	-	-	X	-	-	-	X
24	-	-	-	-	X	-	X
28	X	X	X	X	X	X	-
32	-	-	-	-	-	-	X
40	-	-	-	-	-	-	X
44	X	X	X	X	X	X	-
48	-	-	-	-	-	-	X
52	X	X	X	X	X	X	-
64	-	-	-	-	-	-	X
68	X	X	X	X	X	X	-
84	X	X	X	X	X	X	X
96	-	-	-	-	-	-	X
100	X	X	-	X	X	X	-
124	X	X	-	X	X	X	-
156	X	X	-	X	X	X	-

X = included in the standard.
\- = not included.

for the type E carrier are designed to be compatible with the popular leadless type C package. The 28-pin 0.350- × 0.550-in. outline and the 32-pin 0.550- × 0.450-in. outline are intended for ROMs, PROMs, EPROMs, DRAMs, and static RAMs.

In addition to the type E designs, there are also leadless type F units for memory chip packaging. Although they too have a rectangular outline and will house memory dice, their thinner configurations of 0.285 × 0.350 in. and 0.285 × 0.425 in. allow for mounting on a conventional 300-mil center ceramic-DIP carrier, thereby minimizing chip carrier solderability problems while retaining chip carrier density advantages. These devices are furnished with 18 terminals (Figure 3.11).

The use of chip carriers for the packaging of LSI and VLSI circuits with high terminal counts has several limitations. Therefore, the JEDEC group is planning to standardize on both 0.020-in. and 0.025-in. centerline chip carrier packages (Figure 3.6) for these applications.

While there are numerous types of chip carrier packages and packaged devices, it should be noted that there are two basic categories: plastic and ceramic. These basic material categories have substantially different properties, and some of the associated critical material factors will be discussed later.

In addition, there are several critical electrical properties of chip carrier packages based both on materials and on the configurations of chip carrier packages, as compared to DIPs, for instance. These include electrical loss properties and electromigration. Thermal expansion is the major mechanical property to be considered.

3.2.3 Flatpacks

Flatpacks are the oldest type of integrated circuit package for surface mount applications, with their original use dating back to the early 1960s. The typical dual-row flatpack has leads on both sides of its body on 0.050-in. centers. They are supplied with from 14 to 50 flat ribbon leads. Therefore, the dual-row flatpack's leads are usually formed prior to assembly in a "gullwing" fashion for surface-mount applications.

A new leaded integrated circuit package that is gaining in popularity is the plastic "quad" flatpack (PQFP). The PQFP (Figure 3.12) is a high-density/low-cost JEDEC-approved package for lead counts from 52 to 244 terminals. It utilizes a small 0.025-in. lead spacing to achieve high density and a small footprint. The PQFP also features a gullwing leadform and molded "bumpers" in the corners to protect the leads from mechanical damage.

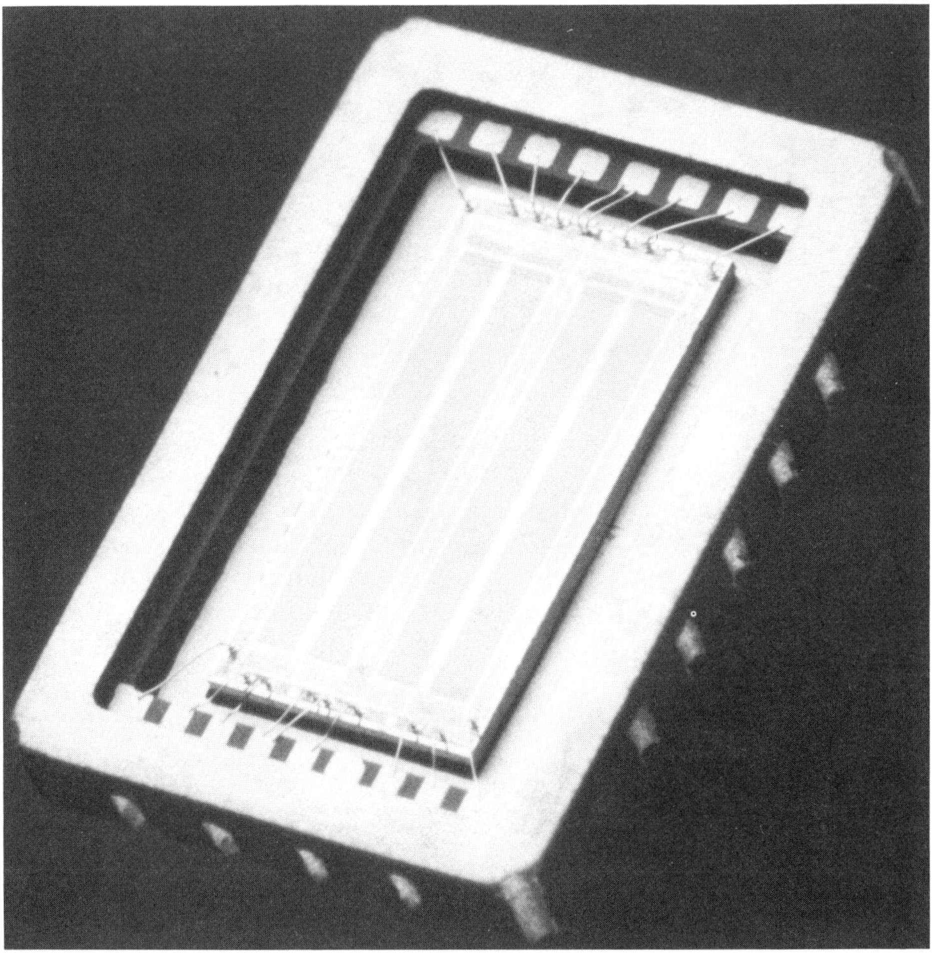

FIGURE 3.11 Leadless type F chip carrier.

3.2.4 Grid Arrays [7]

The higher the I/O lead count becomes, the lower the percentage of the total package area that any given size die cavity will occupy. (For example, the 96-lead, 40-mil center chip carrier is over 1 in. square.) For optimum packaging density, this percentage should

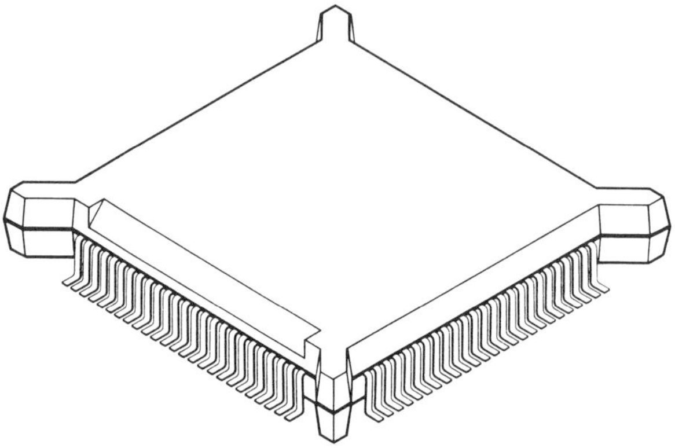

FIGURE 3.12 Plastic quad flatpack (PQFP).

be as high as possible. Therefore, for high-I/O applications, 0.025- and 0.020-in. chip carriers (Figure 3.6) and grid arrays are used. Grid arrays are usually through-hole-mount integrated circuit packages, i.e., pin-grid arrays, with I/O contacts that populate one surface of the package on a 0.100-in. grid. (For special applications, 0.050-in. centers pin-grid arrays and leadless grid arrays are being developed.) The use of a solid grid necessitates placing the die cavity on the side opposite the I/O contacts (Figure 3.13). A double or

FIGURE 3.13 "Cavity-up" pin-grid array package.

multiple concentric row grid permits having die cavity and I/O contacts on the same side with an optional heat sink on the opposite side (Figure 3.14).

Figure 3.15 shows the I/O density achievable with the various packages. As can be seen, for lead count requirements in excess of 100, a 0.100-mil centerline solid grid provides greater I/O density than either 50- or 40-mil centerline chip carriers.

The 25-mil centerline chip carrier is clearly more I/O efficient than is a 0.100-mil grid array. However, a double-row 50-mil grid array is equivalent to a 25-mil centerline chip carrier. The 50-mil solid grid provides the greatest I/O density.

FIGURE 3.14 "Cavity-down" pin-grid array package.

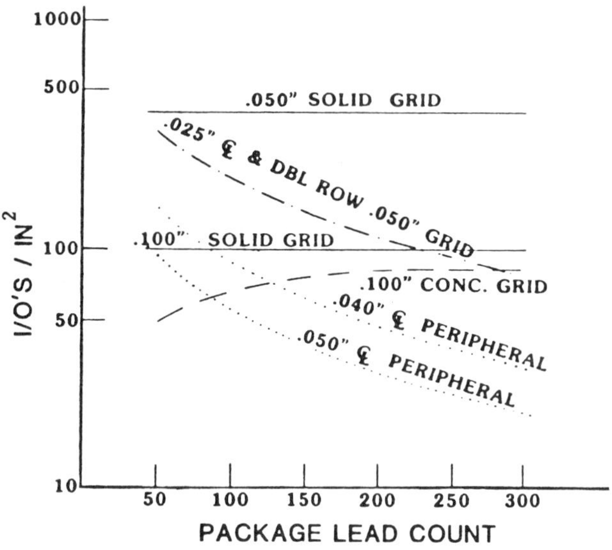

FIGURE 3.15 Integrated circuit package I/O density versus lead count.

With pins brazed on 100-mil centers, pin-grid arrays are a logical carrier for high lead counts intended for through-hole (not surface-mount) board assembly that is compatible with the use of DIPs. This should serve to extend the useful life of present-day technology for some high-lead count applications.

3.2.5 Small-Outline (SO) Packages [6, 8]

With all of the recognition being given to chip carriers, another entry in the microminiature packaging arena, the small-outline (SO) package, is finding applications. Resembling miniature versions of molded plastic DIPs, SO transistors (SOTs) (Figure 3.16) and 0.050-in. center SO integrated circuits (SOICs) with up to 40 terminals (Figure 3.17) are quickly gaining acceptance.

The primary advantage of the SO package is its small size (Figure 3.18). The SO package occupies an area some 30-50% less than an equivalent DIP, with a typical thickness that is 70% less. Consequently, assemblies that incorporate SO devices can serve in applications where space and weight are at a premium (Table 3.6).

FIGURE 3.16 Typical SOT 143 transistor packages. (Source: Siemens Components, Inc.)

3.3 DISCRETE CHIP COMPONENTS [9]

Discrete chip components (Figure 3.19) are basically miniature axial-leaded components minus the leads. They are available in both cylindrical (Figure 3.20) and flat rectangular shapes for resistors (Figure 3.21), capacitors (Figure 3.22), and diodes.

FIGURE 3.17 Small-outline "J-lead" integrated circuit.

Some of the factors promoting interest in chip components are:

Insertion equipment manufacturers are claiming insertion rates in excess of 250 chips in 4 sec (3750/min).

Increased component density and mounting techniques are compatible with chip carrier technology.

Chips often sell in a price range that is comparable to (or less than) that of axial-leaded components.

Since there are no leads, basic circuit reliability is improved.

Chip components are ideal for high-frequency applications.

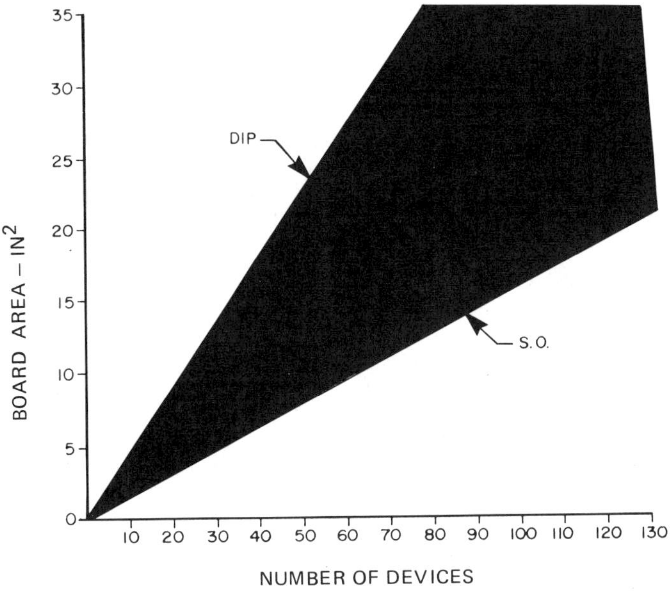

FIGURE 3.18 Board area utilization for DIPs versus small-outline integrated circuits.

TABLE 3.6 Dual-Inline Package and Small-Outline Integrated Circuit Comparison

Item	8-Pin SO	8-Pin DIP	16-Pin SO	16-Pin DIP	28-Pin SO	28-Pin DIP
Body size, L × W, mm^2	20	70	40	140	140	500
	(3.5)		(3.5)		(3.5)	
PC board area, mm^2	31	80	62	175	192	590
	(2.6)		(2.8)		(3.0)	
Body thickness, mm	1.45	3.1	1.45	3.6	2.45	3.9
Height above board, mm	1.75	4.2	1.75	5.1	2.65	5.1
Weight, mg	60	600	130	1200	700	4300
	(10)		(9)		(6)	

Ratios are shown in parentheses ( ).

FIGURE 3.19 Discrete chip components.

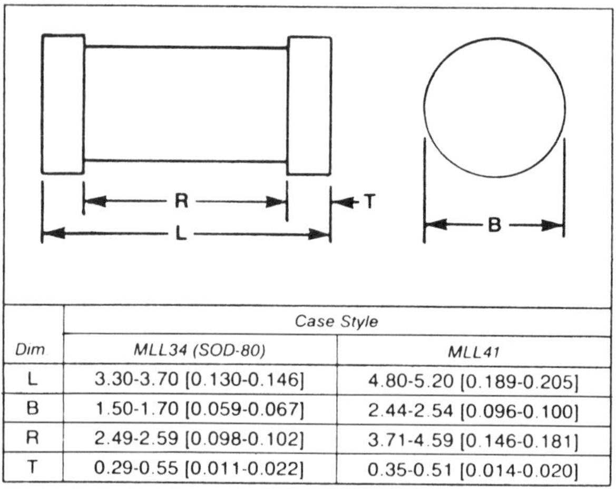

Dim	Case Style	
	MLL34 (SOD-80)	MLL41
L	3.30-3.70 [0.130-0.146]	4.80-5.20 [0.189-0.205]
B	1.50-1.70 [0.059-0.067]	2.44-2.54 [0.096-0.100]
R	2.49-2.59 [0.098-0.102]	3.71-4.59 [0.146-0.181]
T	0.29-0.55 [0.011-0.022]	0.35-0.51 [0.014-0.020]

FIGURE 3.20 Metallized electrode face (MELF) component dimensions, mm [in.]. Note: All dimensions apply after solder coating.

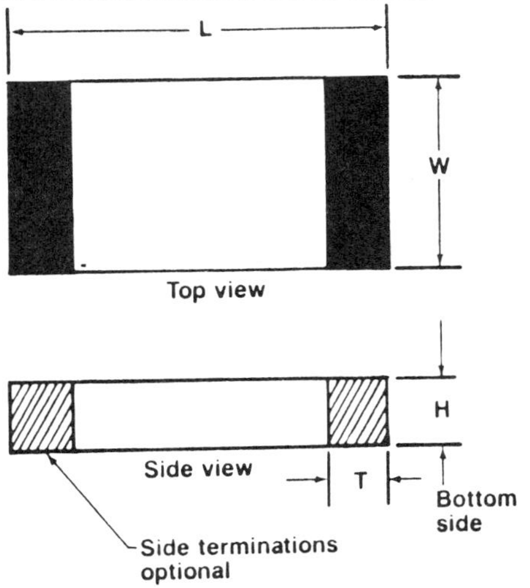

Resistor Number	L	W	H	T[a]
RC0805	1.8-2.2 [.070-.087]	1.0-1.4 [.040-.055]	0.3-0.7 [.012-.028]	0.3-0.6 [.012-.024]
RC1206	3.0-3.4 [.118-.134]	1.4-1.8 [.055-.070]	0.4-0.7 [.016-.028]	0.4-0.7 [.016-.028]
RC1210	3.0-3.4 [.118-.134]	2.3-2.7 [.090-.106]	0.4-0.7 [.016-.028]	0.4-0.7 [.016-.028]

[a] Dimensions have been tailored to facilitate standard land pattern design

FIGURE 3.21 Chip resistor dimensions, mm [in.]. Note: All dimensions apply after solder coating.

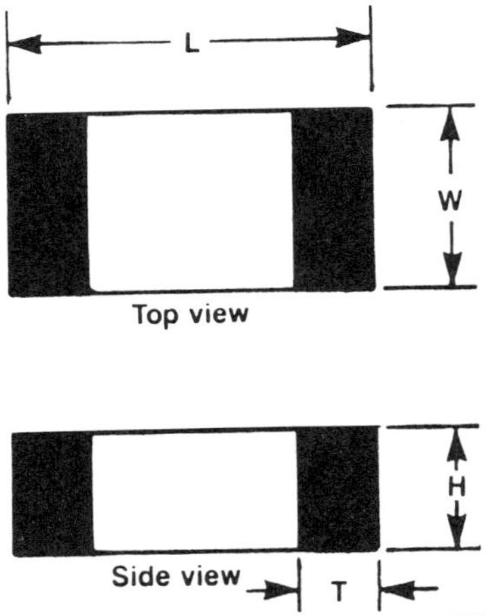

Capacitor Number	L	W	H max	T[a]
CC0805	1.8-2.2 [.070-.087]	1.0-1.4 [.040-.055]	1.3 [.05]	0.3-0.6 [.012-.024]
CC1206	3.0-3.4 [.118-.134]	1.4-1.8 [.055-.070]	1.5 [.06]	0.4-0.7 [.016-.028]
CC1210	3.0-3.4 [.118-.134]	2.3-2.7 [.090-.106]	1.7 [.067]	0.4-0.7 [.016-.028]
CC1812	4.2-4.8 [.166-.190]	3.0-3.4 [.118-.134]	1.7 [.067]	0.4-0.7 [.016-.028]
CC1825	4.2-4.8 [.166-.190]	6.0-6.8 [.236-.268]	1.7 [.067]	0.4-0.7 [.016-.028]

[a] Dimensions have been tailored to facilitate standard land pattern design

FIGURE 3.22 Ceramic capacitor dimensions, mm [in.].

REFERENCES

1. Printed Board Component Mounting, IPC-CM-770, Rev. C, Institute for Interconnecting and Packaging Electronic Circuits, January 1987.

2. E. R. Winkler, "Integrated Circuit Packaging Trends," *Solid State Technology*, June 1982, pp. 94-100.

3. W. L. Morgan, "From VLSI to ULSI," *Semiconductor International*, May 1984.

4. J. Lyman, "Growing Pin Count Is Forcing LSI Package Changes," *Electronics*, March 17, 1977, p. 81.

5. F. C. Rydwansky, Jr., "Chips, DIPs and PWBs: Semiconductor Packaging and Interconnection of the Future," *Circuits Manufacturing*, September 1981, pp. 54-62.

6. J. Tsantes, "Technology Update: Leadless Chip Carriers Revolutionize IC Packaging," *EDN*, May 27, 1981, pp. 49-74.

7. B. M. Hargis and D. J. Westervelt, "High Lead Count Packaging and Interconnection Alternatives," *Electronic Packaging and Production*, September 1981, pp. 82-89.

8. J. Resendes, "Production Techniques with S.O. Packages," *Electronic Packaging and Production*, March 1982, pp. 115-128.

9. B. Winkelmann, "The New Chip Resistors," *Appliance*, June 1981.

4

Surface Mount Electromechanical Components

For the most part, the impact of surface mount technology has been felt at the discrete component level (chip resistors and capacitors) and at the integrated circuit level (chip carriers and small-outline integrated circuits) [1]. However, as the industry becomes more familiar with the technology, other components, such as switches and inductors, will be made available in surface mount form. Not to be left out are electromechanical components, such as printed wiring connectors and sockets.

Developments in this area reflect the general approach being taken throughout the electronics industry with respect to surface mounting. For the majority of users, the full impact of surface mount technology will be realized over several years, during which time there are likely to be three distinct stages of implementation: near term, medium term, and long term.

Near term: Users will retain their existing through-hole mount technology (especially connectors) where it is most cost- and density-effective, and gradually implement surface mount components and technology only in those areas where they are most effective for a few high-density VLSI devices.

Medium term: The switchover to surface mounting will be incorporated to the fullest extent possible, while retaining compatibility with existing through-hole mount connector backplanes and other interconnection hardware. This will minimize the impact on existing equipment designs and allow for the gradual replacement of through-hole mounted assemblies by their surface mount counterparts.

Long term: During this period, every effort will be made to take the fullest cost and density advantages that are possible with surface mount technology, including the use of connectors designed especially for these applications.

The timetable for each of these three stages will vary greatly, depending on the size of the company and the degree to which it is willing to make the investment needed to effectively implement surface mount technology.

4.1 CONNECTORS, GENERAL

In the near-term scenario, the emphasis is on the continued use of through-hole mounted connectors with a higher number of contact positions and lower insertion and withdrawal forces per mated pair of contacts.

4.1.1 Connector Housing Materials

Housing materials (Table 4.1) are usually flame-retardant, glass-filled thermoplastics or thermoset plastics that are able to withstand the high-temperature exposure associated with vapor-phase (condensation) soldering.

4.1.2 Contact Base Materials

For cost effectiveness, the connectors usually have either phosphor bronze or brass contacts (Table 4.2), which are available in a choice of platings, including tin, gold over nickel, and selective gold.

4.1.3 Mechanical Attachment

Attachment options are offered for positive mechanical retention (Table 4.3) with rivets or hardware or for interim mechanical or adhesive attachment during processing.

4.2 RIGID PRINTED-WIRING BOARD CONNECTORS

4.2.1 Horizontal-Mount Connectors [3]

It is the medium-term situation that is receiving the most attention from connector manufacturers. As an entry into the surface mount market, some of them are offering lines of surface-mounted, horizontal board-edge connectors in single- and double-row configurations (Figure 4.1), which mate with conventional square-post headers. These board connectors are usually designed so that the contact floats in the housing, allowing contact between its terminals (legs) and the board's solder lands. This feature compensates for board warpage and irregularities. Single-row surface-mounted connectors are usually supplied with from 2 to 65 positions, and double-row connectors have from 4 to 130 positions.

4.2.2 Eurocard Connectors [4]

For other medium-term applications, some connector manufacturers are taking a slightly different approach. To take full advantage of the popularity of DIN (Deutsches Institut für Normung) Eurocard standardization, they have developed surface mount versions of DIN specification 41612 connectors (Figure 4.2). These connectors provide all the features of surface mount in a true "reverse DIN" connector.

4.2.3 Double-Sided Mount Receptacle Connectors

Others are taking still another approach (Figure 4.3). They also offer a female socket for surface mount applications, however, in this instance, the edge-mounted sockets are tin-lead (solder) plated for ease of termination. They also contact both sides of the board to effectively minimize the use of plated-through holes. Low-cost, high-

TABLE 4.1 Typical Connector Body Insulator Materials [2]

	Units	ASTM test method	Crys- Poly-phenylene sulfide
Mechanical			
Tensile strength	kpsi	D-638	17.5
Elongation at break	%	D-638	0.9%
Flexural modulus	Mpsi	D-790	1.70
Tensile impact strength	ft-lb/in.(E2)	D-1822	9
Izod impact strength, 1/8 in.	ft-lb/in.	D-256	
Notched			1.3
Unnotched			4.5
Thermal			
Coefficient of thermal	in/in F	D-696	1.2E-5
Heat deflection lamp			
at 264 psi	F	D-648	500
	C		260
at 64 psi	F	D-648	
	C		
Limiting oxygen index	%	D-2863	46.5
UL			
Thermal ind., w/impact	C	UL-764B	200
UL94 flammability		UL-94	V-O
Thickness	inches		0.028
Electrical			
Dielectric strength, 1/8 in.	V/mil	D-149	450
Short time			363
Step-by-step			383
Diel. constant, 60 Hz		D-150	3.9
1 MHz			3.8
Dissipation factor, 60 Hz		D-150	0.0006
1 MHz			0.0014
Arc resistance	sec	D-495	38
Volume resistivity	Ω-cm	D-257	4.5E16
Surface resistivity	Ω	D-257	
General			
Specific gravity		D-1505	1.60
Mold shrinkage, MD	mil/in.	D-955	2
TD	mil/in.		10

Thermoplastics				Thermosets	
talline	Amorphous		Cryst./ amorp. blend		
Polyethylene-terephthalate	Poly-etherimide	Poly-ether-sulfone	Modified polyarylether	Diallyl phthalate	Phenolic
22.0	24.5	18.0	15.9	9.0	9.0
2.3%	2.7%	3.0%	1.5%		0.2%
1.50	1.20	0.86	1.34	1.80	1.80
			16		
1.6	2.0	1.5	1.0	1.0	0.9
	8.0	8.0	7.8		
1.1E-5	1.1E-5	1.5E-5	1.5E-5	1.5E-5	1.3E-5
435	410	420	293	450	500
224	210	216	145	232	260
	414		381	450	500
	212		194	232	260
33		40		50	
150	170	180	150	130	150
V-O	V-O	V-O	V-O	V-O	V-O
0.032	0.063	0.017	0.032	0.058	0.067
430		375	455	400	475
			419	400	425
3.6			3.8	4.1	5.6
3.6		3.5	3.7	4.0	5.2
0.0040			0.0089	0.0090	0.0290
0.0100		0.0060	0.0086	0.0130	0.0250
117	85		125	130	180
1.0E15	3.0E16	2.0E16	1.6E16	1.0E15	1.0E9
1.0E14			2.1E16		
1.67	1.51	1.51	1.47	1.90	1.73
2	2	3	2	2.4	1.5
9			5		

TABLE 4.2 Typical Contact Base Materials [2]

Alloy and temper	C72500 Hd. temper	C51000 Hd. temper	C51000 Ex-Hd. temper	C17410
Tensile strength (KSI)	75-90	76-91	88-103	110-130
0.2% yield strength (KSI)	74-79	74-88	85-102	95-125
Elastic modulus (MSI)	20	16	16	19
Electrical conductivity (% LACS)	11	15	15	45-60
Formability (R/T) bad way bend 90°	3.0	2.0	4.0	6.0
Stress relaxation—remaining stress (%) 1000 hr @ 105°C 75% of 2% Y.S.	85	80	75	>95
Stress relaxation—1000 hr @ 200°C	c	c	c	80
Stress relaxation—1000 hr @ 225°C				

[a]Aged 700°F/2 hr.
[b]Bulk conductivity with 25% CU.
[c]Not recommended at 200°C.

reliability versions of these connectors are available in sizes from 36 to 140 contact positions, with either 0.1 in. × 0.1-in. or 0.1 in. × 0.2-in. mounting patterns.

Four-row versions have also been developed. Mechanical stability of the socket is achieved through the use of a slot in the socket that engages the printed wiring board. The design uses minimal board area for both mounting and soldering in order to be compatible with the high-density objectives usually associated with surface mount applications.

4.2.4 Vertical-Mount Connectors [5]

For more versatility, vertically mounted, low-profile headers (Figure 4.4) are available that provide interconnections between printed wiring

C17200 1/4 HM temper	C17200 1/2 HM temper	C17200 HM temper	CU/SS/ CU HM temper	C72900 TMO2 temper	C72900 TM04 temper	C72900[a] TSO2 temper
110-120	120-135	135-150	145-160	105-125	115-135	145-175
90-110	100-125	115-135	135-150	90-110	105-125	135-165
19	19	19	25	19	19	19
20-28	20-28	20-28	22-25[b]	8	8	8
1.0	2.0	3.0	<0.3	Sharp	0.5	0.5 before age
>95	>95	>95	80	>95	>95	>95
—	30	—	70	90	90	90
				80	80	80

boards and flat-ribbon insulation-displacement cables (IDC) using surface attachment. The header pins are 0.025-in. square and are on a 0.1-in. × 0.1-in. board termination grid. They are available in single-row configurations with from 2 to 65 positions and in double-row configurations with from 4 to 130 positions. The flat leads are formed to ensure intimate contact to the board surface. The solder-joint interface between the connector and the board is fully visible for ease of inspection and repair. Keying provisions are also available to mate with conventional female socket keying arrangements.

4.2.5 Double-Sided Mount Plug Connectors [6]

Another type of surface mount connector (Figure 4.5) totally replaces traditional printed wiring board gold-plated tab contacts at a reportedly

TABLE 4.3 Surface Mount Connector Mechanical Attachment Methods [3]

	Nut and bolt	Rivet	Metal clip	Plastic clip	Press fit	Heat stake	Heat stake insert	Press fit insert	Bottom up	Two piece
Alignment during assembly	–	–	–	–	+	+	+	+	+	–
Compresses leads to board	+	+	–	–	+	+	+	+	+	+
Prevents board bow	+	+	+	–	–	+	+	–	+	+
Prevents rocking	+	+	–	–	+	+	+	+	+	+
Prevents unmating stress	+	+	–	–	+	+	+	+	+	+

Permits thermal expansion	+	−	+	+	−	−	+	+	+	+
0.100 ± 0.003 hole	+	+	+	+	+	+	+	+	+	−
Mounting ears required	−	−	−	−	+	+	−	−	+	−
Easily repairable	+	−	+	+	−	+	+	−	+	+
Weakened by heat	+	+	+	−	−	−	−	−	−	−
Noninductive	−	−	−	+	+	+	+	+	+	+
Suitable for hand assembly	+	−	−	+	+	−	−	+	+	+
Automation potential	−	+	−	−	+	+	+	+	+	−
Assembly tooling needed	−	−	−	−	+	−	−	+	−	−
Total +	9	7	5	5	10	10	10	10	12	8

FIGURE 4.1 Horizontal-mount connector. (Source: du Pont Connector Systems)

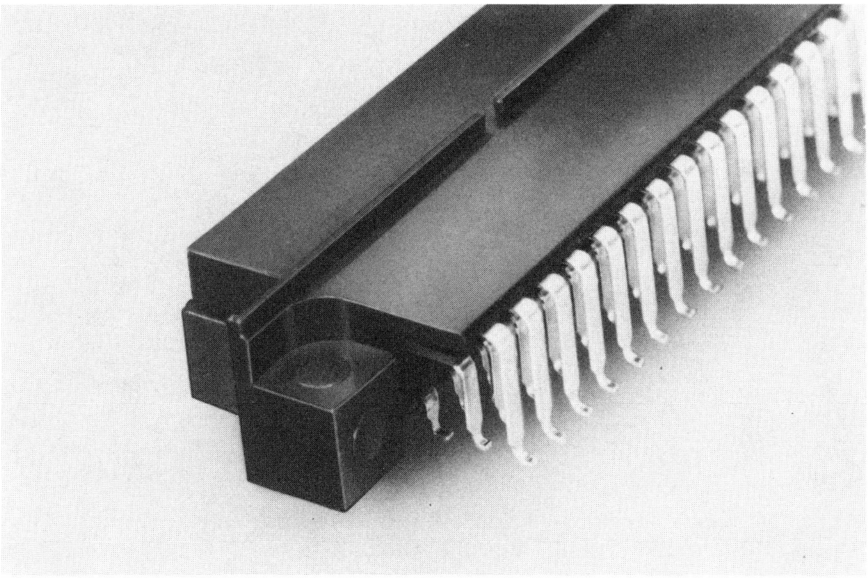

FIGURE 4.2 Surface mount Eurocard connector. (Source: Burndy Corp.)

significant reduction in overall applied cost. The connector is simply installed on the board edge where, after surface mount soldering, it mates with conventional one-part (card-edge) connectors.

4.2.6 Mid-Board Mount Connectors

For non-plug-in applications, a connector with a unique compression terminal design (Figure 4.6) mounts to the board with a snap-action mechanical retention. The mid-board connector thus makes contact to printed wiring board leads while providing a gas-tight connection. It is available for insulation-displacement cabling for rapid mass termination with from 4 through 12 positions for use with 22 through 30 AWG wire.

4.2.7 Zero-Insertion-Force Connectors [7]

New connector types have been developed for long-term equipment designs that are completely oriented toward surface mounting. One such connector type (Figure 4.7) has been designed to satisfy the requirements for high pin counts, zero-insertion-force (ZIF) mating, and termination-to-multilayer printed wiring boards.

FIGURE 4.3 Double-sided mounting receptacle connector. (Source: Thomas & Betts Corp.)

FIGURE 4.4 Vertical-mount IDC shrouded header. (Source: AMP, Inc.)

FIGURE 4.5 Double-sided mounting plug connector. (Source: Augat/Elfab)

The 300-contact version of the connector is 9.3-in. long and has five interchangeable segments of 60 contacts each. Each segment of the connector has an integrally molded locating pin that helps to assure positive registration on the printed wiring board. The contacts are arranged in two rows of 0.050-in. center spacing. Two of these connectors can be mounted end to end to provide a 600-contact capability.

The connector is surface-mounted to solder leads on the board with mounting rails. This permits more space for high-density conductor routing on the board (since no plated-through holes are required for connector terminations), ease in assembling connectors to backplanes, field repairability, and simplified backplane fabrication. A screwdriver is the only tool required to install or remove a connector.

FIGURE 4.6 Mid-board surface mount connector. (Source: Methode Electronics)

4.2.8 Very-High-Density Connectors [7]

A new connector type (Figure 4.8) features very high density with 318 contacts in a 5-in.-long housing. This card-edge connector is possibly the highest-density surface mount connector available. The connector is made up of two sections, each containing 159 contacts. The contacts are located on a 0.050-in. pitch in a two-row staggered layout, thus providing an effective 0.025-in. center-to-center spacing.

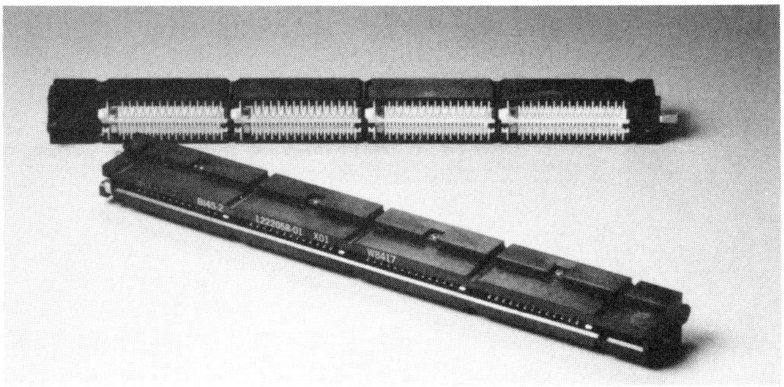

FIGURE 4.7 240-contact ZIF surface mount connector. (Source: Burndy Corp.)

The four rows of contacts are gold-plated to provide optimum compatability with the opposing gold-plated contacts on the plug-in (daughter) board. The slot between the two assembled halves of the connector is adjustable, thereby permitting a plug-in board of any thickness to be inserted between them.

4.3 FLEXIBLE CIRCUIT CONNECTORS

4.3.1 Flexible Circuit/Rigid Board Connectors

A unique approach to providing versatile, high-density, non-plug-in connections to printed wiring boards (Figure 4.9) from flexible

FIGURE 4.8 Very-high-density surface mount connector. (Source: Burndy Corp.)

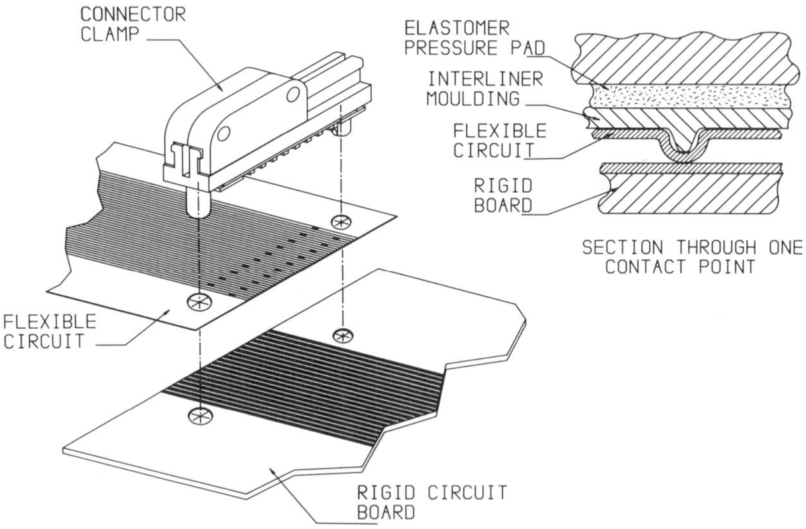

FIGURE 4.9 Flexible circuit/rigid board connector. (Source: Dowty Electronics.)

circuits has custom-made raised pressure peaks. The connector has 40 contacts per inch (0.025-in. centers), which results in densities that are significantly higher than conventional insulation-displacement connectors and in much lower board profiles.

The basic connector (clamp) consists of the connector housing, a sliding cam, a metal pressure spring with cam coupler, and a compliant pad. Two posts project from each end of the housing base to align with matching holes in the printed wiring board. These posts serve to protect the spring from damage, guide the clamp hooks into latched engagement with the board, and force registration and polarization between the clamp and board.

4.3.2 Flexible Circuit/Flexible Circuit Connectors [8]

Another interconnection scheme also takes advantage of raised pressure peaks on flexible circuits. However, with the increased usage of surface mounting, it may be given more consideration than it has in the past. Basically, the interconnection principle consists of depositing a gold-plated bump on the end of a flexible printed circuit conductor (Figure 4.10). The mating surface is a gold-plated lead

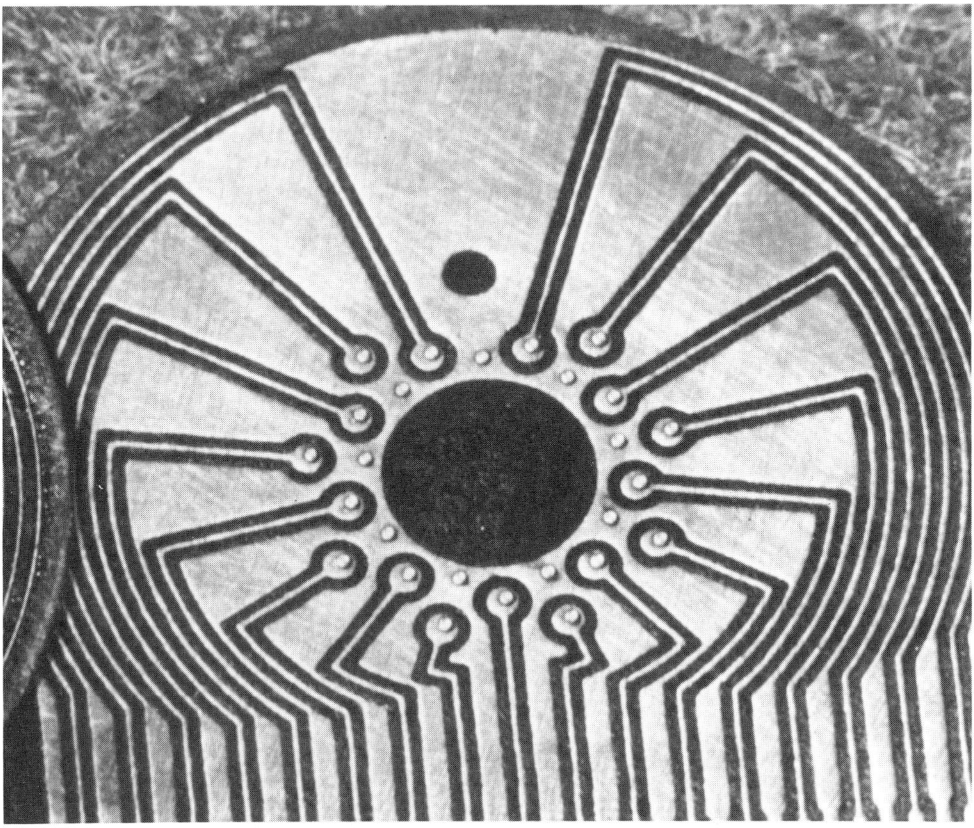

FIGURE 4.10 Flexible circuit/flexible circuit connector. (Source: Hughes Aircraft.)

on the mating circuit. The connector or pressure mechanism consists of two plates that clamp the mating flexible circuits together.

The elimination of pin and socket contacts in this way drastically reduces the connector size and mass, and eliminates contact wear. This, in turn, eliminates the need for soldering, insulation displacement, or crimping, and thereby allows for a mass-termination system. Placing conductors on 0.025-in. centers at the connecting interface allows for extremely high density.

4.4 INTERCONNECTION TAPE [9]

Perhaps the most unique surface-mountable connection scheme does not have a connector or pressure mechanism per se (Figure 4.11). This cost-effective alternative to the other types of connector systems is based on its adaptability to automated assembly, high reliability, and superior inseparable utility. The concept uses a "connection" tape that is both conductor and contact. It has silver paths, each with a conductive adhesive coating, uniformly spaced on a polymer film backing. Interconnections are made by first cutting the tape to the desired length, then attaching it to the edge of the mating board, and, finally, applying heat for 5 sec. The resulting connections are gas-tight and have very low contact resistance.

FIGURE 4.11 Surface mount interconnection tape. (Source: 3M Electronic Products.)

4.5 INTEGRATED CIRCUIT COMPONENT SOCKETS

Sockets for integrated circuits in chip carrier and pin-grid array (PGA) packages play a significant role in near-term surface mount applications. In fact, in some applications, the use of chip carrier sockets of the type shown in Figure 4.12 can allow for the continued use of through-hole technology with sophisticated surface mount VLSI devices in chip carrier packages. As the company matures into medium-term surface mount soldering capabilities, surface mount versions of these sockets (Figure 4.13) can be used. The ultimate

FIGURE 4.12 Through-hole mounting chip carrier socket. (Source: 3M/ Textool.)

FIGURE 4.13 Surface-mounting chip carrier socket. (Source: Augat, Inc.)

long-term application will usually eliminate the socket entirely by soldering the chip carrier directly to the printed wiring board.

At the present time, sockets for integrated circuits in pin-grid array packages are all of the through-hole mount type. Due to the large number of terminals (pins) associated with the use of these components, zero-insertion-force PGA sockets (Figure 4.14) are often used.

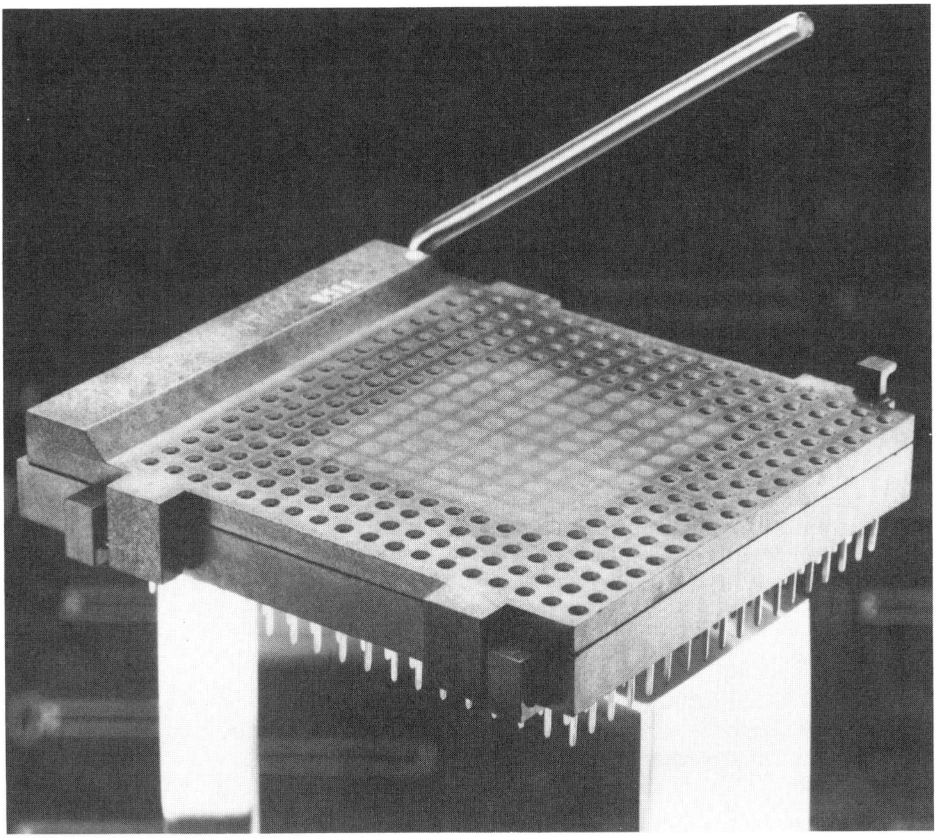

FIGURE 4.14 Through-hole mounting ZIF pin-grid array socket. (Source: AMP, Inc.)

REFERENCES

1. G. L. Ginsberg, "Surface Mounting Impacts Printed Wiring Connectors," *Electronic Packaging and Production*, November 1984, pp. 96-99.
2. "Design and Application Guidelines for Surface Mount Connectors," IPC-C-406, Institute for Interconnecting and Packaging Electronic Circuits, 1988.
3. D. Brearley, Jr., "The Connector/PCB Interface Key to Success in Surface Mounting of Connectors," Proceedings, National Electronic Packaging Conference '85, pp. 820-826.
4. M. Lazar, "Surface-Mounted Connectors," *Design News*, September 17, 1984.
5. J. Rose, Ed., "Surface Mount Connectors: Where the Industry Stands Today," *Connection Technology*, November 1986, pp. 35-39.
6. H. Weaver, "Surface Mount Connector Offers Form, Fit and Function Replacement for Gold Plated Fingers," Proceedings, Connectors and Interconnection Technology Symposium, November 14-16, 1983, pp. 348-358.
7. N. A. Conlon, "High Density Backplane Connectors," *Connection Technology*, July 1985, pp. 35-37.
8. M. J. Howett, "An Advanced High Density Interconnection System for Military/Space Applications," Proceedings, Connectors and Interconnection Technology Symposium, September 19-21, 1984.
9. G. C. Quinn, "Ingenuity Is Alive and Thriving in Circuit Interconnect Technology," *Circuits Manufacturing*, June 1984, pp. 44, 46, 52.

5

Packaging and Interconnecting Structures

The advances in semiconductor design and fabrication described in Chapter 2 have made it easier to incorporate very large functions into a single component package and even larger functions onto a single printed board while retaining cost effectiveness. At the same time, improvements in semiconductor processing have dramatically increased circuit speeds. For example, switching rates that were once the province of products requiring the most sophisticated ECL are now finding their way into products ranging from consumer products to supercomputers [1-3].

As a result of this rapid advance, printed board technology finds itself confronted with the dual need to manage the packaging and interconnecting of components with as many as 320 terminations while simultaneously controlling the high-speed behavior of the printed board in order to minimize degradation in circuit performance. To complicate matters further, close attention must be given to costs, manufacturability, testability, and repairability.

Unfortunately, the tools that have traditionally worked well to design, manufacture, assemble, and test printed boards are beginning to show their age when applied to these more demanding applications. All too often the solutions turn out to be more expensive than is generally deemed to be cost-effective and/or unrealistically long to implement. Therefore it is becoming critical to examine the issues involved in order to develop an overall balance among cost, complexity, and performance in order to arrive at a successful product.

Certain technical dynamics within the integrated circuit industry can be better understood by realizing that the complexity of leading-edge devices has doubled when compared to the previous year (Figure 5.1). These increases in chip complexity are primarily the result of increasing device functional density and increases in chip size. (Integrated circuit density doubles approximately every two years.)

The increase in functional density has been the direct result of decreasing semiconductor line widths from about 25 μm in 1960 to about 1.5 μm today. Shrinking feature geometries have been accompanied by significant decreases in cost, since the sizes of transistor functions have shrunk accordingly. Cost reductions have been the result of the concurrent use of better material, reduced labor content, and increased capital-usage efficiencies—in each instance, related to more transistor functions per unit of manufacturing cost.

The printed board analogy to the increases in integrated circuit complexity is illustrated in Figure 5.1, which shows the emergence of ever-more complex printed boards and their year of introduction. Unfortunately, here the degree of complexity doubles only every two and one-half years.

Underlying this disparity has been much slower decreases in printed board feature size (i.e., line widths, lead sizes, and hole sizes) that have not been compensated for enough by an increasing number of board layers. Putting this in terms of feature size relationships, in 1965 printed board feature size was about 30 times larger than the comparable semiconductor value; this ratio increased to 50:1 in 1975, and to nearly 100:1 in 1985.

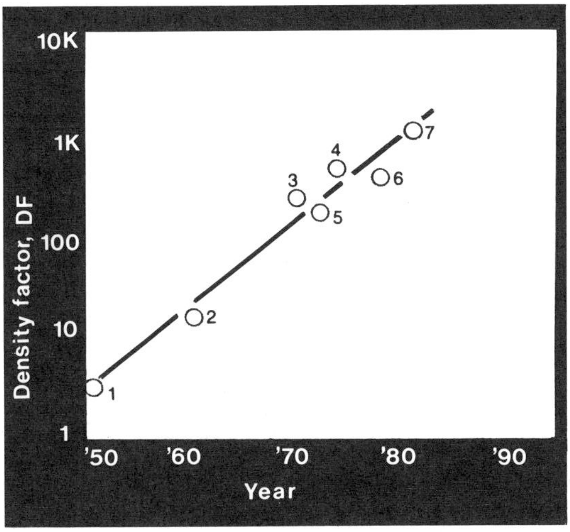

FIGURE 5.1 Printed board complexity. (1) Early single-sided board: 1 layer, 0 LPC, assumed 250-mil grid. DF = 4. (2) PTH board: 2 layers, 0 LPC, assumed 125-mil grid. DF = 16. (3) Military board: 25 layers, 0 LPC, 125-mil grid. DF = 200. (4) Commercial board: 19 layers, 1 LPC, 100-mil grid. DF = 380. (5) IBM ELSI board: 4 layers, 3 LPC, 100-mil grid. DF = 160. (6) IBM Advanced ELSI board: 8 layers, 3 LPC, 100-mil grid. DF = 320. (7) IBM TCM board: 18 layers, 4 LPC, 98.4-mil grid. DF = 965.

Also, if the trend in Figure 5.1 continues, printed boards introduced commercially in 1990 will have density factors of nearly 5000. This is equivalent to a 25-layer board, with four conductor lines per channel, built on a 25-mil (0.025-in.) grid. Perhaps more to the point, this is an indication of the level of future printed board (or any other advanced second-level interconnecting structure) interconnectivity that will be required.

This demand for increasing functional density at the interconnecting substrate level, combined with other concerns about improved thermal management and higher circuit operating speeds, brings into consideration the use of combinations of high-density multilayer boards, sophisticated constraining/supporting cores, improved dielectric materials, and new fabrication procedures. Because of their multifunctional use, these new interconnecting substrates are generically referred to as "packaging and interconnecting structures" (P&I structures).

5.1 GENERAL SELECTION CRITERIA [5]

The selection of materials for conventional (multilayer) boards and the newly emerging composite structures is important (and complicated) for thermal, mechanical, and electrical system reliability reasons. Each candidate material or combination of materials has a set of properties with particular advantages and disadvantages (Table 5.1). It is probable, however, that no single material will satisfy all design needs or applications. Therefore, a compromise of material properties should be sought that offers the best "tailoring" to suit end-product requirements cost-effectively.

Packaging and interconnecting structures vary from basic printed wiring boards to very sophisticated constraining-core structures. However, there are some selection criteria that are common to all of them. To aid in the selection process, Table 5.2 lists design parameters and material properties that affect system performance regardless of the P&I structure type. Also, Table 5.3 lists the properties of the most common materials that are candidates for these applications.

In general, packaging and interconnecting structures will fit into one of four basic categories of construction: organic base material, nonorganic base material, supporting plane, and constraining core. Thermal management and solder joint reliability concerns often preclude the use of conventional printed board materials and manufacturing processes in most high-reliability applications. Therefore, to fully realize the packaging advantages of chip carriers, these devices are often surface-mounted on sophisticated P&I structures.

When the chip carrier is surface-mounted, solder acts as both the electrical and mechanical attachment medium. Mounted to traditional printed board materials, such as glass epoxy, the thermal expansion mismatch between the chip carrier and the board can cause solder-joint stress failures when a wide operational temperature range is encountered, e.g., in military, automotive, and telecommunications applications. This problem becomes more apparent as component input/output pin count increases due to the use of larger package sizes and the resultant higher stresses. Potentially damaging microcracks can typically begin to appear at solder joints within 100 to 200 thermal cycles from -55 to +125°C when leadless chip carriers with pin counts of 28 or more are utilized. In addition, the higher packaging densities achievable with chip carriers generate more heat per unit area, which must be dissipated in order to prevent integrated circuit failures. Unfortunately, conventional organic base board materials are thermal insulators that require heat sinks and sophisticated cooling mechanisms to achieve reliable chip carrier operation, with resultant penalties in increased system weight and size.

Thermal expansion mismatch is the absolute difference in the thermal expansion of two components, usually a chip carrier and the P&I structure, due to the influence of heat dissipation of an operating circuit and/or changes in ambient temperature. Other things being equal, differences in coefficients of thermal expansion (CTE) become the underlying element of this problem (Table 5.3), with the ceramic in the chip carrier and the printed board coefficients being from 5 to 7 versus from 12 to 18 ppm/°C, respectively. Therefore, it is apparent that a P&I structure having the best characteristics of both conventional boards and ceramics would be desirable.

Existing chip carrier P&I structures bypass some of these problems by the use of good thermal management in conjunction with leaded chip carriers, sockets, and/or ceramic motherboards. Although these approaches can offer reliable surface mounting of chip carriers, they do not always allow the full potential of chip carrier packaging to be achieved.

The use of constraining metal-core P&I structures, such as those based on the use of porcelain-enamel or epoxy-coated steel substrates, appeared to offer significant improvements by virtue of their unlimited size, mechanical ruggedness, thermal conductivity, and inherent ground plane. Unfortunately, the thermal expansion coefficient of steel still appears to be too high.

Although attempts are still being made to surface mount chip carriers using conventional organic-base board materials, the P&I structures and substrates described in the following sections are being given serious consideration. It is quite possible that each of them will find a niche when used in conjunction with the thermal management of a particular set of packaging parameters.

5.2 ORGANIC-BASE P&I STRUCTURES [6]

5.2.1 Conventional Printed-Wiring Boards

The "king" of laminates for organic-base printed wiring boards is still the FR-4 (GF) materials. However, laminate manufacturers offer a variety of alternative materials to meet higher-performance multilayer board needs, such as polyimide (GI), multifunctional epoxies (FR-5), bismaleimide-triazine (BT), and other engineered resin systems. Driven heavily by military requirements for improved reliability, in-service thermal performance, and field repairability, the polyimides have evolved as the most significant new material in the high-performance laminate marketplace.

TABLE 5.1 Packaging and Interconnecting Structure Comparison [5]

Type	Major advantages	Major disadvantages	Comments
Organic base substrate			
Epoxy fiberglass	Substrate size, weight, reworkable, dielectric properties, conventional board processing	Thermal conductivity, X, Y and Z axis CTE	Because of its high X-Y plane CTE, should be limited to environments and applications with small changes in temperature and/or small packages
Polyimide fiberglass	Same as epoxy fiberglass plus high-temperature Z axis CTE, substrate size, weight, reworkable, dielectric properties	Thermal conductivity, X and Y axis CTE, moisture absorption	Same as epoxy fiberglass
Epoxy aramid fiber	Same as epoxy fiberglass, X-Y axis CTE, substrate size, lightest weight, reworkable, dielectric properties	Thermal conductivity, X and Y axis CTE, resin microcracking, Z axis CTE, water absorption	Volume fraction of fiber can be controlled to tailor X-Y CTE; resin selection critical to reducing resin microcracks
Polyimide aramid fiber	Same as epoxy aramid fiber, Z axis CTE, substrate size, weight, reworkable, dielectric properties	Thermal conductivity, X and Y axis CTE, resin microcracking, water absorption	Same as epoxy aramid fiber
Polyimide quartz (fused silica)	Same as polyimide aramid fiber, Z-axis CTE, substrate size, weight, reworkable, dielectric properties	Thermal conductivity, X and Y axis CTE, Z axis CTE, drilling, availability, cost, low resin content required	Volume fraction of fiber can be controlled to tailor X-Y CTE; drill wearout higher than with fiberglass

Fiberglass/aramid composite fiber	Same as polyimide aramid fiber, no surface microcracks, Z axis CTE, substrate size, weight, reworkable, dielectric properties	Thermal conductivity, X and Y axis CTE, water absorption, process solution entrapment	Resin microcracks confined to internal layers and cannot damage external circuitry
Fiberglass/Teflon laminates	Dielectric constant, high temperature	Same as epoxy fiberglass, low-temperature stability, thermal conductivity, X and Y axis CTE	Suitable for high-speed logic applications, same as epoxy fiberglass
Flexible dielectric	Light weight, minimal concern to CTE, configuration flexibility	Size	Rigid-flexible boards offer trade-off compromises
Thermoplastic	3-D configurations, low high-volume cost	High-injection-molding setup costs	Relatively new for these applications
Nonorganic base			
Alumina (ceramic)	CTE, thermal conductivity, conventional thick-film or thin-film processing, integrated resistors	Substrate size, rework limitations, weight, cost, brittle, dielectric constant	Most widely used for hybrid circuit technology
Supporting plane			
Printed board bonded to plane support (metal or nonmetal)	Substrate size, reworkability, dielectric properties, conventional board processing, X-Y axis CTE, stiffness, shielding, cooling	Weight	The thickness/CTE of the metal core can be varied along with the board thickness, to tailor the overall CTE of the composite

(continued)

TABLE 5.1 (Continued)

Type	Major advantages	Major disadvantages	Comments
[Supporting plane]			
Sequential processed board with supporting plane core	Same as board bonded to supporting plane	Weight	Same as board bonded to supporting plane
Discrete wire	High-speed interconnections; good thermal and electrical features	Licensed process; requires special equipment	Same as board bonded to low-expansion metal support plane
Constraining core			
Porcelainized copper-clad invar	Same as alumina	Reworkability, compatible thick-film materials	Thick-film materials still under development
Printed board bonded with constraining metal core	Same as board bonded to supporting plane	Weight, internal layer registration	Same as board bonded to supporting plane
Printed board bonded to low-expansion graphite fiber core	Same as board bonded to low-expansion metal cores, stiffness, thermal conductivity, low weight	Cost	The thickness of the graphite and board can be varied to tailor the overall CTE of the composite
Compliant layer structures	Substrate size, dielectric properties, X-Y axis CTE	Z axis CTE, thermal conductivity	Compliant layer absorbs difference in CTE between ceramic package and substrate

TABLE 5.2 P&I Structure Selection Considerations [5]

Design parameters	Material properties								
	Transistion tempera- ture	Coefficient of thermal expansion	Thermal conduc- tivity	Tensile modulus	Flexural modulus	Dielectric constant	Volume resistiv- ity	Surface resistiv- ity	Mois- ture absorp- tion
Temperature and power cycling	X	X	X	X					
Vibration				X	X				
Mechanical shock				X	X				
Temperature and humidity	X	X				X	X	X	X
Power density	X		X						
Chip carrier size		X		X					
Circuit density						X	X	X	
Circuit speed						X	X	X	

TABLE 5.3 P&I Structure Material Properties [5]

	Material		
Material	Glass transition temperature (°C)	X-Y Coefficient of thermal expansion (ppm/°C) (note 4)	Thermal conductivity (W/M°C)
Epoxy fiberglass	125	13-18	0.16
Polyimide fiberglass	250	12-16	0.35
Epoxy aramid fiber	125	6-8	0.12
Polyimide aramid fiber	250	3-7	0.15
Polyimide quartz	250	6-8	0.30
Fiberglass/Teflon	75	20	0.26
Thermoplastic resin	190	25-30	
Alumina-beryllia	NA	5-7	44.0
Aluminum (6061 T-6)	NA	23.6	200
Copper (CDA101)	NA	17.3	400
Copper-clad Invar	NA	3-6	150XY/20Z
Graphite (P-100)	160	-1.15	110XY/1.1Z
Graphite (P-75)	160	-0.97	40XY/.7Z

Notes: 1. These materials can be tailored to provide a wide variety and processing methods.
2. The X and Y expansion is controlled by the core material will be the same as the reinforced resin system used.
3. When used, a compliant layer will conform to the CTE of the strain between the component and P&I structure.
4. Figures are below glass transition temperature, and are NA—Not applicable.

properties

X-Y tensile modulus (PSI × 10^6)	Dielectric constant (at 1 MHz)	Volume resistivity (Ω/cm)	Surface resistivity (Ω)	Moisture absorption (%)
2.5	4.8	10^{12}	10^{13}	0.10
2.8	4.8	10^{14}	10^{13}	0.35
4.4	3.9	10^{16}	10^{16}	0.85
4.0	3.6	10^{12}	10^{12}	1.50
	4.0	10^{9}	10^{8}	0.50
0.2	2.3	10^{10}	10^{11}	1.10
3-4	10^{17}	10^{13}	NA	
8.0	10^{14}			
10	NA	10^{6}		NA
17	NA	10^{6}		
17-22	NA	10^{6}		NA
37	NA			
25	N/A			

of material properties based on resins, core materials, core thickness,

and only the Z axis is free to expand unrestrained. Where the T_g

the base material and to the ceramic component, therefore reducing

dependent on method of measurement and percentage of resin content.

Table 5.3 summarizes some of the important properties of the more popular multilayer board dielectric materials. The properties that most significantly affect both multilayer board design and processing developments are glass transition temperature (T_g) and coefficient of thermal expansion (CTE).

The simplest definition of glass transition temperatures is just what its name suggests: the temperature at which a material changes from a hard, brittle, "glasslike" form to a more rubbery, softer consistency. The important thing to remember is that significant changes occur in many properties of the material at the glass transition temperature. These changes can cause some processing, handling, and electrical performance problems.

Most important for our purposes is the change in the resin from a hard, almost brittle material to a softer one at the glass transition temperature. For example, this can cause smearing during the hole-drilling operation, which generates substantial temperatures at the drill's cutting interface with the multilayer board. Also, the board dielectric material may "lose its grip" on the copper foil's bonding surface while at the high temperatures that are common during solder reflow or component repair and replacement.

The coefficient of thermal expansion is a measure of how much a given material expands when it is heated. If a resin is unreinforced, its expansion is essentially isotropic, i.e., the same in the X, Y, and Z directions. Because glass has a lower CTE than the resin, multilayer board material reinforced with glass fabric tends to have its expansion restrained in the X and Y directions and, to a much lesser extent, in the Z direction. Moreover, the Z-direction coefficient of thermal expansion can increase by as much as fivefold above the T_g from the values shown in Table 5.3. This results in a substantially higher strain on plated-through hole (PTH) plating and, with sufficient thermal cycling, in a high PTH failure rate.

The key here is not that polyimide has a better CTE than the epoxies; it does not. But polyimide does have a significantly higher T_g. Therefore, the high Z-direction expansion and resin softening do not both occur with polyimide at the temperatures of most solder reflow operations. Thus, the problems normally associated with "pad lifting" and PTH damage during processing are minimized, and board repairability is enhanced.

Where lifted pads are not allowed and where there is extensive solder touch-up, the use of BT-type resins often results in an acceptable product. Also, in these specific cases, the BT-type resins are often less expensive and easier to process than the polyimide materials.

5.2.2 Special Reinforcement Materials

With the increasing emphasis on surface mount technology, new multilayer board reinforcement materials are being developed. For CTE-sensitive applications, the most interesting developments have been with respect to the use of quartz fabric and aramid fiber materials.

Quartz Fabric Materials [7]

Multilayer boards with improved CTE characteristics are made from a quartz fabric that has been impregnated with either epoxy or polyimide (Table 5.1). Such boards may be processed in the same shop as a conventional multilayer board with essentially the same processes. However, because of the hardness of the fused silice, the tough carbide bits for drilling holes wear out much faster, which can present production problems. Also, quartz fabric materials are expensive, sometimes as much as several times the cost of epoxy glass. Yet it is quite possible that the positive aspects of using quartz fabric materials will outweigh the negative ones for some applications.

Aramid-Fiber Materials [8]

Aramid fiber (du Pont's Kevlar) is a high-strength, high-modulus, low-density reinforcing material. When it is impregnated with either epoxy or polyimide resins, the resulting fabrics have enhanced mechanical and thermal properties.

The major concerns when using these materials in multilayer printed board applications are delamination and proper drilling—delamination, because the aramid fiber has a significantly lower CTE than the other dielectric materials in the board; and proper drilling, because of the fiber's high strength. However, in the latter case there are indications that solid carbide drills can be used successfully for this purpose along with a proper set of drilling parameters and the correct entry and exit materials.

Copper plating of the holes in the aramid fiber laminate does not appear to present any significant problems. Procedures have been developed for electroplating these materials that have resulted in copper depositions that are virtually identical in appearance to those on conventional multilayer boards.

5.3 NONORGANIC (CERAMIC-BASED) P&I STRUCTURES [9]

Before the development of chip carriers, hybrid circuit manufacturers tended to use bare chip-and-wire techniques for IC mounting and interconnection. However, the disadvantage of this approach is that bare chips, once mounted, cannot be pretested or easily reworked. With chip carriers, which can be pretested and more easily reworked, there is no such restriction.

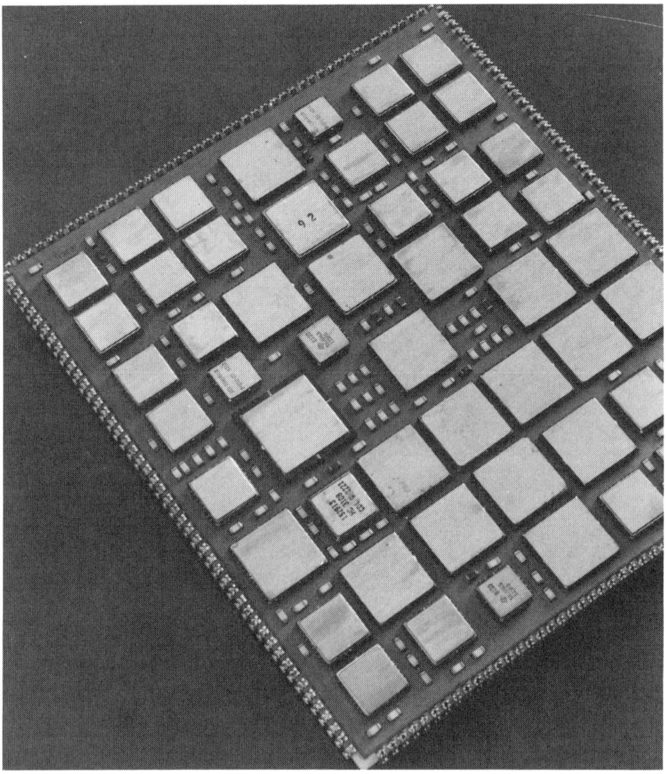

FIGURE 5.2 Alumina P&I structure assembly. (Source: Trocor, Inc.)

By moving to large substrates (Figure 5.2), the role of alumina and thick-film technology has been extended to do the tasks normally assumed by multilayer printed boards.

Since the whole structure is fabricated on ceramic substrates, the overall interconnection density achievable is that of 700 connections per square inch, as compared with 30 to 60 for conventional boards. Also, conductor widths and spaces can be 10 mils; the dielectric layer can have holes or vias that can be matched to these parameters.

5.4 CONSTRAINING-CORE P&I STRUCTURES

Constraining-core substrate materials, such as copper-Invar-copper or copper graphite have been developed for use in composite multilayer printed boards for use primarily in surface mount/military applications. The purpose of the constraining core is to "tailor" the multilayer boards' overall thermal properties.

Compared to glass epoxy at a few dollars per pound, the materials required for constraining-core structures might appear to be prohibitively expensive. On the other hand, they are used when the resulting properties cannot be achieved as cost-effectively by any other method.

5.4.1 Copper-Invar-Copper Materials

Composite multilayer printed wiring boards are being made in the form of packaging-and-interconnecting structures with thin, 0.004- to 0.010-in. (0.1- to 0.25-mm), copper-clad Invar as electrically functional ground and power planes (Figure 5.3). After the planes have been predrilled, they are located (ideally) in a symmetrical arrangement within the "lay-up" and subsequently laminated as an integral part of the multilayer P&I structure. The overall CTE of the structure can be tailored by varying the composition and thickness of the planes.

5.4.2 Graphite Materials [10]

Graphite materials are available with a wide range of thermal and mechanical characteristics. The form may be a continuous fiber with many filaments in a one-ply construction, or a woven cloth. Depending on the specific need, a particular graphite-resin system is used.

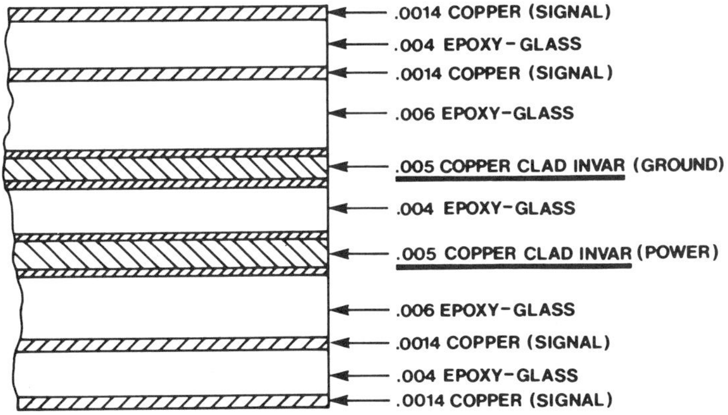

FIGURE 5.3 Multilayer P&I structure with copper-clad Invar power and ground plane constraining cores (cross section).

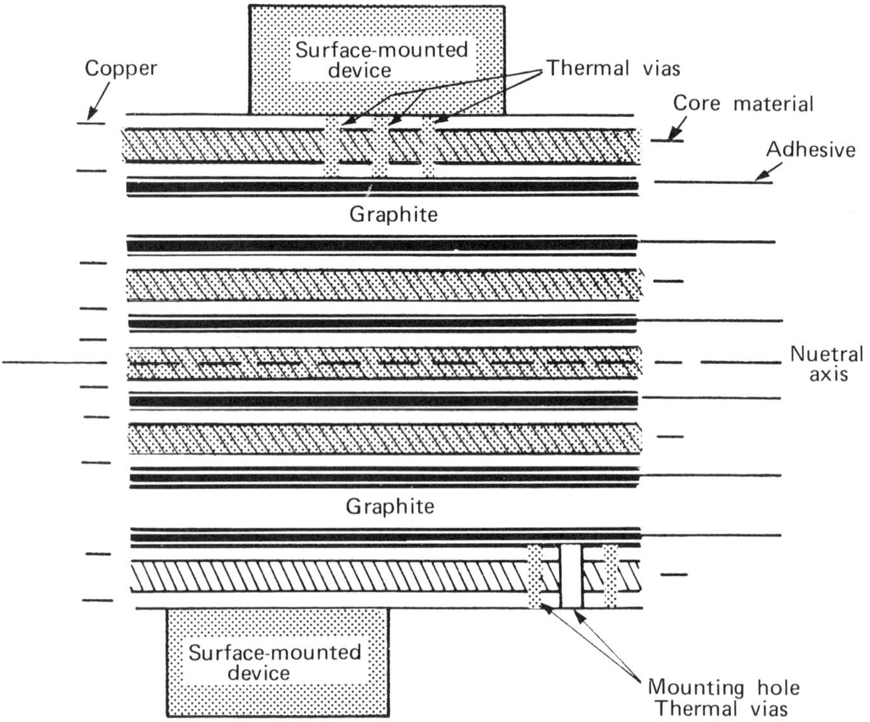

FIGURE 5.4 Graphite constraining-core P&I structure (cross section).

In terms of weight needed to achieve desired thermal properties, graphite is very attractive for use as a restraining core(s) in a multilayer P&I structure (Figure 5.4). Considering the specific conductivity, or ratio of a material's thermal conductivity to its density, graphite exhibits an advantage over the use of copper-Invar-copper. However, care must be used in maintaining registration during board fabrication because of the great mismatch in thermal expansion between graphite and the circuit layers. As with the use of aramid fibers, substantial dimensional changes are experienced at processing temperatures.

Thermal mismatch of the P&I structure's material elements also places a great demand on adhesion. This includes adhesion of conductive (copper) circuitry to core laminates and adhesion of graphite layers to circuit and filler layers, because substantial sheer stresses are generated within the laminated structure during "cooldown" from processing temperatures and during thermal cycling.

5.5 SUPPORTING-PLANE P&I STRUCTURES

5.5.1 Copper-Clad Invar Structures [11]

A clad-metal supporting-plane P&I structure has been developed that combines the metal-core board advantages with a CTE match similar to that of a ceramic chip carrier. This structure is based on the use of a substrate that is a metallurgically bonded composite of copper surfaces with a core of 64% iron/36% nickel alloy (Invar).

By varying the ratio of high-thermal-expansion copper to low-thermal-expansion Invar, the clad copper-Invar-copper material's CTE can be adjusted from 3 to 10 ppm/°C, with 16% copper/68% Invar/16% copper yielding a CTE of 6.4 ppm/°C. The latter consists of approximately 5 mils of copper on both sides of a typical 30-mil-thick clad metal substrate and provides excellent lateral thermal conductivity to facilitate heat transfer through the structure.

Copper-clad Invar can be punched, drilled, sheared, bent, or drawn to practically any desired configuration. Thus, it can serve as its own chassis or structural member (Figure 5.5).

With surface mounting, plated-through holes are no longer needed to make connections from one side of the board to another. Instead, vias need only interconnect between layers of the multilayer board, i.e., they may be "blind" or "buried" vias. This allows for conductor routing on the other layers of the board in areas that would otherwise be blocked by plated-through holes. Since such holes do not have to go through the clad metal core, the clad metal core need be only the mechanical member of the P&I structure, and thin laminate sheets attached to it can serve as the interconnection and component mounting members.

The use of thin laminates, 0.003 to 0.005 in. thick, offers advantages with respect to achieving reduced via diameters. In turn, the smaller the via diameter, the higher the potential interconnection density with more conductors routed between via holes.

Another advantage is realized when polyimide glass laminates are used. The laminate's lower thermal expansion rate and modulus, in relation to epoxy glass, facilitates the control of the CTE of the overall P&I structure, thereby allowing for reduced clad metal content (Figure 5.6).

5.5.2 Laminated-Package P&I Structures [12]

The laminated-package P&I structure is very similar to the clad metal technique. It involves the fabrication of a rigid circuit board by adhesively bonding a thin circuit to a punched and dielectrically coated metal support plane (Figure 5.7). By this construction, the circuit interconnection functions are separated from the physical support, ground plane, and heat sink functions.

FIGURE 5.5 Copper-clad Invar supporting-plane P&I structure. (Source: Cirtel Circuits.)

The use of a thin circuit allows for the low-cost generation of small plated-through holes. These small holes, in turn, make it possible to implement an interconnection capability in a double-sided board that is similar to that of a multilayer board.

Since this approach uses only one face of the support plane, two independent (or connected) circuits could bond to either face of the same support plane.

Fabricating a support plane involves punching an array of holes, then blanking the board from the sheet. Following punching, the support plane undergoes grit blasting, to remove burrs, and then straightening. An electrostatic powder coating, with an appropriate electrical-grade powder, completes the support plane.

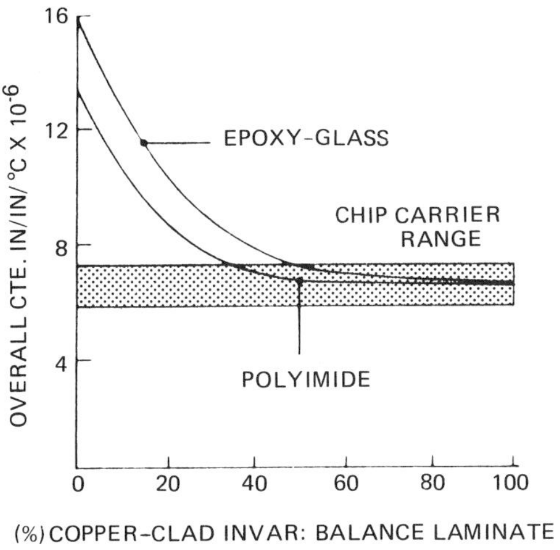

FIGURE 5.6 Influence of laminate types on overall CTE. (Source: Texas Instruments.)

A variety of laminates can then be used for the thin circuit, including epoxy glass and polyimide, both clad with thin copper. Thin-clad epoxy glass, 0.005-in. thick, has the stability to provide the dimensional tolerances required. Conventional processes produce the copper circuitry, which then undergoes testing before cover coating.

The cost of a completed thin circuit compares roughly to the cost of a conventional rigid epoxy glass circuit with equivalent line and space constraints, but with larger through holes. The added cost of the support plane and lamination produces a finished P&I structure that is somewhat more expensive than a double-sided board of comparable density, but significantly less expensive than a multilayer board.

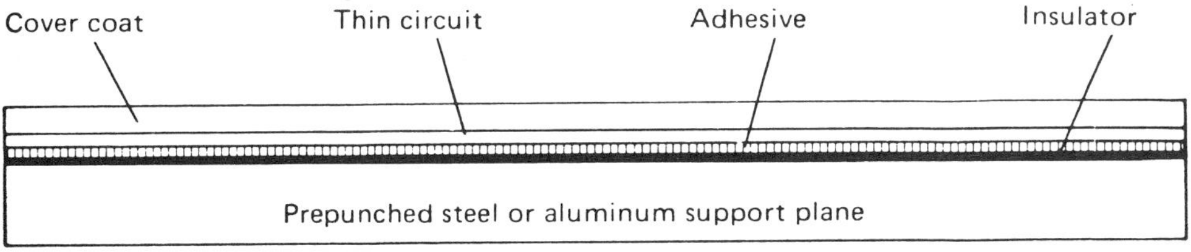

FIGURE 5.7 LAMPAC P&I structure (cross section).

5.5.3 Discrete-Wire P&I Structures [13, 14]

High-density interconnection systems designed specifically for surface-mounted components can be in the form of discrete-wire P&I structures (Kollmorgen Corporation's Microwire).

The discrete wiring matrix (Figure 5.8) requires no signal artwork, and engineering changes can be implemented rapidly. The wiring is encapsulated in a compliant resin to absorb local stress and provide good vibration dampening. Electrical access to the conductors is by 10-mil-diameter copper vias. The small via size can be accommodated in the component-attachment lead, thus eliminating the need for fan-out patterns with chip carriers down to 25-mil centers. Therefore, very high packaging densities can be achieved.

These P&I structures are available as daughter boards (Figure 5.9), large stand-alone panels of 50 in.^{2} or more, or as thin unsupported circuits for bonding to a custom heat sink. They are suited for use with chip carriers containing VLSI devices in military, computer, and communications applications that require high-density packaging with high-speed electrical characteristics in compact/rugged systems, especially those with short development times.

Discrete-wire P&I structures such as these are built up from metal supporting planes that offer good heat dissipation and a coefficient of thermal expansion match with leadless chip carriers. The interconnections are made by 2.5-mil-diameter insulated copper conductors, which are laid down precisely by numerically controlled machines on a nominal 12.5-mil grid. This geometry ensures a thin

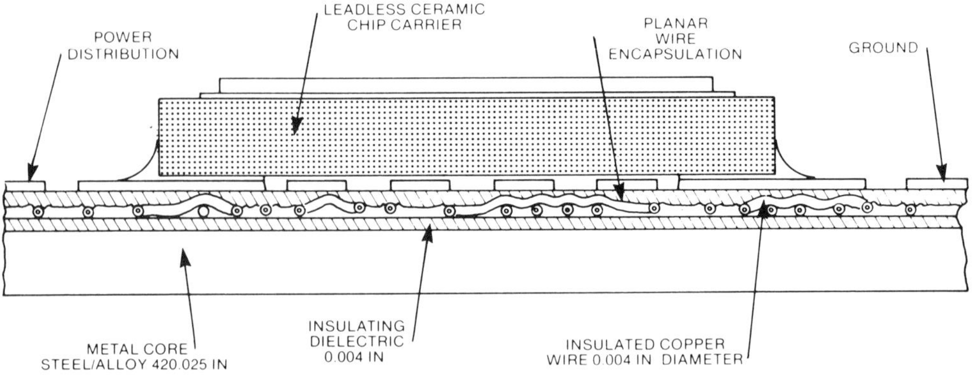

FIGURE 5.8 Discrete-wire P&I structure (cross section). (Source: Kollmorgen Corp.)

FIGURE 5.9 Discrete-wire P&I structure assembly. (Source: Kollmorgen Corp.)

circuit with excellent high-speed electrical characteristics and an interconnection density that is normally associated with thick-film technology.

Another discrete-wire technique (Kollmorgen Corporation's CoAxe) provides a coaxial transmission line. It takes the form of shielded leads embedded in a copper-plated ground plane (Figure 5.10) to minimize return currents and to provide a metallized surface that can be used as a heat sink. Conventional printed board substrates can be used.

The shield is in the form of a continuous copper plating with a thickness of 10 μm (0.4 in.). This is equivalent to several GaAs "skin" depths and maintain shield integrity and provide the desired electrical characteristics.

To achieve maximum speed, expanded Teflon, which at 1.3 has the lowest dielectric constant available, is used for the wire insulation. The inner conductor signal wire is 40 AWG.

The distance of the shield from the conductor determines the impedance. The result is a microcoaxial wire that has an impedance of

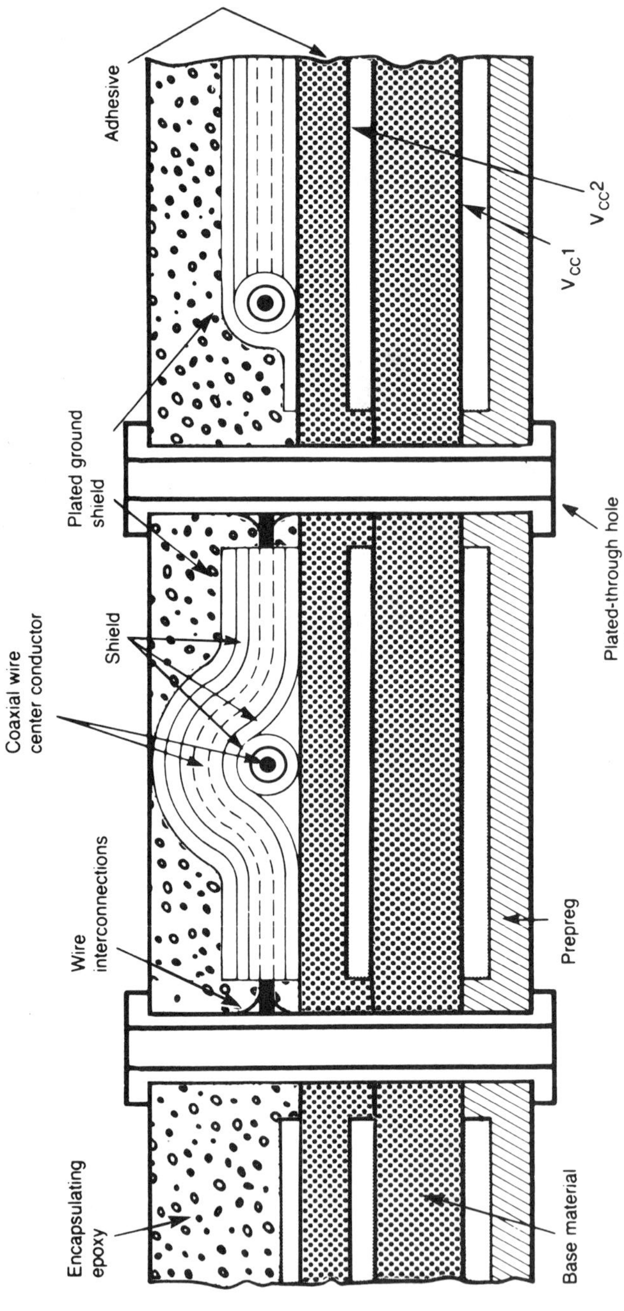

FIGURE 5.10 Discrete-wire transmission line (cross section).

50 Ω. The resulting structure provides signal propagation velocities that are 87% of the speed of light, which is a 30% improvement over the speeds achievable with Teflon multilayer boards.

5.5.4 Multiplanar P&I Structures [15]

Multiplanar P&I structures (Exacta Circuits, Ltd., Chipstrate) were also developed specifically for use with chip carriers and surface mounting (Figure 5.11). These structures feature a cushioning effect by the use of a novel elastomeric first layer that minimizes the risk of solder joint failure due to differential thermal expansion between the substrate and large surface-mounted chip carriers. In this structure, interconnection density is greatly increased by a combination of fine lines and spaces and either extremely small vias or plated pillars. Lastly, thermal dissipation is enhanced by constructing integral metal layers to conduct heat away.

FIGURE 5.11 Multiplanar P&I structure. (Source: Exacta Circuits, Ltd.)

The manufacturing process closely resembles that for conventional additive-circuit printed boards. The conductor pattern is formed by the electrodeposition of ductile copper, with the supporting substrate being either a dielectric insulator, such as glass epoxy, or a conductor, such as copper or aluminum that has been coated with the elastomer. Examples of the elastomer are polyimides, epoxy, modified epoxy, nitrile-rubber blends with phenolics, and acrylics.

The thermally conducting planes are fabricated so as to form clear apertures to allow for the through-hole interconnections. The planes are then treated to promote adhesion and included with other layers in the lamination and bonding of the multilayer structure.

5.5.5 Sequentially Laminated P&I Structures [16, 17]

Another approach to these complex packaging requirements is a combination of precision lithography, additive copper plating, and multilayer polyimide lamination (Augat, Inc., Microtec). Figure 5.12 shows the process from the initial photomask operation through to a completed substrate that features fully shielded conductors, stripline configurations, and thermal vias.

Large power and ground planes in the multilayer structure minimize crosstalk, control signal-line characteristic impedance, and minimize power and ground impedances. The low relative dielectric constant of the polyimide used (3.8) keeps the distributed capacitance of signal lines low, while retaining the benefits of very small dimensions between the signal traces and ground elements.

The end product has lines and spaces as small as 0.002-in. and vias as small as 0.004 in. square. Changing the relationship between the signal lines and the power planes provides layout flexibility to attain the exact characteristic impedance desired. Although the impedance is typically 50 ± 2 Ω, 25- and 75-Ω lines have been produced.

A further advantage is that vias can be "blind"; i.e., they do not have to go completely through the substrate. In addition, vias can be "blind," i.e., not connected to an outside surface of the substrate. In these instances, conductors can be routed above and below them on the other layers of the structure. These advantages make the process highly suited for use with high-speed circuitry where short line lengths are required.

The total number of layers that can typically be fabricated with this process is six, with a normal structure thickness of from 18 to 20 mils. However, by using the high-density features of the process, these parameters are adequate for most applications.

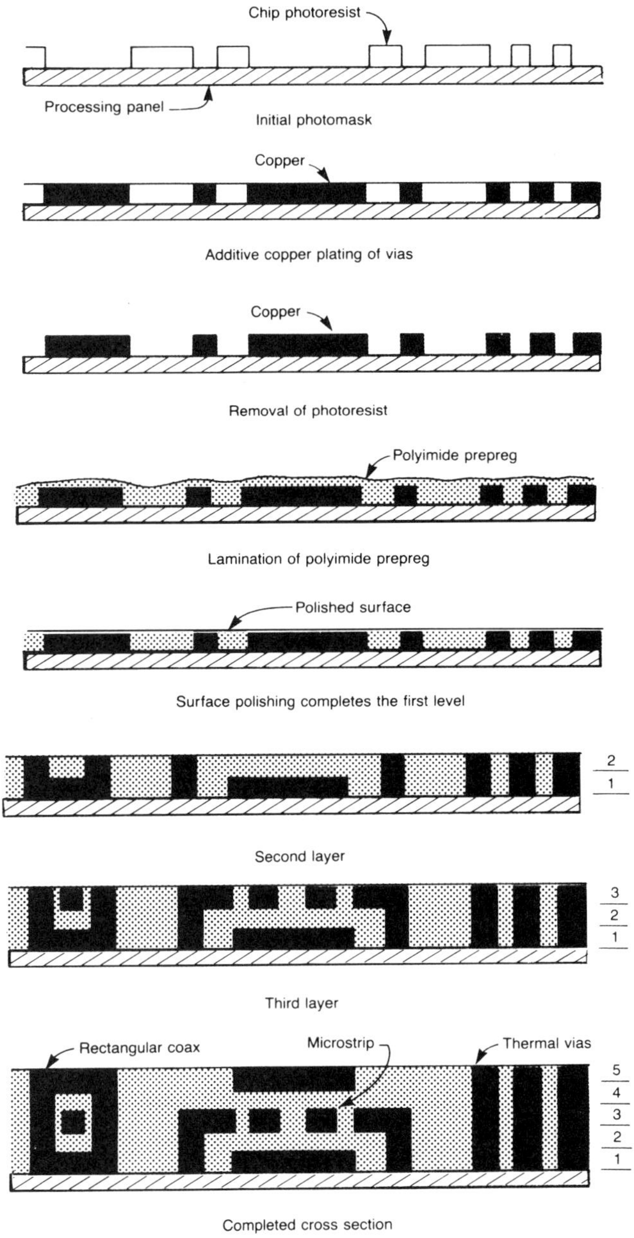

FIGURE 5.12 Additive multilayer transmission line (cross section).

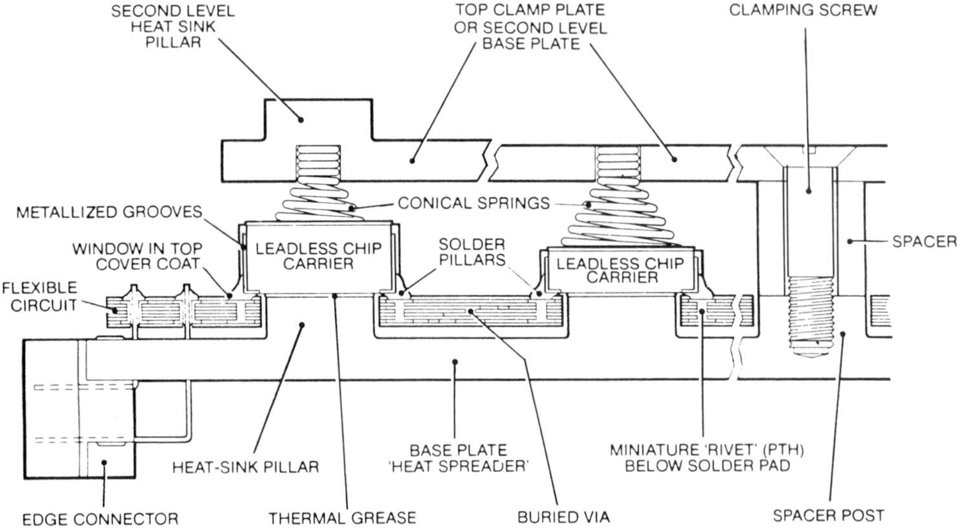

FIGURE 5.13 Flexible printed circuit P&I structure assembly.

5.5.6 Flexible-Circuit P&I Structures [18]

Much has been written about the various P&I structures that are being considered for chip carrier applications. One new candidate (Figure 5.13) involves the use of conventional fine-line polyimide flexible printed wiring. Assemblies of this nature can be constructed in multilayer form while retaining a low modulus feature that reduces residual strain at the soldered joints. Furthermore, the thin printed wiring laminate enables very fine holes to be produced by laser drilling, which can be plated-through or filled with solid copper as required.

To retain inherent flexibility while dissipating heat away from the solder joint, cutouts in the flexible circuit accommodate pillars from the metal heat-sink baseplate. Although this concept appears to be heavy and cumbersome, if the heat-sink baseplates are fabricated from thin sheets of aluminum, the resulting total weight per unit area of the combined circuit and heat-sink assembly might be less than that of alternative constructions.

The development of concepts such as this illustrate how far afield thermal management thinking can go. As previously mentioned, in all probability P&I structures will have to be "tailored" to satisfy each particular application, with no one approach being clearly predominant over the others.

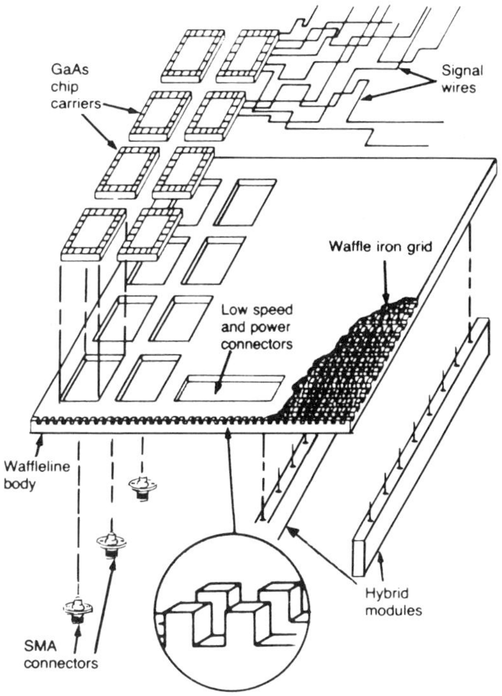

FIGURE 5.14 Rectangular-cavity coaxial system components.

5.5.7 Rectangular-Cavity Coaxial Structures [19]

The outer conductor or shield of a coaxial structure may also be rectangular. (In fact, that is basically what is achieved with the additive multilayer coaxial structures described in Section 5.5.5). This is the general concept used in a unique quasi-coaxial structure that has been developed expressly for use with GaAs circuits (Figure 5.14). In this approach, rectangular channels are milled in an aluminum substrate (Harris Corporation's Waffeline). Teflon-insulated 30-AWG wires are orthogonally (X-Y) routed in the appropriate channels, then a metal foil is placed over the top to complete the series of rectangular coaxial paths. However, any one outer shield is periodically interrupted by crossing channels.

The signal wires are at a height that allows them to lay right on top of one of the pads of a cavity-up chip carrier holding the GaAs device. Then, with the aid of a microscope, the center conductor of the wire is manually soldered to the chip carrier pads. (Automatic wire routing followed by laser soldering is being considered for production volumes.)

The main advantages of this packaging are its high signal isolation and bandwidth, which allow fast rise times. Electrically, it is an extremely low-dispersion transmission medium, with propagation speeds of approximately 80% of the speed of light. Also, the channel depth is such as to maintain a 50-Ω impedance throughout the structure and still permit wire crossovers.

REFERENCES

1. G. L. Ginsberg, "Packaging and Interconnecting Structures for Chip Carrier/Surface-Mounting Technology," Proceedings, Printed Circuit World Convention III, May 22-25, 1984, IPC.
2. L. W. Ritchey, "The Impact of VLSI on PCB Design," *VLSI Systems Design*, August 1986, pp. 86-88.
3. W. S. Fujitsubo, "Lithography Impact on Printed Wiring Boards," *Solid State Technology*, June 1986, pp. 161-164.
4. W. Loeb, "Printed Circuits Mirror Integrated Circuits," *Electronic Packaging and Production*, February 1986, pp. 168-170.
5. Guidelines for Component Packaging and Interconnection with Emphasis on Surface Mounting, IPC-SM-780, Institute for Interconnecting and Packaging Electronic Circuits, 1988.
6. C. Guiles, "Polyimide PWBs—Materials and Processes," *PC Fab*, June 1985, pp. 47-59.
7. "Quartz Fiber Suits Big Ceramic Carriers," *Electronics*, October 20, 1982.
8. C. T. Brooks, "Kevlar as the Reinforcing Fiber in Printed Board Materials," *PC Fab*, February 1982.
9. K. Smith, "Huge Alumina Substrates Shrink Avionics Computer," *Electronics*, January 13, 1982, pp. 81-82.
10. J. Liebowitz, W. Winters, and J. Kolkin, "Graphite Layers in SMT Boards Control Thermal Mismatch," *Electronic Packaging and Production*, June 1985, pp. 86-88.
11. F. J. Dance and J. L. Wallace, "Clad Metal Circuit Board Substrates for Direct Mounting of Ceramic Chip Carriers," Proceedings, International Electronic Packaging Conference, November 1981.

12. V. L. Brown, "LAMPAC—An Alternative to MLBs," *Circuits Manufacturing*, October 1978, pp. 22-26.

13. G. Messner and C. Lassen, "Substrates for Surface Mounted Components," *Circuit World*, August 1981, pp. 10-14.

14. L. Schieber, "Microcoaxial Board Interconnect," *VLSI Systems Design*, March 1986, pp. 57-60.

15. M. El Rafaie, "Chip-Package Substrate Cushions Dense, High-Speed Circuitries," *Electronics*, July 14, 1982, pp. 135-141.

16. S. Lebow, "A Method of Manufacturing High Density Fine Line Printed Circuit Multilayer Substrates Which Can Be Thermally Conductive," *Transactions, IEEE CHM&T*, 1980.

17. L. Nelson, "Success of GaAs Semiconductor Hinges on Packaging," *Electronic Packaging and Production*, June 1985, pp. 48-51.

18. P. L. Kirby, "Substrates for Chip Carriers—A Flexible Approach," VLSI Packaging Workshop, September 12-13, 1983.

19. R. Davis, "Board and Microwave Chip Carriers for GaAs Systems," *VLSI Systems Design*, June 1986, pp. 125-130.

6

Electrical Design Considerations

The demand for increased data processing in modern electronic systems has resulted in new high-speed digital processing technology [1-3]. Thus, more capable architectures have been developed, and major advances have been made in high-speed solid-state logic devices. This is reflected in system clock speeds that are rapidly increasing to greater than 100 MHz, and which require integrated circuits with rise times that are frequently less than 1 nsec. Since these priorities are often in direct conflict with producibility constraints, it is important to know first how this affects the design engineer.

6.1 THROUGH-HOLE VERSUS SURFACE MOUNT INTEGRATED CIRCUITS [4, 5]

In general, the through-hole mount dual-inline package (DIP) does not perform electrically as a comparable surface mount chip carrier as input/output (I/O) lead counts increase (Table 6.1). This is

TABLE 6.1 Dual-Inline Package and Chip Carrier Electrical Characteristics

Parameter	Lead count	Chip carrier Longest lead	Chip carrier Shortest lead	DIP Longest lead	DIP Shortest lead
Trace inductance (H)	16	1.13	0.73	6.40	1.62
	28	1.80	1.15	14.77	1.62
	40	2.90	1.87	24.94	1.62
	64	6.44	4.21	49.14	2.34
Line-to-line capacitance (pF)	16	0.13	0.09	0.74	0.25
	28	0.19	0.13	1.48	0.25
	40	0.27	0.19	2.13	0.25
	64	0.52	0.36	4.12	0.33
Trace resistance (Ω)	16	0.114	0.108	0.242	—
	24	0.139	0.136	0.319	—
	40	0.147	0.109	0.644	—
	64	0.222	0.222	1.000	—

in part because of differences in lead and conductor lengths. The longest trace within a 64-lead DIP is almost eight times that of the corresponding trace within a 64-lead chip carrier. The ratio of the longest-to-shortest DIP trace is 12.5:1, while the corresponding ratio for the chip carrier is 1.4:1. Long traces mean increased lead resistance and inductance. Unequal trace lengths affect system and device performance by restricting power and ground capabilities. Long side-to-side conductor traces result in significant lead-to-lead capacitance, which can affect some devices.

One of the electronic application areas for which chip carriers offer significant advantage is that of high-speed circuitry. This rapidly growing circuit design area may well reach the point at which the packaging becomes the limiting factor in the success of the application. For instance, the circuit speeds gained with advanced semiconductors might well be lost if limiting propagation delays exist in the package or the P&I structure onto which the packages are mounted. Typical propagation delays for 64-lead devices are 0.3 nsec for a DIP, 0.1 nsec for a flatpack, and 0.05 nsec for a chip carrier; typical P&I structure delays are 0.13 nsec/in. When maximum system delay times are on the order of 0.5 to 0.3 nsec, these individual elements of propagation delay can seriously affect packaging density and power density requirements (Figure 6.1).

6.2 CHARACTERISTIC IMPEDANCE [5, 6]

The design engineer's task is to see where compromises can be made in each set of requirements, starting with an understanding of the

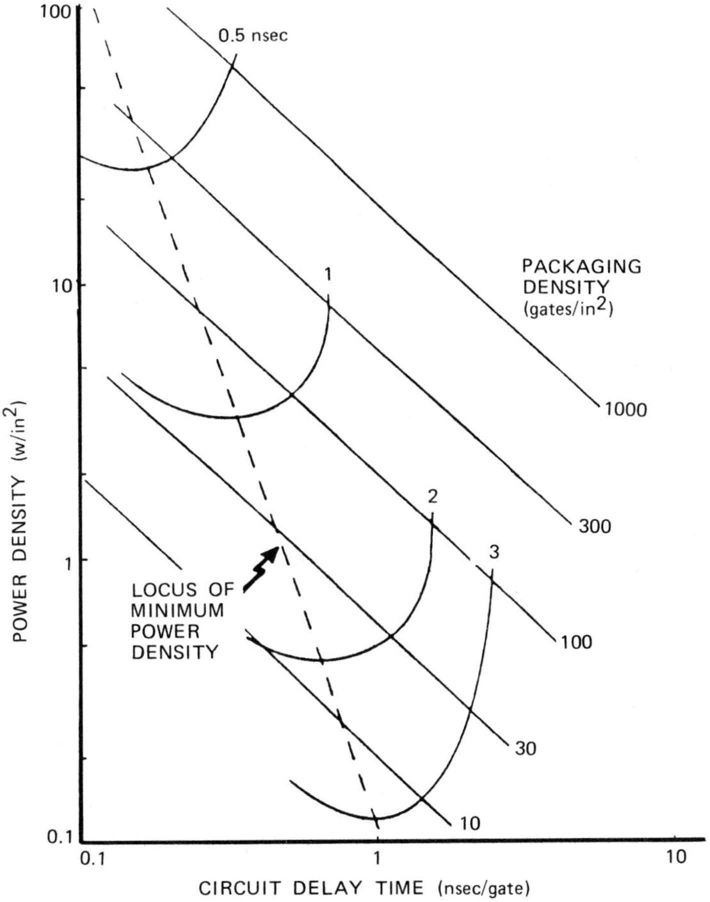

FIGURE 6.1 Factors affecting system delay time.

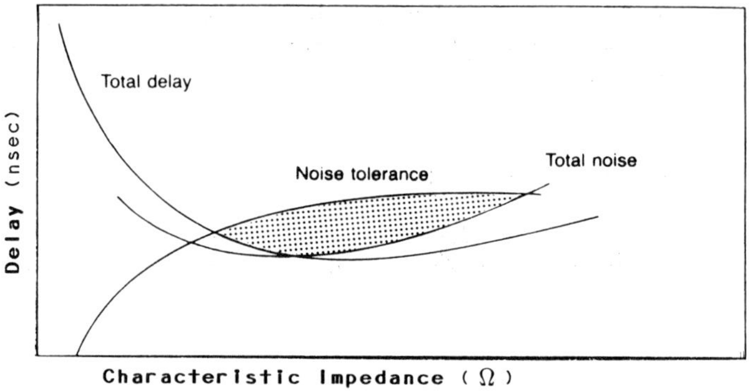

FIGURE 6.2 Best system performance "design window."

key parameter that interrelates most aspects of system performance; the characteristic impedance (Z_0). A range of characteristic impedance values ("design window") exists for any electronic system and is a function of both electrical and mechanical requirements, such as the example shown in Figure 6.2. This "window" represents the optimum compromise among noise, delay, crosstalk, and producibility constraints. Once system characteristic impedance has been determined, design rules can be generated that not only help to ensure acceptable electrical performance, but also help to achieve cost-effective producibility. Component packages and higher-level interconnections, such as printed wiring, are the intervening elements that are crucial in getting the most out of circuit performance at the system level (Figure 6.3).

A variety of transmission-line configurations have been developed for use at very high (microwave) frequencies. These are shown in Figure 6.4. Unfortunately, in both printed board and semiconductor technologies, the conductors basically have a rectangular cross section. This presents a difficult problem to analyze because of the fringing effect or concentration of the electromagnetic fields in the corners. Also, rectangular cross sections require additional consideration to the conductor's w/h aspect ratio in order to determine the effective field interactions.

Specific standard transmission-line characteristic impedance values are 50 ± 2 Ω, 75 ± 3 Ω, and 95 ± 5 Ω. These nominal values and tolerances are very hard to hold with current printed board technology. In fact, it may be very difficult even to determine variations of ±5% and less with conventional printed-board inspection tools.

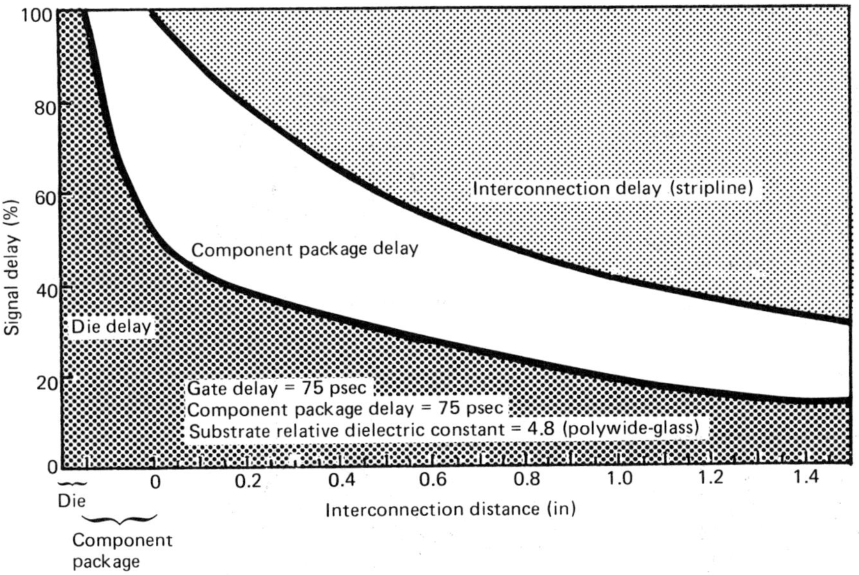

FIGURE 6.3 Transmission delay.

In addition to the problem of achieving a certain absolute characteristic impedance value and reproducing it within allowable limits, another equally difficult task is matching impedances among sets of conductors. This must be done without creating the undesirable effects that result from signal reflections; it must also be done without using an excessive amount of board real estate. One solution is to have "curves" and "waves" that adjust the conductor's length while minimizing sharp turns that cause reflections, which in turn cause changes in impedance and distortion and attenuation of electrical signals. Otherwise, additional layers will have to be added to the board to allow for more tailored routing of conductors.

There are two basic printed wiring configurations that can be used to implement designs concerned with characteristic impedance and other critical electrical parameters: microstrip and stripline. The microstrip configuration (Figure 6.5) is basically for a double-sided board (or the outer layers of a multilayer board). The stripline configuration is, obviously, that of a multilayer board (Figure 6.6).

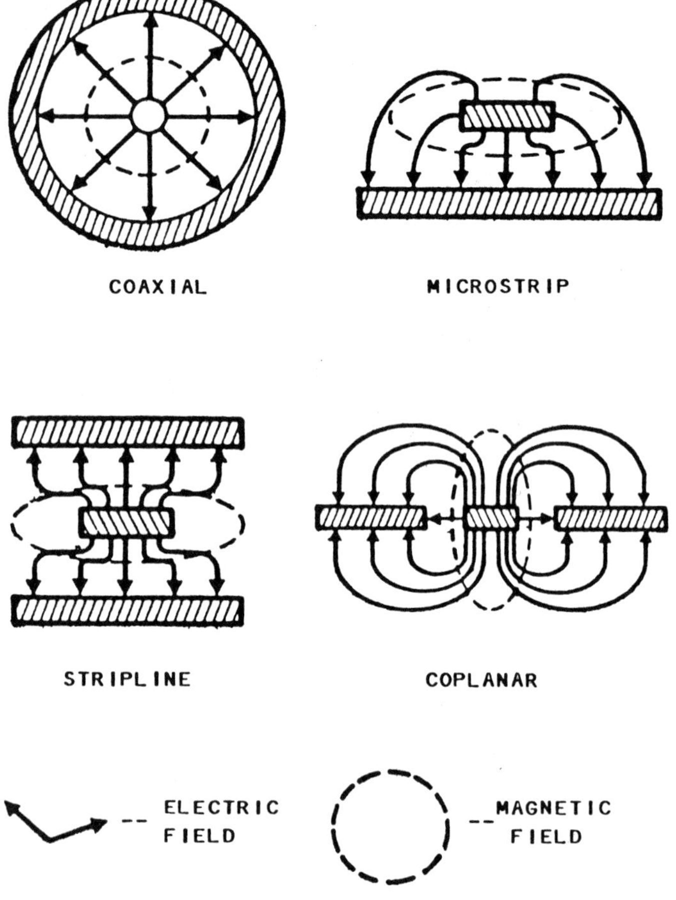

FIGURE 6.4 Basic transmission line configurations.

The electrical characteristics of the two printed wiring configurations are a complex flunction of the board material's dielectric constant, dielectric thickness, conductor width and conductor thickness. For the purposes of this work, it is not important to understand the magnitude of these complexities. What is important to understand is that printed wiring used in these applications is concerned with much more than just connectivity and that is highly sensitive to changes in values, especially those resulting from manufacturing tolerances.

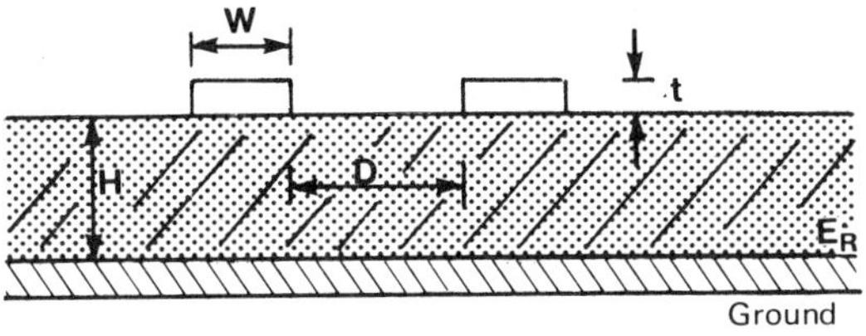

$$Z_O = \frac{87}{\sqrt{E_R + 1.41}} \ln\left(\frac{5.98H}{0.8W + t}\right) \Omega$$

$$T_{PD} = 0.08471\sqrt{0.475E_R + 0.67} \text{ nsec/in}$$

$$C_O = \frac{T_{PD}}{Z_O} \text{ pF/in} \quad T_{PD} \text{ in nsec/in} \quad Z_O \text{ in } K\Omega$$

$$X_{TALK} \propto \frac{1}{D}, H$$

FIGURE 6.5 Microstrip relationships.

Characteristic impedance matching/control is one of the electrical parameters imposed on designers and fabricators of interconnecting networks that consist of integrated circuit packages, wire bonds, and printed boards. Designing the interconnecting network with a characteristic impedance that is equal to the component "load" impedance

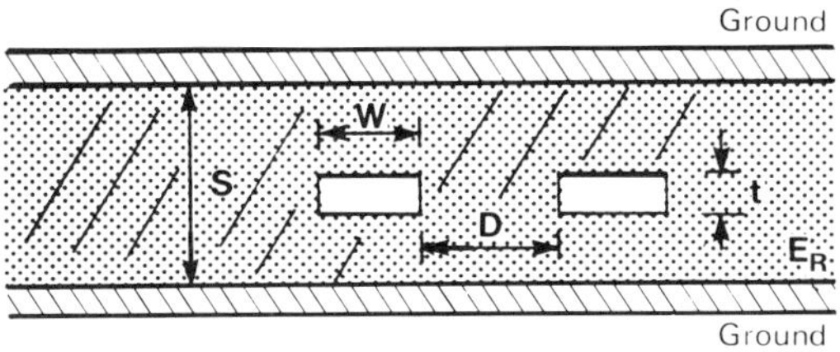

$$Z_O = \frac{60}{\sqrt{E_R}} \ln\left(\frac{4S}{0.67\Pi W}\left(0.8 + \frac{t}{W}\right)\right) \Omega$$

$$T_{PD} = 0.08471\sqrt{E_R} \text{ nsec/in}$$

$$C_O = \frac{T_{PD}}{Z_O} \text{ pF/in} \quad T_{PD} \text{ in nsec/in} \quad Z_O \text{ in } K\Omega$$

$$X_{TALK} \propto \frac{1}{D}, S$$

FIGURE 6.6 Stripline relationships.

makes it possible to transport total power from the source to its destination without reflections or distortions. In addition, the propagation delay due to the length of travel of the signal along the conductor will be of short duration.

The equations that are used to determine characteristic impedance show a somewhat complicated relationship among conductor width (w), conductor thickness (h), the spacing between coplanar conductors, and the dielectric constant and thickness of the insulating material. However, assuming that the resistive losses are very small compared to the reactive impedance and that the leakage current through the dielectric is also very small, the characteristic impedance of the interconnecting network is proportional to $(L/C)^{1/2}$. The values for L (self-inductance) and C (capacitance) in printed boards are determined by the materials used and by the geometries of the conductors (Figure 6.7).

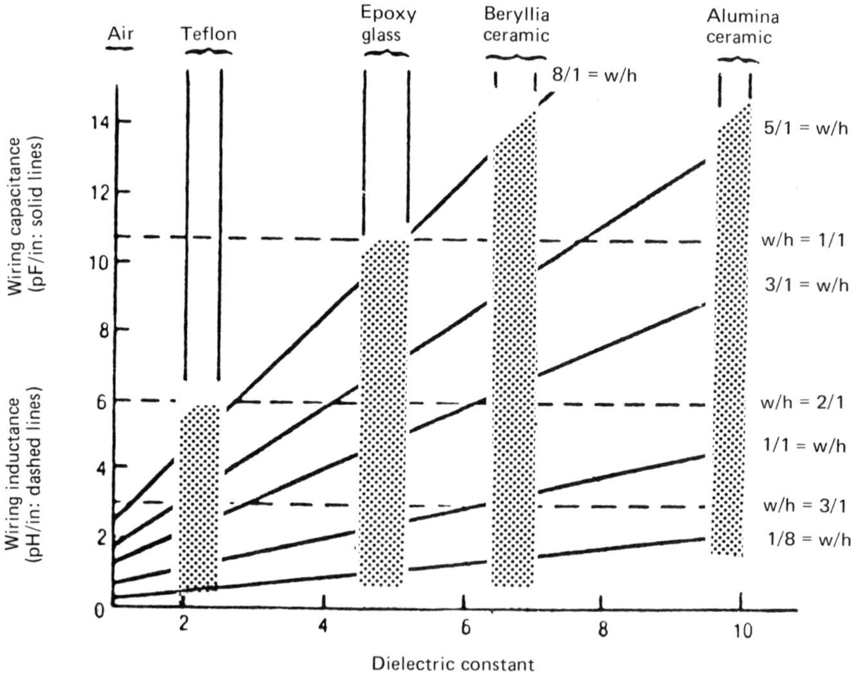

FIGURE 6.7 Printed board inductance and capacitance (for different geometries and dielectrics).

6.3 CROSSTALK ISOLATION

Signals at high frequencies on closely spaced conductors (and the subsequent electromagnetic signal phenomena) give rise to problems associated with isolating signal lines from inductive and capacitive coupling. This induction of unwanted signals into other conductor lines from active signal lines is called "crosstalk."

There are five parameters that affect the degree of crosstalk:

1. The distance between conductors (the farther apart, the better)
2. The dielectric medium (the higher the dielectric constant, the better)
3. The presence of ground shields (the more, the better)
4. The signal rise time (the slower, the better)
5. The length of the coupled lines (the shorter, the better)

Unfortunately, the action taken to reduce crosstalk is often in conflict with the implementation of high-density interconnections, maintaining high propagation velocities, and controlling impedance. For example, the lower the dielectric constant, the higher is the signal propagation speed; the higher the dielectric constant, the better is the electromagnetic isolation.

The presence of ground shields enhances the degree of signal isolation and permits closer control of impedance values. However, the use of extra ground shields and ground planes complicates the fabrication of the printed board.

Another approach to reducing interlayer coupling is to route signal lines orthogonally to each other between adjacent signal layers. Intralayer reductions can be achieved by using interdigitated guard lines between signal lines on the same layer of the printed board. However, both these approaches increase conductor routing pattern complexity accompanied by requirements for higher resolution and better alignment accuracies between layers of the printed board. This, in turn, tends to reduce manufacturability and increases costs.

6.4 MATERIAL DIELECTRIC PROPERTIES [7, 8]

Transmission line properties must be considered when the travel time of the pulse to and then back from a point of reflection, or twice the interconnection length between devices, is less than the rise time. Thus, for different logic types there are different critical interconnection lengths, which when exceeded require the interconnection to be treated as a transmission line.

This critical length also depends on the dielectric material through which the signal propagates. For example, with polyimide, with a dielectric constant of 4.0, as the dielectrical material, the rise time and critical interconnection length for typical logic families are TTL: 8.0 nsec/24.0 in.; Schottky high-speed TTL: 3.0 nsec/9.0 in.; 100k high-speed ECL: 0.6 nsec/1.8 in.; and GaAs: 0.1 nsec/0.3 in.

The critical interconnection length is so short for some of these logic families, such as for GaAs, that it becomes a factor even within the integrated circuit package containing a single chip. For these applications custom packaging and interconnecting structures (Chapter 5) have been developed with discrete wiring, additive multilayer, and rectangular coaxial transmission-line configurations.

The electrical properties of the printed-board laminate materials become extremely critical with the use of high-speed devices, especially GaAs. The values of these properties control a host of other electrical parameters, such as signal propagation delay, characteristic impedance, signal rise time, and overall system losses. To make matters worse, the material's dielectric constant and dissipation factor begin to vary significantly with increases in frequency and changes in temperature.

As previously mentioned, as signal speeds approach the gigaherz frequency range, the electrical energy no longer resides within the conductors. As signal frequency increases, the phenomenon known as "skin" or surface effect occurs whereby the flow of current concentrates around the periphery of the conductor. Therefore, the electromagnetic waves are propagated primarily in the dielectric material surrounding the conductors. Consequently, it is the dielectric constant of the insulating material that determines the propagation velocity of the signal.

Electrical signals propagate at the speed of light if the dielectric constant of the dielectric material happens to be 1, the same as that of air. However, the signal propagation speed for dielectric constants greater than 1.0 is inversely proportional to the square root of the relative dielectric constant; e.g., for a polyimide material with a dielectric constant of 4.0, the speed would be one-half of what it would be in air.

To illustrate this point, Figure 6.8 plots the system speed of a 1-GHz GaAs device versus interconnection path length in various dielectric materials. For a 3-in. interconnecting path length, the speed has been reduced 25% for PTFE (i.e., 750 MHz), 30% for polyimide, and over 45% for ceramic. As the path lengths get longer, the degradation becomes even worse. (Ironically, if wafer-scale integration is used, the actual system speed is decreased even more, as silicon has a dielectric constant of about 15.) Consequently, the interconnection's total length must be kept very short in order to preserve most of the inherent switching speeds of the newer-generation devices.

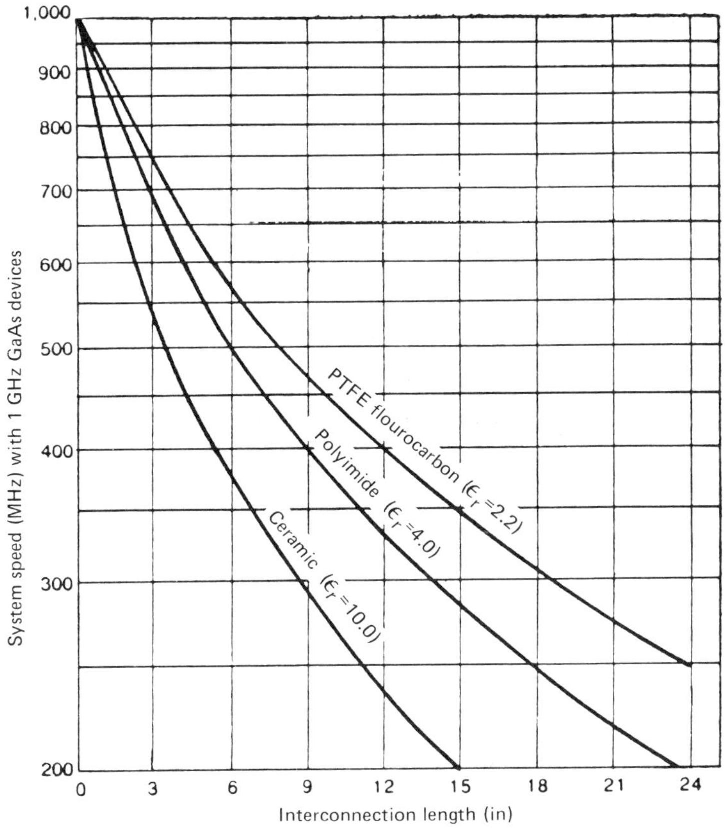

FIGURE 6.8 Switching speed versus interconnection length.

6.5 POWER DISTRIBUTION [7, 9]

Within the world of modern digital circuitry, most discussions have centered primarily on the theme of transmission line theory. Although not discussed as openly, the requirements for increasing power distribution bear equal consideration by the system designer. On the surface, the treatment of these two dominant concerns, speed and power, may appear to be distinct issues that must be accommodated separately. They are, however, closely intertwined physical relationships. More specifically, although select facets of design that accommodate one aspect of power may be totally independent of speed, there are others that negatively impact it. The converse relationship is, of course, also true.

The power and ground planes of the more demanding circuit logic families, such as ECL and GaAs, usually most be designed using heavy copper (2 oz and greater) to provide the general electrical characteristics of increased ampacity, reduced resistance, and depreciated component (inductive and capacitive) parasitics.

More specifically, the heavy ground planes are used for:

Noise immunity: A homogeneous/noninterrupted path improves laminar electron flow and, consequently, reduces signal transmission noise.

Energy dissipation: Some of the high-speed logic families consume power at an extremely high rate. This energy must, of course, be dissipated without adversely affecting the operating environment limits.

Current surges: Because of the switching speed and possibilities of spontaneity of switching states, excessive current surges are likely. These surges must be restrained in the system by inductive (parasitic) elimination and restraint of impedance flow.

State level: Owing to the reduced switching state levels and bias voltages, the system is extremely sensitive to voltage/ground level variations.

In addition to complicating the fabrication of printed boards, the use of heavy copper planes brings concerns as to which type of copper foil should be used, because of the phenomenon of "skin effect." The skin-effect depth is defined as that distance from the surface at which 80% of conduction occurs. This depth varies with frequency (about 2 μm for ECL circuit speeds and 1 μm or less for GaAs circuit speeds), so the conductive path mostly follows the contour of the foil.

As a result of its fabrication process, electrodeposited (ED) foil has both a smooth and a matte side. It is supplied with a treatment on the matte side to enhance bonding of the surface to the laminate. The treatment essentially provides an additional roughness texture consisting of minute projections of copper and copper oxide that are deposited most heavily on the peaks of the tooth (Figure 6.9). Over these projections is deposited a second metallic layer of brass to provide stain-proofing.

Unlike ED copper, foil made by rolling copper ingot to progressively smaller thicknesses has uniformly smooth surfaces on both sides. It too must have a surface treatment to enhance its bonding to a laminate. For this purpose, very small metallic nodules are uniformly deposited to mechanically lock the resin to the otherwise smooth foil surface (Figure 6.10). The treatment further provides the wrought copper with a surface-stabilizing agent that inhibits oxidation and tarnish.

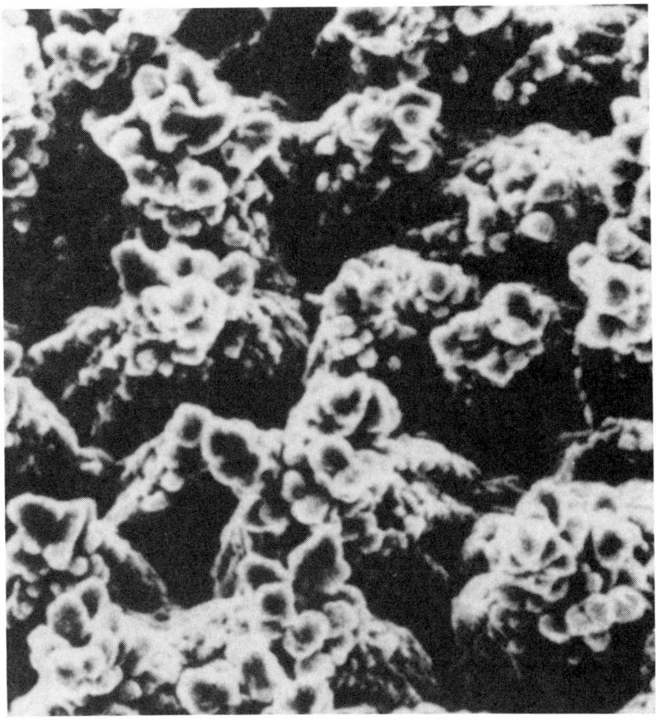

FIGURE 6.9 Electrodeposited (ED) copper foil, 2000X.

Since the conductive path basically follows the irregular contours of the treated foil surface, wrought copper foil does not display the same large increase in conductor resistance at high frequencies as does ED copper. Thus, it is better suited for these applications.

6.6 PLANAR RESISTOR TECHNOLOGY [10]

It is especially effective for some computer circuit designs, such as those using emitter-coupled logic (ECL), to incorporate a large number of terminating resistors into the assembly. The use of planar resistors has proven to be a viable alternative to discrete resistors and resistor networks for several of these applications.

Planar resistor technology is a thin-film system that utilizes conventional subtractive printed-board fabrication methods. The laminate

FIGURE 6.10 Wrought copper foil, 3000X.

system consists of an insulating substrate and, rather than a single conductive layer, has a two-layer bifunctional cladding. The lower layer, immediately against the base laminate surface, is an electrically resistive material. The upper layer is copper. The two layers of this bifunctional cladding are in contact with each other over their entire area. By means of selective etching, the two layers can be etched differently so that separate conductors and resistors are formed.

The advantages of using this type of resistive system include:

Increased circuit density (fewer discrete components)
Improved reliability (fewer solder joints)
Improved electrical properties (the resistors are closer to the integrated circuits)

Therefore, the use of this type of resistive system depends to a great degree on the number of resistors being implemented (and the cost savings this provides) and the level of sophistication of the end-product application (see Section 9.8).

REFERENCES

1. C. Harper and W. Staley, "Interconnection Challenges with High Speed Digital Processing Systems," *Electronic Packaging and Production*, April 1985, pp. 58-62.
2. M. Saubert and D. Snyder, "Packaging Engineers Face Conflicting Demands," *Electronic Packaging and Production*, June 1985, pp. 152-153.
3. R. Pound, "Packaging Links Fast GaAs Dice to High-Speed Systems," *Electronic Packaging and Production*, August 1985, pp. 71-74.
4. G. A. Katronge and G. Northover, "Chip Carriers for Computer applications," *Circuits Manufacturing*, February 1981, pp. 69-74.
5. C. A. Harper and W. W. Staley, "Some Critical Materials Factors in the Application of Leadless Chip Carrier Packages," *Electronic Packaging and Production*, August 1981, pp. 137-140.
6. V. Milutnovic, "Issues of Importance in Designing GaAs Microcomputer Systems," *Computer*, October 1986, pp. 45-53.
7. I. Hutkin, "Foiled by Gallium Arsenide," *Circuits Manufacturing*, June 1986, pp. 53-58.
8. R. Pound, "Maintain the Speed of GaAs in Digital Systems," *Electronic Packaging and Production*, October 1985, pp. 50-52.
9. R. Coulombe and M. Pullam, "Heavy Copper Inners for High Speed (ECL) Logic Circuits, *PC Design*, March 1985, pp. 3-11.
10. B. Mahler and P. Schroeder, "Planar Resistor Technology for High-Speed Multilayer Boards," *Electronic Packaging and Production*, January 1986, pp. 150-154.

7

Thermal Management

While the power consumed per integrated circuit (IC) function is decreasing, the number of functions per chip is increasing. This has led to increased power consumption per device. Opinion is divided on what power the VLSI of the 1990s will dissipate, but a norm of a few watts per device package is probable [1-3]. The greatest impact on thermal management will result from having such devices mounted very close together. This means that with one 68-pin chip carrier per square inch, the power density will be on the order of a few watts per square inch.

If the junction temperatures of the IC devices are to be held to a reasonable level, about 100°C, by using convection cooling or forced-air cooling in the worst case, it will certainly be necessary to have a good thermal path from the chip to the P&I structure or to its own heat sink. The latter will necessitate having the heat

FIGURE 7.1 Chip carriers with individual heat sinks. (Source: Unisys Corp.)

sink as an integral part of an individual chip carrier (Figure 7.1) and/or having chip carriers with integral heat sinks mounted together on a mother carrier.

The heat concentrations encountered in today's "cutting-edge technology," as exemplified by multichip modules (Chapter 13) pose a significant challenge to the packaging engineer with respect to thermal management. Table 7.1 reveals that at the bare chip level within the package, these heat fluxes are in the range of 5 to 40 W/cm^2 (31 to 250 W/in.^{2}). The upper end of these heat densities is comparable to the thermal loading experienced by space reentry vehicles, and at the lower end to heat fluxes imposed on rocket motor cases.

These thermal management requirements and the critical need to reduce off-chip time delays, as well as the need to provide significantly higher system mean-time-between-failures (MTBF) values at substantially lower costs, have focused efforts on the use of multichip modules. Thus, the thermal management associated with multichip modules is, by necessity, at the forefront of thermal control technology.

7.1 POWER CYCLING [4]

Most thermal analysis heretofore has been concerned with solder joint reliability under temperature-cycling testing. Thus, initial design

TABLE 7.1 Single-Chip and Multichip Module Thermal Parameters [3]

Technology	Chip size (mm)	Maximum power dissipation (W)	Maximum chip flux (W/cm^2)	Thermal resistance (K/W)
Single-chip modules				
Mitsubishi Alumina HTCP	8 × 8	4	6.25	5.2
Mitsubishi SiC	8 × 8	4	6.25	4.7
Hewlett-Packard Finstrate	6.3 × 6.3	4	10.10	<8.7
Hitachi S-810	1.9 × 4	1	13.1	7.0
Fujitsu M-380	4.5 × 4.5	3	14.8	8.0
Fujitsu M-780 air-cooled	9.3 × 9.3	6.5	7.5	3.5
Fujitsu M-780 water-cooled	9.3 × 9.3	9.5	11.0	2.5
Burroughs PGA	4.5 × 4.5	5	24.3	12
Motorola MCA-2	7 × 7	12	24.5	3.3
Sperry Compact HX	5 × 5	10	40	4.9
Multichip modules				
Mitsubishi HTCM	8 × 8	4	6.25	7.3
NEC SX liquid-cooled module	8 × 8	>5.4	>8.4	5 (water-cooled)
Hitachi RAM	1.9 × 4	1	13.1	34.7
IBM 4381	4.6 × 4.6	3.8	17.0	17.0
IBM 3090 TCM	4.85 × 4.85	7	29.8	8.7 (water-cooled)
NTT grooved substrate	8 × 8	15.1	23.6	3.3 (water-cooled)

considerations focused on matching the coefficients of linear thermal expansion of the leadless chip carrier and the P&I structure on which it is mounted.

However, temperature cycling will cause behavior in some of the constituent materials that is not representative of normal operating conditions. This might lead to misleading or wrong conclusions for applications where cycling stresses are not imposed primarily by such temperature extremes but by power cycling (power on/off). In reality, functioning assemblies do not heat up and cool with the chip carrier and P&I structure at the same temperature. Rather, when the circuit is turned on, the chip gets hot first and then its chip carrier package expands. As shown in Figure 7.2, this produces a compressive strain in the solder joints, with the hot (expanding) chip carrier on one side and the still cold board on the other side. Only later does the board temperature rise and reduce the solder joint strain. Conversely, when the circuit is turned off, the chip carrier cools down and contracts. Then the strain becomes one of tension while the board is expanding, as it serves its role as a heat sink for the heat dissipated by the chip carrier.

This localized solder joint analysis is further complicated by the fact that such events take place randomly over the surface of the assembly and at varying magnitudes, depending on the thermal dissipation of the various chip carriers in the circuit. Thus, the ideal P&I structure is one that has a thermal expansion coefficient tailored to satisfy the specific requirements of its assembly.

Therefore, thermal management with respect to selecting the appropriate P&I structure for an application is much more complicated

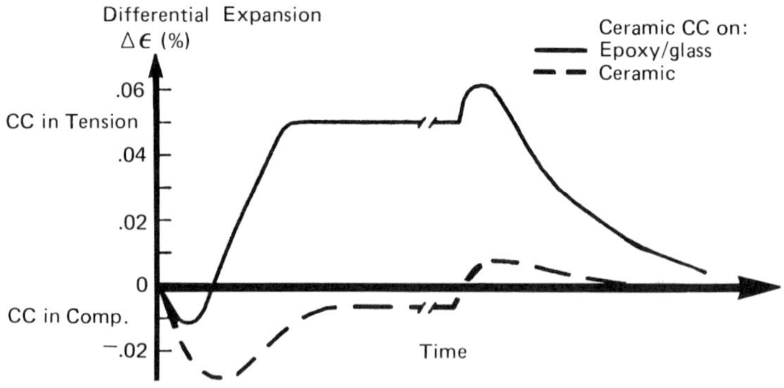

FIGURE 7.2 Differential expansion of chip carriers.

than just matching coefficients of linear thermal expansion, as originally thought. This will probably bring about a reconsideration of P&I structure design criteria, with some leading candidates falling back and some trailing candidates coming to the forefront.

7.2 FORCED-AIR COOLING

7.2.1 IBM Series 4381 Computers [5]

In addition to the multichip modules described in Chapter 13 for the Hitachi and NEC computers, air-cooled multichip modules are also used in the IBM 4381 series computers. These (MCM 641) modules are made up of a complex arrangement of circuit chips on top of a multilayer ceramic substrate.

The MCM 641 module has a 64-mm (2.5-in.) square substrate with a maximum of 36 chips. These chips dissipate up to 90 W at the module level and 3 W at the chip level. The module is capped with a serpentine heat sink (Figure 7.3a), which is used to remove the power generated by these powerful devices. The air-cooled heat sink provides the thermal bridge to connect the circuit chips to the surrounding environment (Figure 7.3b).

7.2.2 Mitsubishi High Thermal Conduction Module

Mitsubishi's concern with high thermal resistance of flip-chip bonded devices has led to the development of a packaging technology that relies on heat transfer from both the top and solder-bump sides of each chip in order to achieve a relatively low junction-to-ambient thermal resistance. The Mitsubishi high-thermal-conduction multichip module (HTCM, Figure 7.4), contains nine 3000-gate ECL chips. The chips are each 8 mm (0.3 in.) square with 223 I/Os and a maximum dissipation of 4 W. The chips are solder-bumped to the substrate and soldered on top to a 13-mm (0.5-in.) square × 0.25-mm (0.1 in.) thick copper plate.

In operation, heat released by the chips flows to the heat sink along two parallel paths, one passing through the solder bumps and the substrate and then onto the cap and heat sink through the module seal, and the other going directly to the cap and heat sink via the thermal spreader and the gas gap.

It must be noted, however, that in order to achieve the thermal performance listed in Table 7.1, the HTCM must remain hermetically sealed, and the gas gap dimension must be kept to approximately 25 μm (1 mil), and the quality of the mating surfaces and/or the thickness of the bonding layers must be carefully controlled.

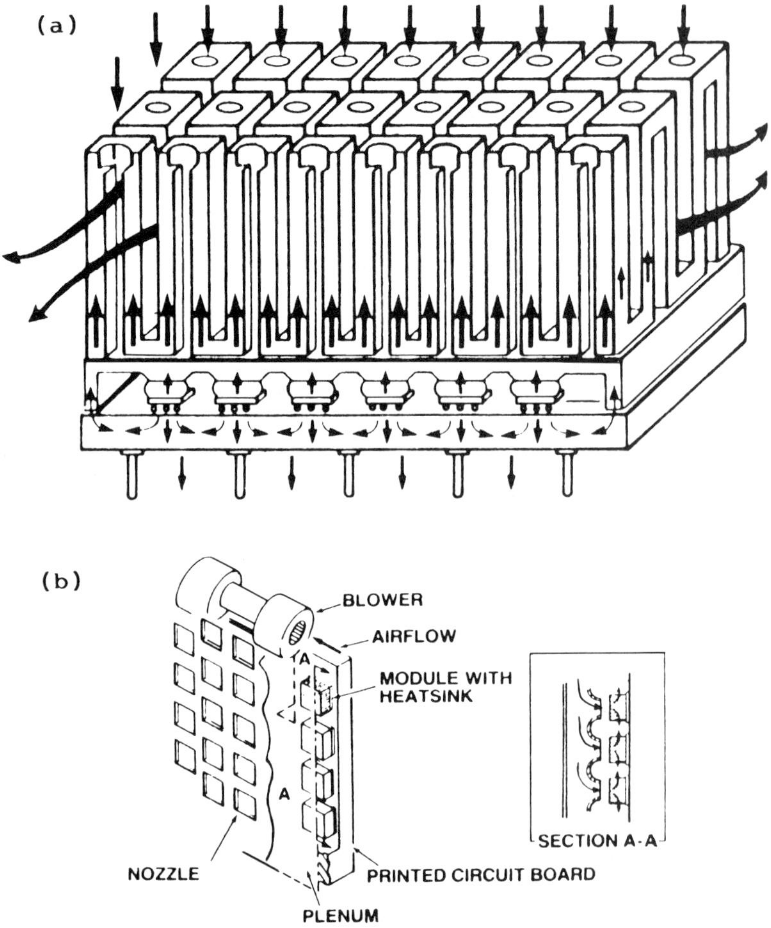

FIGURE 7.3 Air-cooled IBM 4381 multichip module (from ref. 3).

7.3 LIQUID COOLING

With their multitude of high-speed, high-power chip carriers, the highly sophisticated electronics systems (computers) dissipate vast amounts of power. Such a system may have to dissipate over 30 kW of power, have several thousand 5-W devices, and have more than 1 kW in individual assemblies. In these instances, the industry is in a transitional stage, gradually changing from forced-air cooling to liquid cooling.

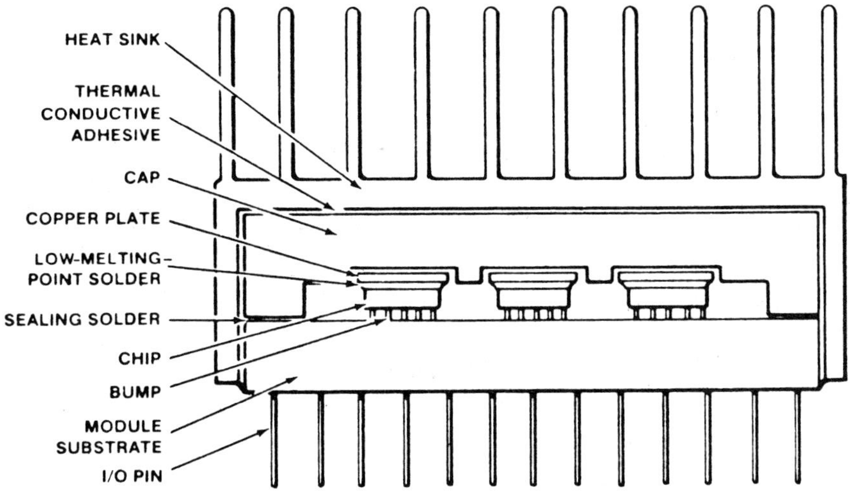

FIGURE 7.4 Mitsubishi high thermal conduction module (from ref. 3).

7.3.1 Sperry Unisys 1100/90 Computer

Liquid cooling, with its compressors, manifolds, and tubing, is a much more convoluted matter than air cooling, but above certain heat loads it is often the only way to go. In the example shown in Figure 7.5, each chip carrier dissipates about 5 W.

The technology is based on water-cooling pairs of board assemblies with leaded chip carriers separated by an aluminum cold plate. The outer layer of the cold plate is a plastic membrane of Tedlar compliant film enclosing a thermal conductive paste.

The boards are pressed against the membranes, forcing the chip carriers against the film, and the thermal paste conducts the heat of the chip carriers to circulating water running through the cold plate. The water flows from the cold plate to a cooling unit, which serves as a heat exchanger between the coolant and a customer-supplied chilled-water system. The cooling system accommodates 26 board pairs, a heat load of up to 10 kW.

7.3.2 Nippon Telegraph and Telephone Modules [6]

Although quite suitable for two-dimensional packaging, conventional liquid-cooling systems are not well suited to three-dimensional or high-density stack packaging. This is because the application of these systems results in complicated and bulky modules, as represented by the

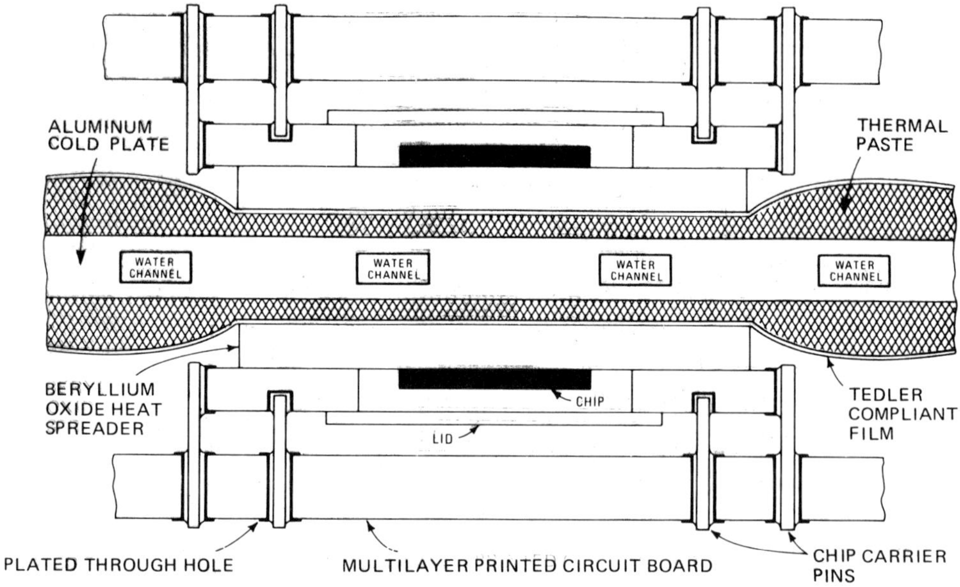

FIGURE 7.5 Liquid-cooling elements of Sperry (Unisys) 1100/90 computer.

IBM and NEC thermal conduction modules described in Chapter 13. Therefore, other companies have found it essential to develop liquid-cooling packages having a higher cooling capability and better suited for high-density packaging.

Nippon Telegraph and Telephone (NTT) recently developed a new indirect liquid-cooling multichip module that is part of a high-density stacked packging system. The fabricated package dimensions are 85 mm (3.3 in.) wide × 105 mm (4.1-in.) long onto which a 5 × 5 array of 10-mm (0.4-in.) square chips are mounted.

In NTT's multichip module, very fine coolant channels are formed into the lower portion and between the buried vias of a six-layer alumina substrate (Figure 7.6a). Interconnection pads are fabricated on the front and rear surfaces of the substrate to permit connection of the substrate to the printed wiring boards located on three sides of the package (Figure 7.6b).

In the heat transfer process the coolant is supplied from the inlet port and distributed to each of the twenty-nine 800-μm (32-mil)

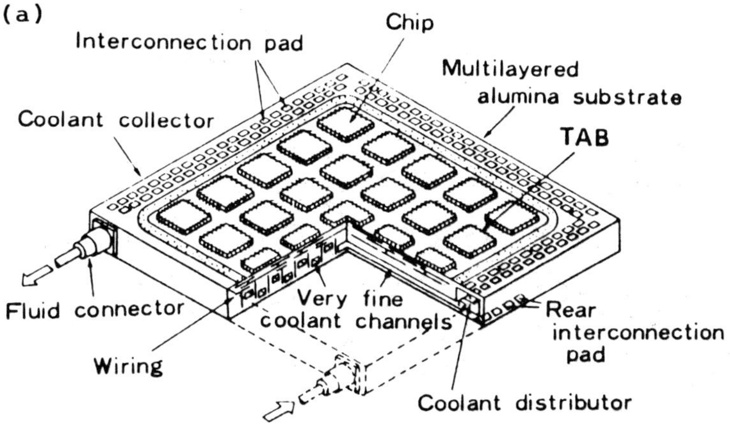

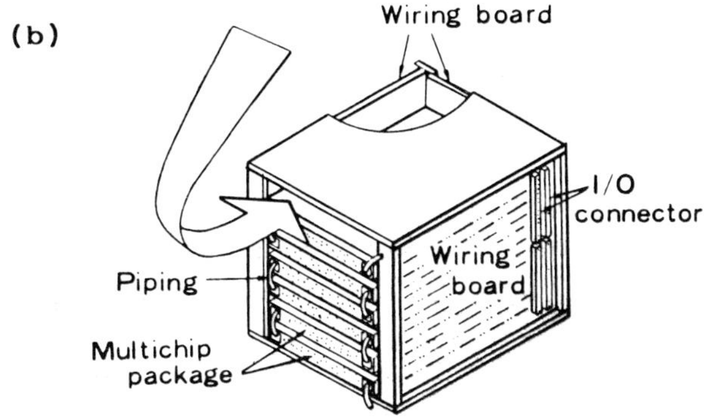

FIGURE 7.6 NTT's stacked indirect liquid-cooling multichip module package (from ref. 3).

wide and 400-μm (16-mil) wide coolant channels by the coolant distributor, thereby ensuring a uniform velocity profile. The heat generated in the chips is conducted to the coolant channels by way of the alumina substrate, where it is transferred to the liquid flow.

The warmed coolant is subsequently connected after flowing through the coolant channels. It then flows out through the coolant outlet port. The upper and lower stack 900-I/O packages are connected in series by piping.

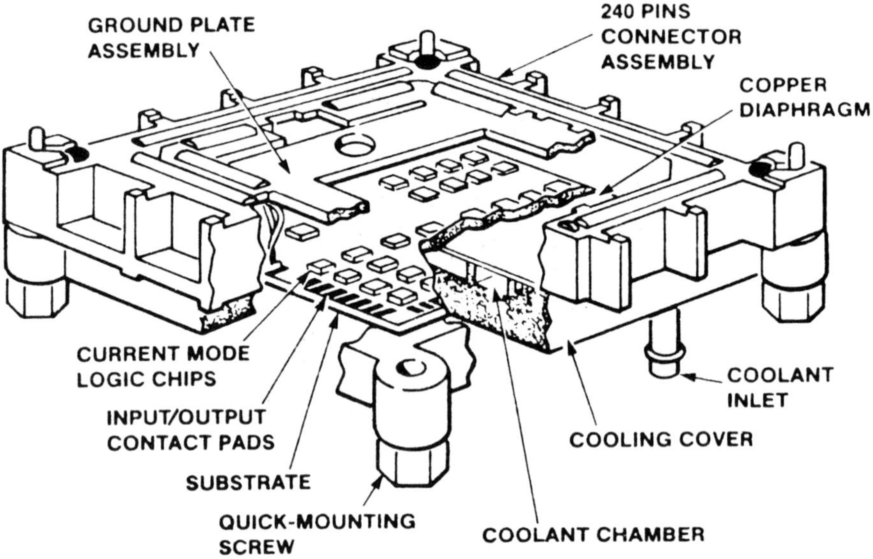

FIGURE 7.7 Honeywell silent liquid integral-cooler multichip module (from ref. 3).

7.3.3 Honeywell Silent Liquid Integral Cooler Module

As shown in Figure 7.7, cooling of the Honeywell DPS-88 computer described in Section 2.5 is provided at the multichip module level by means of a silent liquid integral cooler (SLIC) cover, which serves as a liquid-cooled cold plate. A flexible copper diaphragm, which conforms to the back surface of the substrate on one side and is wetted by the circulating water or by a water-cooled plastic membrane on its other side, constitutes the bottom of that cover. Each multichip module is reported to dissipate 60 W, for an average chip dissipation of approximately 0.55 W.

7.4 SUMMARY AND CONCLUSIONS

Many other sophisticated thermal management schemes exist; some are based on the cooling of the P&I structure instead of cooling the chip carriers directly, others are based on the use of forced-air cooling, and still others utilize a high degree of conduction cooling. However, no matter what the ultimate scheme chosen, all such designs require thermal management that takes into account the thermal properties of both the chip carrier and the P&I structure. Therefore, careful consideration must be given to these elements of the packaging system when a high degree of heat dissipation is required.

REFERENCES

1. C. L. Lassen, "Wanted: A New Interconnection Technology," *Electronics*, September 27, 1979, pp. 113-121.
2. J. Lyman, "Supercomputers Demand Innovation in Packaging and Cooling," *Electronics*, September 22, 1982, pp. 136-143.
3. A. Bar-Cohen, "Thermal Management of Air and Liquid Cooled Multi-Chip Modules," IPC-TP-632, IPC Technical Paper, April 1987.
4. W. Engelmaier, "Effects of Power Cycling on Leadless Chip Carrier Mounting Reliability and Technology," Proceedings, International Electronic Packaging Conference, November 1982.
5. N. J. Timko, Jr., "Air-Jet Impingement Keeps Computer Cool," *Electronic Packaging and Production*, May 1987, pp. 33-34.
6. T. Kishimoto and T. Ohsaki, "VLSI Packaging Technique Using Liquid-Cooled Channels," Proceedings, Electronic Components Conference, May 1986.

8

Printed-Board Design Considerations

The continuing trend toward finer board lines and spaces (Figure 8.1) and the availability of thinner copper foils, preferably 1/2 oz, seem to be sufficient for current requirements for surface-mounting chip carriers [1]. For example, approximately thirty five 84-pin chip carriers can be interconnected with two signal layers on an 8-in. × 13-in. board with a conventional 0.034-in.-diameter vias, 0.005-in. conductor widths, and 0.007-in. spacing between conductors (Figure 8.2). However, as the technology continues to mature, higher component densities will be required, and these land pattern features will not be sufficient. The board designer will then be forced to push for fine via holes with land diameters of about 0.020 in. along with finer line widths and spaces. As shown in Figure 8.2, this will

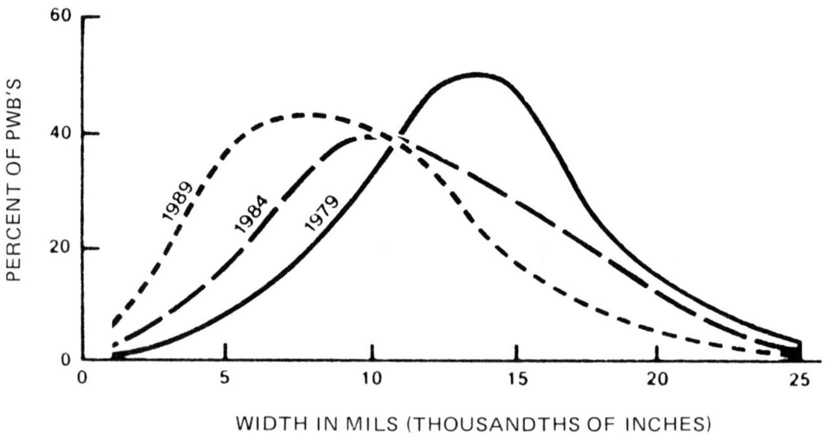

FIGURE 8.1 Printed-wiring conductor width and spacing trends (from ref. 2).

provide density improvements on the order of 35% using 0.050-in.-pitch chip carriers.

Looking further into the future, fine-pitch chip carriers with 0.025-in. centers will begin to appear that will provide component density improvements of well over 100% of present-day capabilities if the associated improvements in fine via size, smaller line widths, and smaller spacing can be achieved. How successfully subtractive circuit technology can be improved to meet these requirements will to a great extent determine the importance of additive circuit technology for future chip carrier/surface-mounting applications.

8.1 BOARD SIZE/QUANTITY

The envisioned improvements in land pattern feature size will have a significant effect on the board area/number of layers required to interconnect a given circuit function. However, this does not mean that boards will get smaller and the number of layers will decrease. The driving force will continue to be to put more and more circuit functions on a board. Therefore, board fabrication, assembly, and testing capabilities/limitations will probably keep board size and the number of layers where they are today.

A more important question is whether or not the number of boards used will decrease. The answer to this question will depend on the

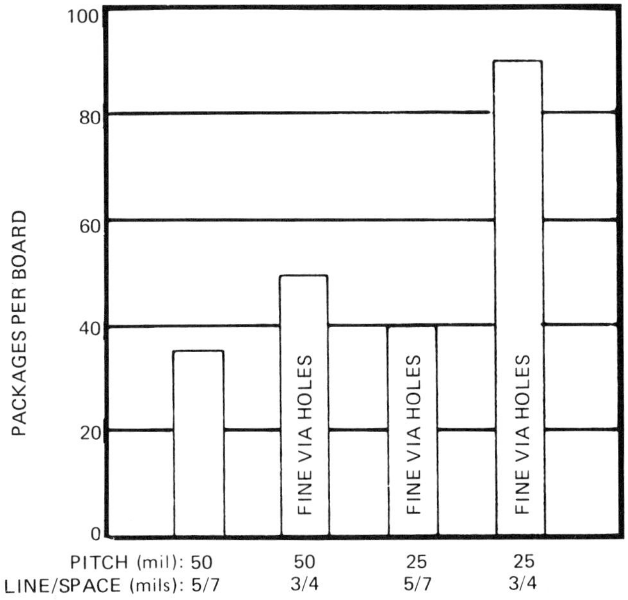

FIGURE 8.2 Printed-wiring design trade-offs (from ref. 3).

availability of fine-pitch chip carriers, the relative size of component packages, capability improvements, etc. If past history is any guide, equipment complexity increases will compensate for function size decreases. Thus, whether or not board usage will be affected by this new technology will probably depend more on marketing forces than on technical decisions.

8.2 SPACE EFFECTIVENESS [2, 4]

The space effectiveness (die size/package size) of a DIP is inversely proportional to the number of pins it has. For example, an 18-pin DIP requires about 0.28 in.^{2} to support a die of 0.039 in.^{2}, or is 14% efficient as a package. In a 40-pin DIP, the die size increases to 0.05 in.^{2} with a 1.25-in.^{2} package and the efficiency drops to only 4%. In comparison, an 18-pin rectangular chip carrier requires about 0.1-in.^{2} to support a maximum die area of 0.03 in.^{2}, for a space efficiency of 30%--that is, over twice the value for an 18-pin DIP.

As the I/O goes to 40 pins, a leadless chip carrier that uses 0.48 in.^{2} will accept a die that is 0.25 in.^{2}, giving a slightly reduced efficiency of 26%. Thus, leadless carriers retain a relatively high space efficiency as pin count increases, while being more space-effective overall.

Comparison of the space required for the various chip carriers shows that for the relatively lower pin counts the fine-pitch chip carriers require less space (Figure 3.16). However, as the pin counts increase, the advantages of using the fine-pitch chip carriers decreases, especially when compared to the use of pin-grid arrays.

8.3 LAND PATTERNS [4, 5]

None of the advantages of using chip carriers and surface mounting would have any real significance without the ability to properly interconnect the devices. Fortunately, the use of chip carriers and surface mounting allows for the use of small hole sizes, dictated only by the substrate thickness and/or the through-hole technology being used. Thus, land patterns ("footprints") can be arranged to produce configurations with a routing capability commensurate with the increased packaging density.

The holes need not be large; 20-mil-diameter holes in 30-mil square lands is sufficient. In turn, the reduced land size produces larger between-land spacings, which can accommodate three conductors instead of two or two conductors instead of one for conservative tolerances and double-sided boards (Figure 8.3). In either case, the use of surface mounting can significantly increase routing density.

The grid array packages utilize 100-mil terminal spacing. This dictates the next level of interconnection in that hole sizes are fixed by the pin diameter. Annular ring requirements establish the land size. This in turn establishes the size of the available wiring channel.

High-density land patterns are also possible for 0.025-in. (Figure 8.4) and 0.020-in. (Figure 8.5) "fine-pitch" chip carriers. However, the benefits of using these patterns can be obtained only if a finer hole grid is used on the next interconnection level.

Design guidelines for all the major types of surface-mount components are provided in IPC-SM-782 (Appendix C) (available from the Institute for Interconnecting and Packaging Electronic Circuits, 7380 North Lincoln Avenue, Lincolnwood, IL 60646).

8.3.1 Resistors and Capacitors

Land patterns for rectangular chip components, such as resistors and capacitors, should be designed to provide the mounting surface for an appropriate solder pedestal (Figure 8.6). Manufacturing allowances, phototool tolerances, etc., should also be taken into account.

A common land pattern design guideline for both reflow-soldered and wave-soldered assemblies for resistors is

$$\text{Land width (X)} = W_{max} - K$$

$$\text{Land length (Y)} = H_{max} + 2T_{max} + K$$

$$\text{Gap between lands (A)} = L_{max} - 2T_{max} - K$$

where

W = component body width
H = component body height
L = component body length
T = solderable termination thickness
K = constant of 0.25 mm (0.010 in.)

When all the numbers derived from this equation are rounded to the nearest sensible land size, taking into account that board processing tolerances can reduce or enlarge the conductive pattern, land patterns for resistors can be developed like those shown in Figure 8.7. (Details concerning the physical dimensions of the components used in the illustrations in this chapter can be obtained from the Electronic Industries Association, 2001 Eye Street NW, Washington, DC 20006.)

For capacitors, the equation is modified slightly due to the height of the metallization. For most capacitors, except tantalum, the land length formula becomes

$$\text{Land length (Y)} = H_{max} + T_{min} - K$$

Determining land width and gap is the same as for resistors. Typical land patterns for capacitors are shown in Figure 8.8.

When tubular metal-electrode-face (MELF) components are used, cutouts in the rectangular areas of the land pattern (Figure 8.9) have been found to be helpful in holding the component in place during an automatic reflow soldering process. They may be omitted if other processes are used.

8.3.2 Transistors

The land pattern configurations shown in Figures 8.10 and 8.11 provide nominal dimensions for lands used to mount small-outline transistor (SOT) types SOT-23 and SOT-89, respectively.

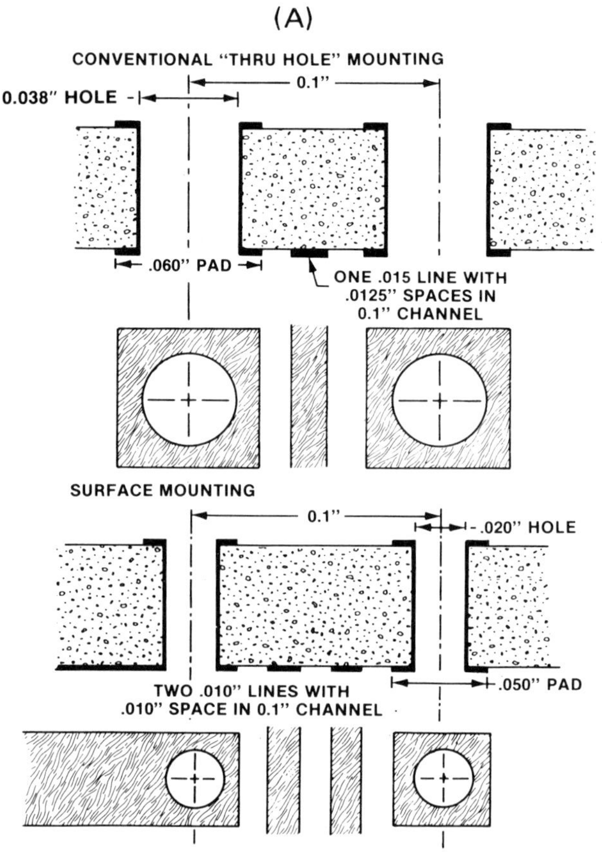

FIGURE 8.3 (A) Chip carrier land pattern routing densities. Low-density conservative tolerances.

8.3.3 Small-Outline Integrated Circuits

A common land pattern design guideline for reflow-soldered, small-outline integrated circuit (SOIC) packages with outward-formed, gull-wing-shape, leads is

Land width = 0.63 mm (0.025 in.)

Land length = 2.0 mm (0.080 in.)

Gap between lands (A) = F - K

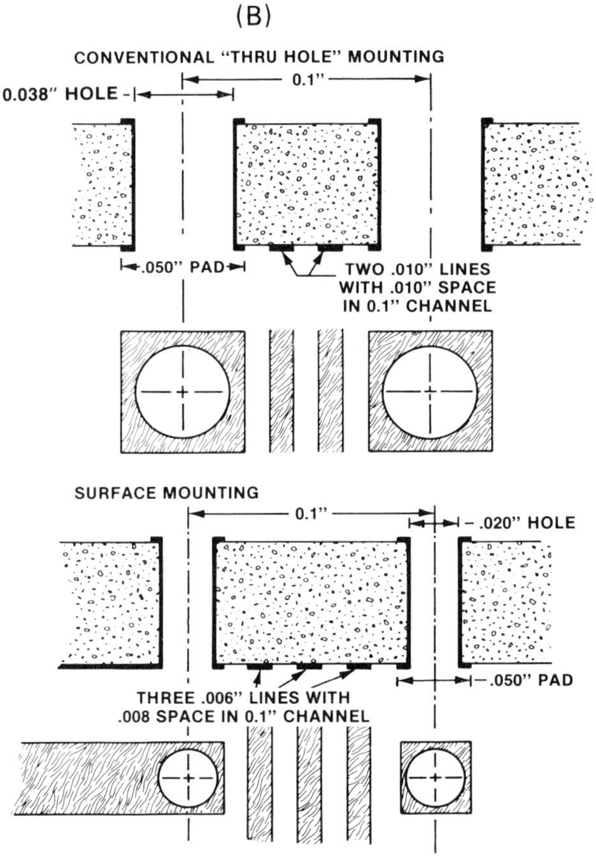

FIGURE 8.3 (B) High-density stringent tolerances.

where

F = SOIC body width
K = constant of 0.25 mm (0.010 in.)

Use of this equation in the same manner as the previous equation was used for resistors yields land patterns such like that shown in Figure 8.12. When the land pattern is modified to accommodate SOICs with inward-formed, J-shape leads, Figure 8.13 can be used.

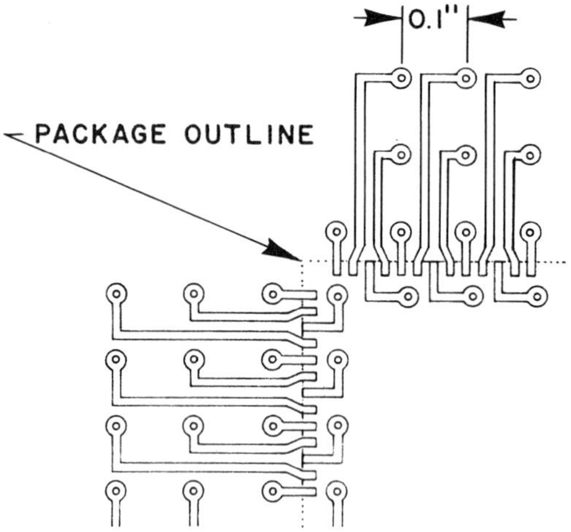

FIGURE 8.4 0.025-in.-pitch chip carrier land pattern detail.

8.3.4 Plastic-Leaded Chip Carriers

The land pattern design guidelines for plastic-leaded chip carriers (PLCCs) are based on

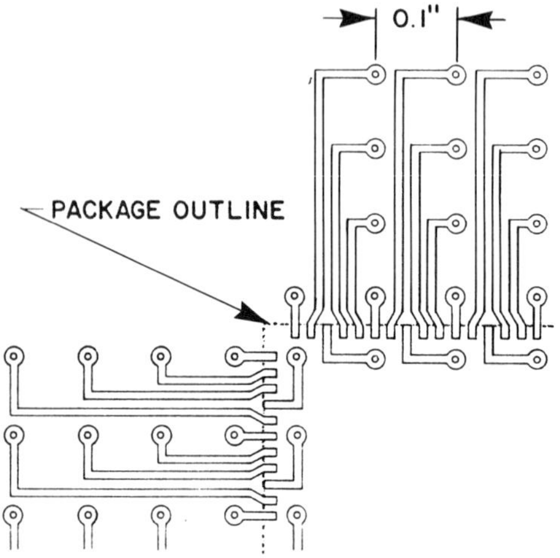

FIGURE 8.5 0.020-in.-pitch chip carrier land pattern detail.

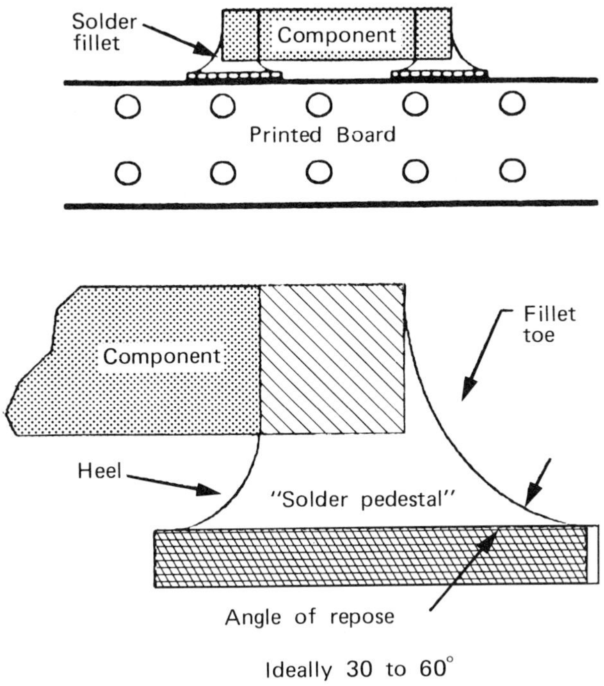

FIGURE 8.6 Solder fillet formation (IPC-SM-782).

Land width = 0.63 mm (0.025 in.)

Land length = 2.0 mm (0.080 in.)

$$B = C + K$$

where

B = overall land pattern reference dimension
C = maximum PLCC body dimension
K = constant of 0.75 mm (0.030 in.)

Based on this, land patterns for square-outline JEDEC (Joint Electronic Device Engineering Council) postmolded type A PLCCs should be similar to those shown in Figure 8.14.

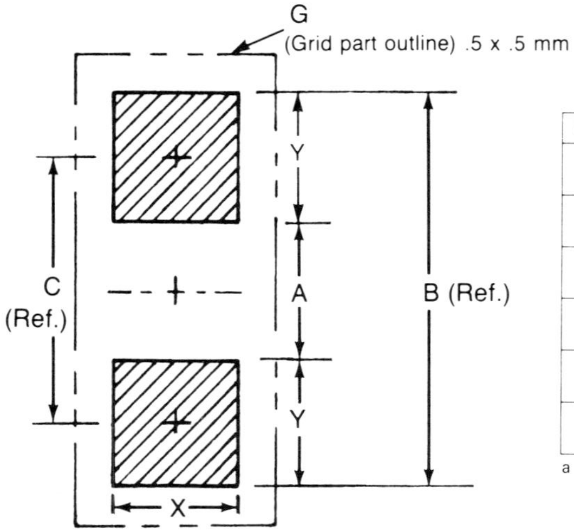

Code Letter	RC 0805	RC 1206 [a]	RC 1210
A	0.8 [.032]	1.8 [.070]	1.8 [.070]
B	3.8 [.150]	5.0 [.200]	5.0 [.200]
C	2.3 [.090]	3.4 [.134]	3.4 [.134]
X	1.4 [.055]	1.6 [.063]	2.6 [.102]
Y	1.5 [.060]	1.6 [.063]	1.6 [.063]
Grid part outline	4 x 8	4 x 12	6 x 12

[a] Example shown in graphics

FIGURE 8.7 Land patterns for rectangular chip resistors, in. [mm] (IPC-SM-782).

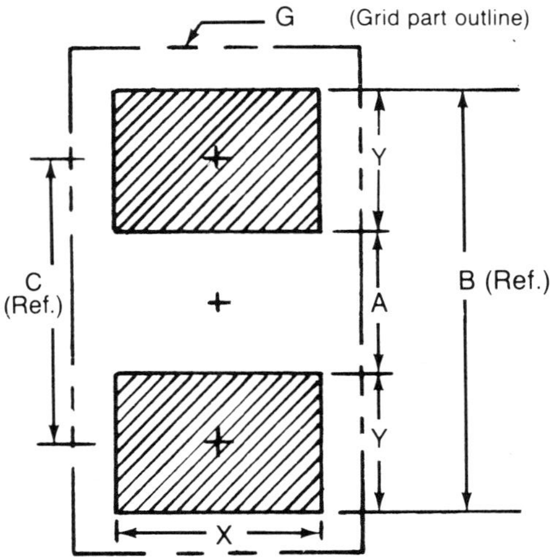

Code Letter	CC 0805	CC 1206	CC 1210[a]	CC 1812	CC 1825
A	0.8 [.032]	1.8 [.070]	1.8 [.070]	3.2 [.126]	3.2 [.126]
B	3.8 [.150]	5.0 [.200]	5.4 [.213]	6.8 [.268]	6.8 [.268]
C	2.3 [.090]	3.4 [.134]	3.6 [.142]	5.0 [.200]	5.0 [.200]
X	1.4 [.055]	1.6 [.063]	2.6 [.102]	3.2 [.126]	6.6 [.260]
Y	1.5 [.060]	1.6 [.063]	1.8 [.070]	1.8 [.070]	1.8 [.070]
Grid part outline	4 x 8	4 x 12	6 x 12	8 x 16	14 x 16

[a] Example shown in graphics

FIGURE 8.8 Land patterns for rectangular chip capacitors, in. [mm] (IPC-SM-782).

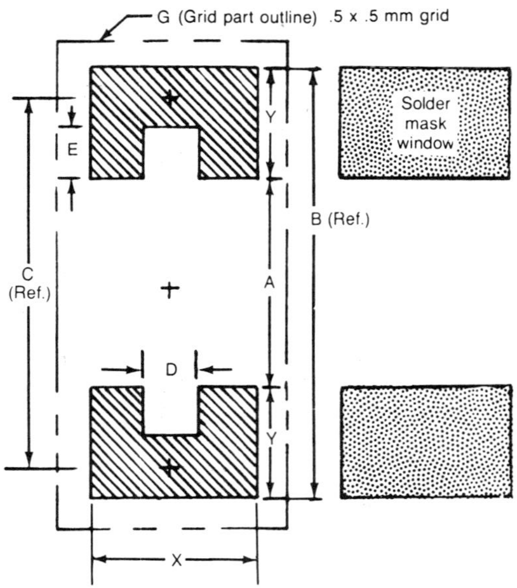

Code Letter	MELF 1/4 W	MLL34	SOD80	MLL41[a]
A	4.4 [.174]	2.2 [.087]	2.2 [.087]	3.4 [.134]
B	8.4 [.330]	5.0 [.200]	5.4 [.213]	7.0 [.276]
C	6.4 [.252]	3.6 [.142]	3.8 [.150]	5.2 [.205]
D[b]	0.3 [.012]	0.3 [.012]	0.3 [.012]	0.3 [.012]
E[b]	0.8 [.032]	0.7 [.028]	0.8 [.032]	1.0 [.040]
X	2.5 [.100]	2.0 [.080]	2.0 [.080]	2.8 [.110]
Y	2.0 [.080]	1.4 [.055]	1.6 [.063]	1.8 [.070]
G	6 x 18	6 x 12	6 x 12	6 x 16

[a] Example shown in graphics

[b] The notch shown by dimensions "D" and "E" are intended to reduce skewing and are optional. The depth of this notch may be determined by the following:

$$E = Y - \left(\frac{B - L_{MAX}}{2}\right)$$

Where "L" equals the maximum body length.

FIGURE 8.9 Land pattern for tubular (MELF) components, in. [mm] (IPC-SM-782).

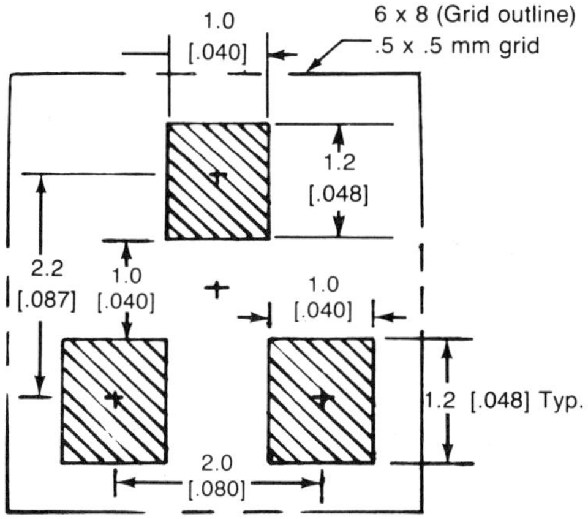

FIGURE 8.10 Land pattern for SOT-23 transistor, in. [mm] (IPC-SM-782).

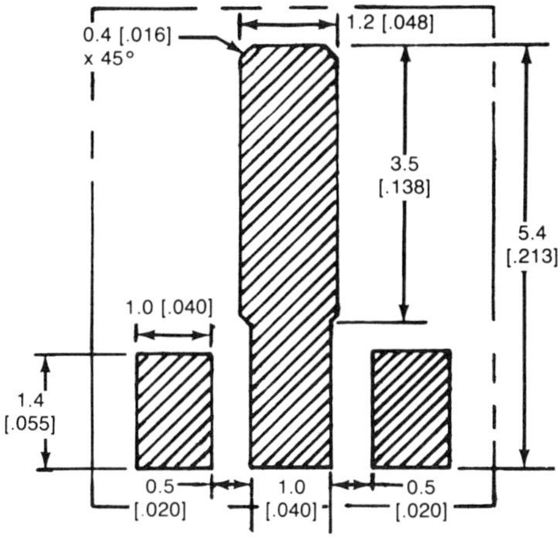

FIGURE 8.11 Land pattern for SOT-89 transistor, in. [mm] (IPC-SM-782).

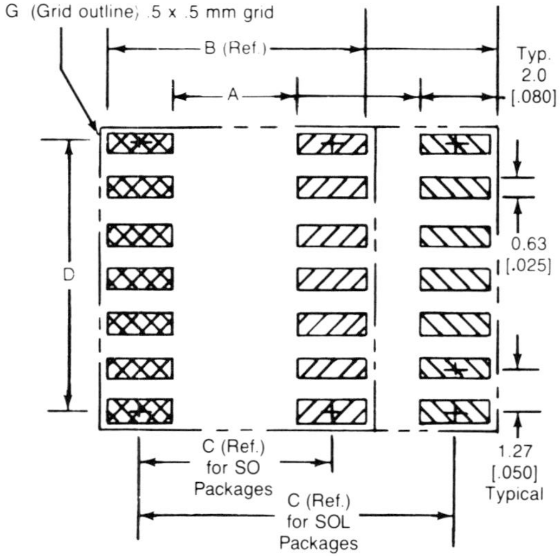

SOIC Lead Count	A	B	C	D	G
SO-8	3.6 [.140]	7.6 [.300]	5.6 [.220]	3.8 [.150]	16 x 10
SO-14[a]	3.6 [.140]	7.6 [.300]	5.6 [.220]	7.6 [.300]	16 x 18
SO-16	3.6 [.140]	7.6 [.300]	5.6 [.220]	8.9 [.350]	16 x 20
SOL-14[a]	7.6 [.300]	11.6 [.460]	9.6 [.380]	7.6 [.300]	24 x 18
SOL-16	7.6 [.300]	11.6 [.460]	9.6 [.380]	8.9 [.350]	24 x 20
SOL-20	7.6 [.300]	11.6 [.460]	9.6 [.380]	11.4 [.450]	24 x 26
SOL-24	7.6 [.300]	11.6 [.460]	9.6 [.380]	14.0 [.550]	24 x 30
SOL-28	7.6 [.300]	11.6 [.460]	9.6 [.380]	16.5 [.650]	24 x 36

[a] Example shown in graphics

FIGURE 8.12 Land pattern for gullwing small-outline integrated circuits, in. [mm] (IPC-SM-782).

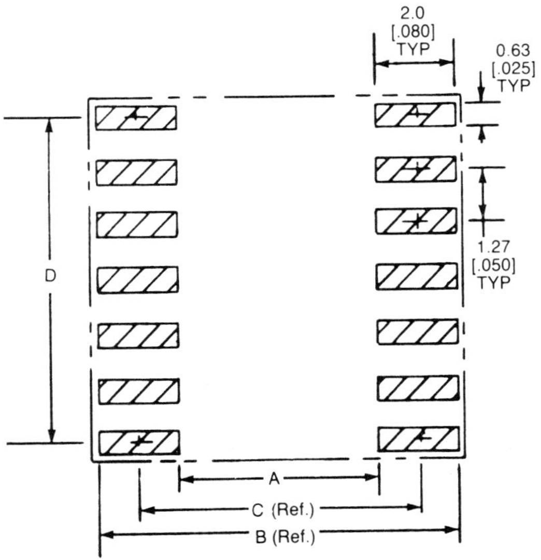

Package Lead Count	A	B	C	D	G
14[a]	4.9 [.190]	8.9 [.350]	6.9 [.270]	7.6 [.300]	18 x 18
16	4.9 [.190]	8.9 [.350]	6.9 [.270]	8.9 [.350]	18 x 20
18	4.9 [.190]	8.9 [.350]	6.9 [.270]	10.2 [.400]	18 x 24
20	4.9 [.190]	8.9 [.350]	6.9 [.270]	11.4 [.450]	18 x 26
22	4.9 [.190]	8.9 [.350]	6.9 [.270]	12.7 [.500]	18 x 28
24	4.9 [.190]	8.9 [.350]	6.9 [.270]	14.0 [.550]	18 x 30
26	4.9 [.190]	8.9 [.350]	6.9 [.270]	15.2 [.600]	18 x 34
28	4.9 [.190]	8.9 [.350]	6.9 [.270]	16.5 [.650]	18 x 36

[a] Example shown in graphics

FIGURE 8.13 Land pattern for J-lead, small-outline integrated circuits, in. [mm] (IPC-SM-782).

8.3.5 Leadless Ceramic Chip Carriers

Leadless ceramic chip carriers (LCCCs) are smaller than comparable terminal count PLCCs, about 1.0 mm (0.040 in.) smaller on a side. However, the land patterns for LCCCs are almost the same as those for PLCCs except that the value for K is different. When the packaging and interconnecting (P&I) structure's coefficient of thermal expansion (CTE) closely matches that of the LCCC, the value for K should be 1.75 mm (0.070 in.). If the CTE is not a good match, such as with glass epoxy substrates, the value of K should be 2.25 mm (0.100 in.) to give further land extension and larger land design (Figure 8.15). However, care should be taken to assure that the land is not too long, as this might allow the solder to "slump" and make the solder joint weaker, with a resulting shortening of product life cycle.

8.4 SPACING

8.4.1 Component Spacing

The land pattern design guidelines discussed so far are important for the reliability of surface mount assemblies. However, the designer should not lose sight of manufacturability, testability, and repairability considerations.

A minimum spacing between components is required to satisfy all of these requirements. However, there is no limit on maximum component spacing; the larger, the better. Unfortunately, some designs necessitate having the components positioned as close as possible. This is not generally a good practice. Based on experience, the minimum component spacing shown in Figure 8.16 should be maintained for good producibility.

8.4.2 Land Spacing

The spacing between lands (terminal areas) for plated-through holes will vary depending on design/manufacturing requirements and, to some extent, on the method of artwork generation, i.e., manual versus photoplotter generation. Figure 8.17 shows some typical recommended values. Figure 8.18 shows various configurations of lands for optimum conductor routing and testability.

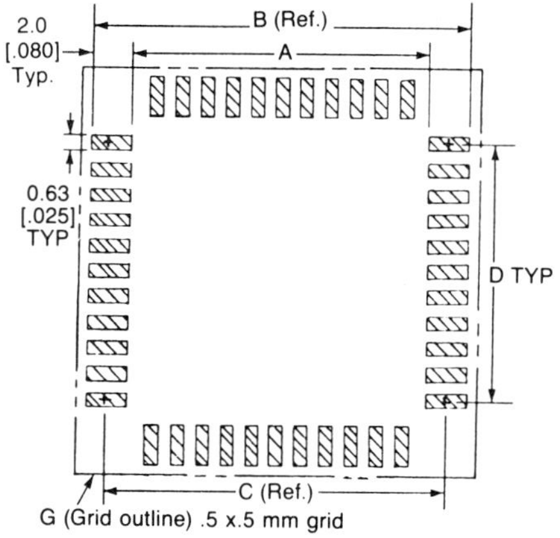

Type A Pin Count	A	B	C	D	G
28	10.0 [.390]	14.0 [.550]	12.0 [.470]	7.6 [.300]	30 x 30
44[a]	15.0 [.590]	19.0 [.750]	17.0 [.670]	12.7 [.500]	40 x 40
52	17.6 [.690]	21.6 [.850]	19.6 [.770]	15.2 [.600]	46 x 46
68	22.6 [.890]	26.6 [1.050]	24.6 [.970]	20.3 [.800]	58 x 56

[a] Example shown in graphics

FIGURE 8.14 Land pattern for JEDEC type-A plastic-leaded chip carriers, in. [mm] (IPC-SM-782).

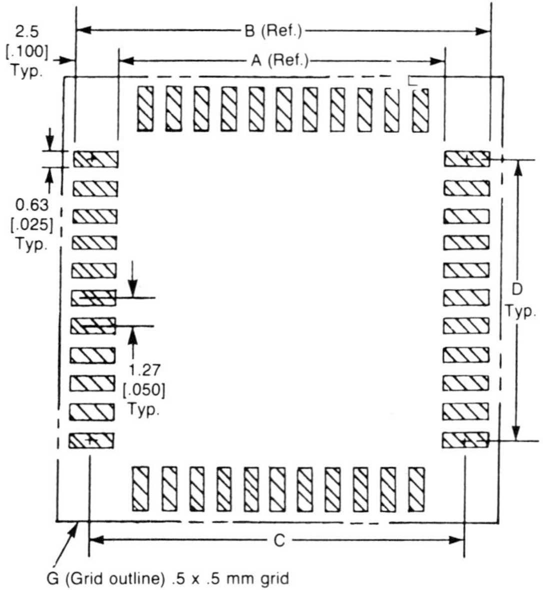

Pin Count	A	B	C	D	G
16	4.8 [.185]	9.8 [.385]	7.3 [.285]	3.8 [.150]	20 x 20
20	6.1 [.245]	11.1 [.435]	8.6 [.335]	5.08 [.200]	24 x 24
24	7.4 [.290]	12.4 [.490]	9.9 [.390]	6.35 [.250]	26 x 26
28	8.6 [.335]	13.6 [.535]	11.1 [.435]	7.62 [.300]	28 x 28
44[a]	13.8 [.540]	18.8 [.740]	16.3 [.640]	12.7 [.500]	38 x 38
52	16.3 [.640]	21.3 [.840]	18.8 [.740]	15.24 [.600]	44 x 44
68	21.4 [.840]	26.4 [1.040]	23.9 [.940]	20.3 [.800]	54 x 54
84	26.5 [1.040]	31.5 [1.240]	29.0 [1.140]	25.4 [1.000]	64 x 64
100	31.5 [1.240]	36.5 [1.440]	34.0 [1.340]	30.5 [1.200]	74 x 74
124	39.2 [1.540]	44.2 [1.740]	41.7 [1.640]	38.1 [1.500]	90 x 90
156	49.3 [1.940]	54.3 [2.140]	51.8 [2.040]	48.3 [1.900]	110 x 110

[a] Example shown in graphics

FIGURE 8.15 Land pattern for standard leadless chip carriers, in. [mm] (IPC-SM-782).

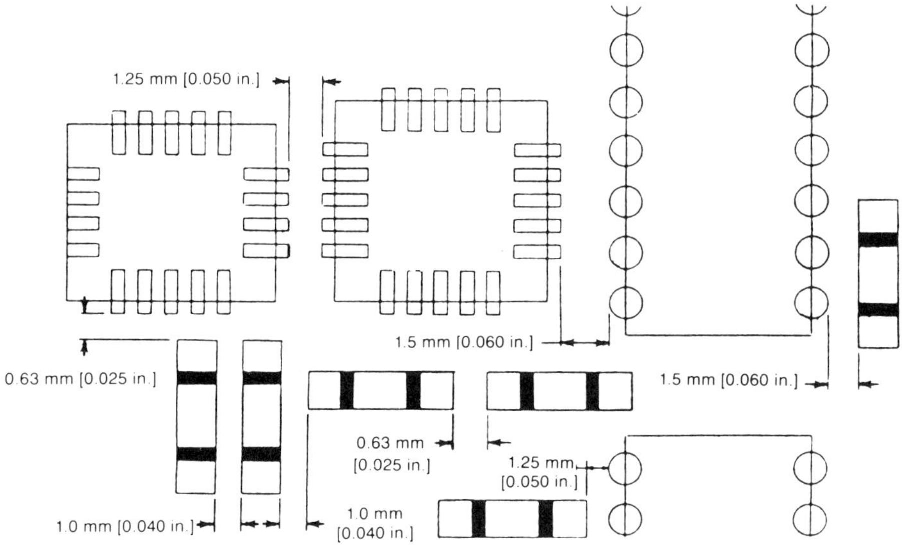

FIGURE 8.16 Recommended minimum land-to-land spacing (IPC-SM-782).

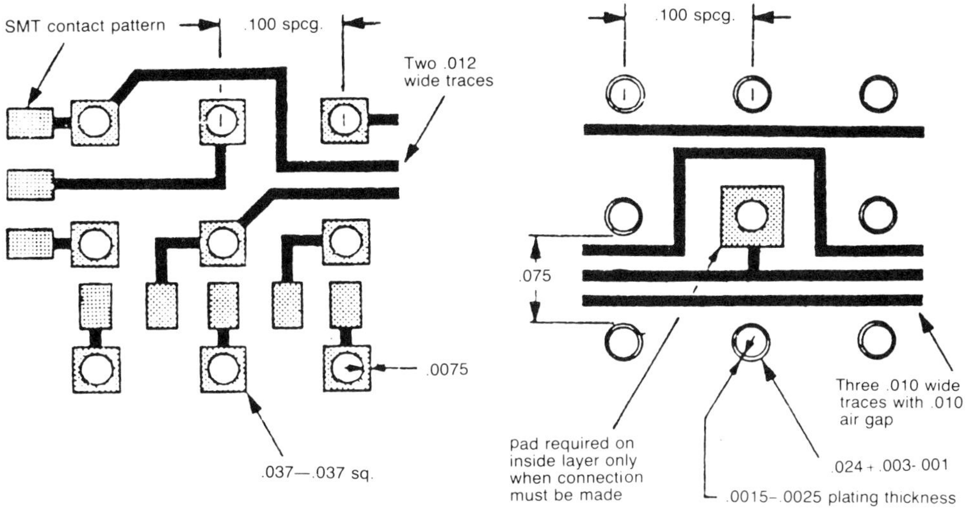

FIGURE 8.17 Recommended minimum land-to-conductor spacing (IPC-SM-782).

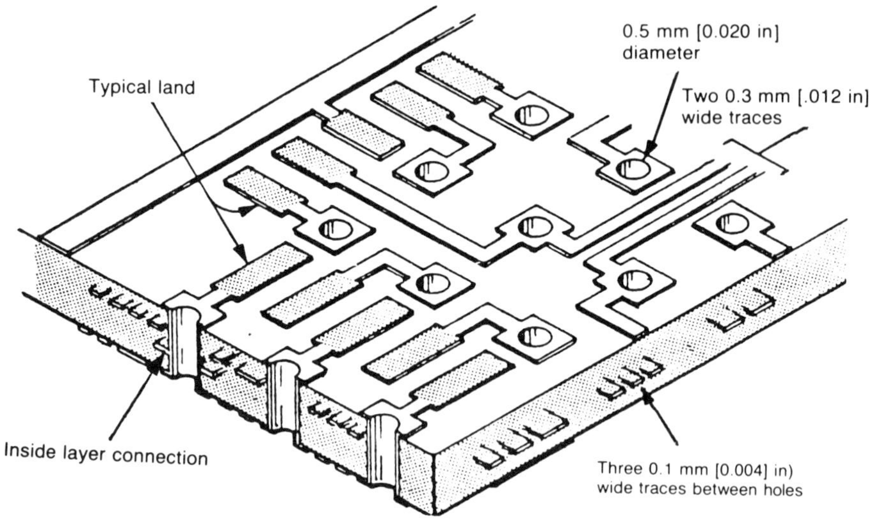

FIGURE 8.18 Land pattern-to-via relationships for 2.54-mm [0.100-in.] grid (IPC-SM-782).

REFERENCES

1. Surface Mount Land Patterns (Configurations and Design Rules), ANSI/IPC-SM-782, Institute for Interconnecting and Packaging Electronic Circuits, March 1987.
2. F. J. Dance, "Can PWBs Meet the Surface Mount Challenge?," *Electri-Onics*, May 1983.
3. N. A. Teneketges and W. H. Knausenberger, "Fine Pitch Chip Carrier Routability Study," Proceedings, First Annual International Electronic Packaging Society Conference, November 1981.
4. D. I. Amey, "Integrated Circuit Package Selection: Pin Grid Array vs. Chip Carriers," Proceedings, First International Electronics Packaging Society Conference, November, 1981.
5. D. I. Amey and J. Balde, "New Chip Carrier Concepts Will Impact LSI Based Designs," *EDN*, September 1978, p. 119.

9

Multilayer Board Fabrication

Many new developments are taking place in multilayer printed-wiring boards [1]. These activities are in response to the increased demands for surface mount products that cost-effectively provide high-density interconnections that are accurately predictable and controlled within specific value "windows." As in all rapidly advancing technologies, certain key elements provide the impetus for this progress. For example, high-density conductor patterns and high-aspect-ratio plated-through holes are becoming more readily achievable with advancements and refinements in drilling, imaging, etching, and plating technology. Thus, the new multilayer boards are becoming more than just a passive means for obtaining interconnections. Except for integrated circuits, they are becoming the most important "component" of critical high-performance electronic assemblies.

9.1 DIMENSIONAL STABILITY

As both the silicon- and GaAs-based technologies drive device speeds into the gigahertz frequency range and drive rise times down to fractions of a nanosecond, the interconnecting role of printed boards assumes a new, dimensionally critical posture. Because the spaces between conductors now become signal paths for electric and magnetic waves to propagate, printed boards must be designed to channel electromagnetic waves in the form of transmission lines rather than just as voltages and currents in conductors.

Control of the physical dimensions of the conductors and spacing between conductors determines the circuit's impedance and crosstalk characteristics. The tolerances of these dimensions ultimately determine the very-high-speed system's performance. Thus, these dimensional tolerances place additional stringent requirements on board fabrication equipment and process controls, with respect to both minimizing the magnitude of tolerances and increasing the degree of repeatability (and dimensional stability) of board features.

9.2 GENERAL CONSIDERATIONS [2]

Despite their advantages, multilayer boards are not simple to make. In fact, fabricating them is a good example of the difficulties that arise when the need to shrink dimensions runs up against physical processing constraints, especially when yields, and ultimately costs, are involved.

The success of a multilayer board design depends as much on the expertise of the board manufacturer as it does on the care taken by the designer. Therefore, there is a continual effort to improve multilayer board manufacturing processes in order both to increase circuit density and reduce end-product costs. The following describes a few recent developments in multilayer board processing that help achieve these goals.

9.3 BURIED VIAS [3]

The use of multilayer boards with "buried" vias is one of the approaches currently being explored in order to provide higher current densities.

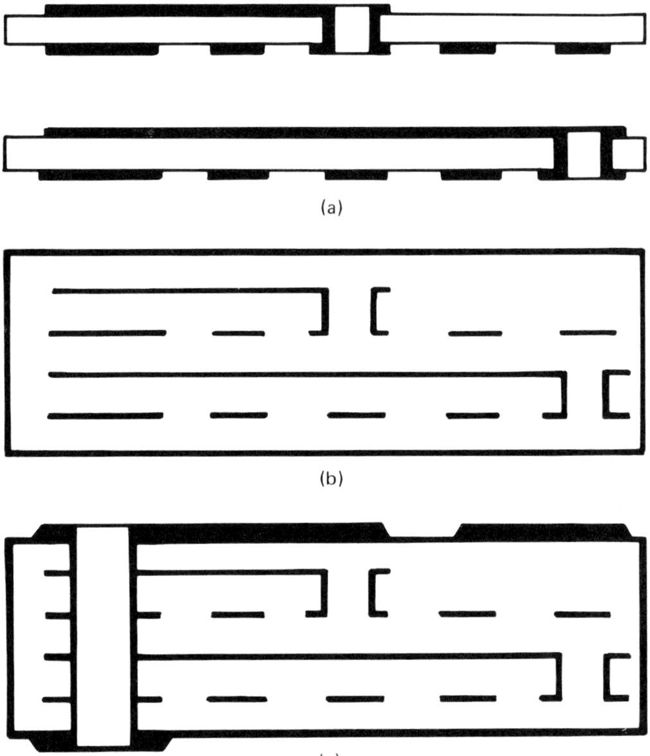

(a)

(b)

(c)

FIGURE 9.1 Multilayer board "buried via" fabrication: (a) internal via holes drilled and plated through; (b) internal layers laminated and through holes drilled; (c) holes plated through and external layers etched.

The main advantage of buried via construction (Figure 9.1) is that it allows connections to be made between internal board layers without connections to external layers. (When the connecting via goes between an internal layer and only one of the outside layers, the via is referred to as a "blind" via.)

Buried via multilayer board fabrication uses the same equipment, materials, and processes required to fabricate conventional boards. The major significant difference is that inner layers are made in pairs connected by small plated-through holes, which subsequently become encapsulated within the board during the multilayer lamination process. Subsequent board processing is the same as with conventional multilayer boards.

Clearly, buried-via board fabrication requires extra processing steps. In many designs, however, the additional cost of these steps is largely offset by a reduction in the number of layers required in the buried-via multilayer board design.

Because the aspect ratio is very low for these holes, they are easily plated. Also, buried via holes can be fabricated with smaller lands than through-via holes. This is possible because less tolerance is needed for registration and drilling a double-sided board than for internal lands in a multilayer board. Annular ring-size requirements are also generally less for lands on internal layers of finished boards than they are for exposed external lands.

9.4 DRILLING SMALL HOLES [4]

The trend toward increased packaging density has resulted in a shifting of commonly used hole drilling diameters downward toward 0.020 in. and less, with corresponding increases in board thickness-to-hole diameter aspect ratios (Figure 9.2). This creates several problems for the multilayer board fabricator, including:

Increased breakage of drill bits, with increased scrap rates
Lower productivity, due to the use of slower "hit" rates and lower stack heights
Higher per-hole drilling costs

In addition, through-hole plating problems are significantly increased.

9.4.1 The Drill Bit

Small-diameter drills cost more and are less readily available than larger ones. Also, they break more easily. Significantly, this breakage is a linear function of drill length and a cubed function of drill diameter. However, there are a few things that drill bit manufacturers can do to compensate for this reduced lack of strength:

1. *Change the material composition*, e.g., increase the cobalt content from 6% to 12%. This increases the strength, although it decreases wear resistance.
2. *Reduce the flute length* to increase resistance to side forces. This, however, reduces the practical drilling depth while still allowing room for chip removal.
3. *Modify geometry*, specifically in three basic areas: (a) eliminating the relief on the margin of the drill; (b) thickening the web of the drill bit; and (c) using single-flute drills.

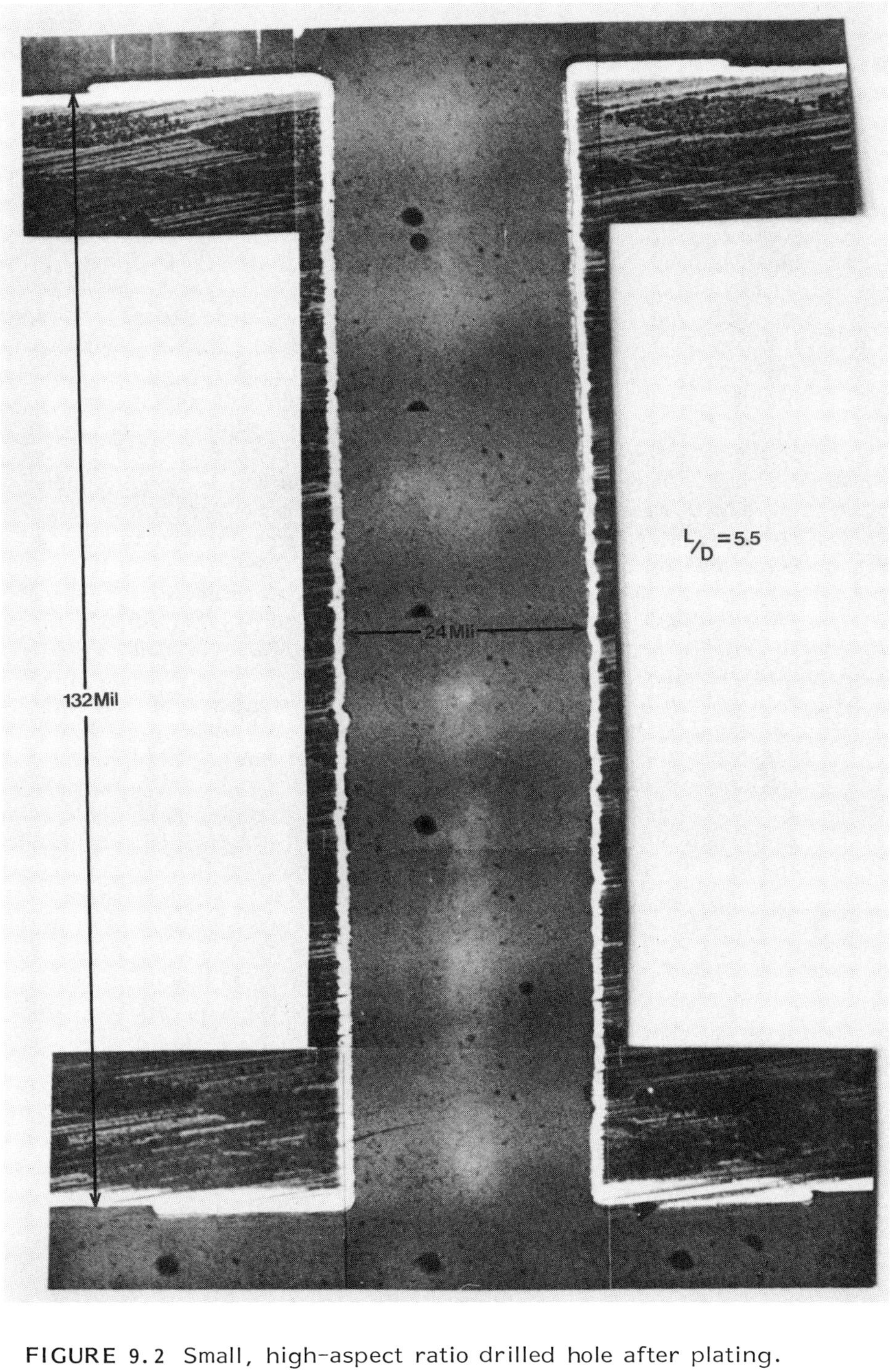

FIGURE 9.2 Small, high-aspect ratio drilled hole after plating.

9.4.2 Drilling Process Variables

There are certain process variables that the board fabricator can control. The ones that are somewhat different when drilling small/deep holes are as follows:

1. *Rotational speed* for drilling holes under 0.020 in. in diameter is the "maximum" practically achievable, presently from 80,000 to 90,000 rpm.
2. *Insertion rate* should be kept relatively low, i.e., from 30 to 160 in./min, to minimize breakage. The success of this will vary with the density and thickness of the board being drilled. Also, the quality of the drill bit has an effect. Therefore, as with most process controls, trial-and-error experimentation is necessary in order to obtain optimum results.
3. *Retraction rate* on most drilling machines is conventionally 1000 in./min to minimize smear, to minimize the time the bit is in the hole, and to maximize productivity. With small-hole drilling the relatively low surface speeds and low stack heights make these factors less critical. Thus, by reducing the retraction rates to from 100 to 200 in./min, the acceleration curve is softened and the shock the drill sees during withdrawal from the hole is reduced.
4. *Process modifications* [5] that are being considered include (a) two-stage "pilot drilling" with a short fluted drill (Figure 9.3), which creates a hole that guides the longer through-hole drill; (b) step "peck drilling" (figure 9.4), where the drill initially enters the material partially, withdraws, then enters the hole again to advance farther into the material, and then withdraws again in order to enter for the last time and finally drill through the material; and (c) the use of "etched vias," which have the drill entry point etched away from the surface foil to guide the drill into the material.
5. *Other variables* that should be evaluated are (a) the entry material, (b) stack height, (c) back-up material, and (d) the drill's pressure foot.

9.4.3 The Drilling Machine

The drilling machine has a great deal to do with successful drilling of small-diameter holes. Therefore, there are some aspects of the machine that require special attention.

1. *Spindle runout* leads to cutters turning eccentrically and can cause drill breakage. Therefore, spindle runout should be minimized, i.e., less than 0.0005 in. TIR.

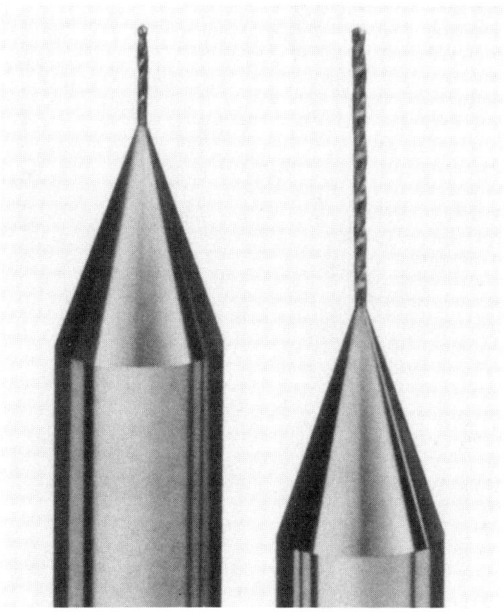

FIGURE 9.3 Short-flute (pilot) and long-flute (through-hole) bits.

2. *Z-axis lost motion* can contribute to drill breakage when the machine reverses its Z-axis travel and begins to withdraw the drill from the hole. Some lost motion is unavoidable and acceptable. However, as with spindle runout, lost motion can be measured and corrected for if excessive, with more than 0.005 in. being a cause for concern.
3. *Computer controls* can be added to drilling machines to provide special functions that address the requirements of small-hole drilling. Depending on their sophistication, such machines can (a) display servo tuning curves and specific response information for performance monitoring; (b) program retraction rates, to allow for "softer" drill retraction; or (c) schedule gain variables, to control the duration and speed of drilling with maximum system stability.

9.4.4 Laser Drilling

In order to overcome many of the problems mentioned above, small-hole drilling may require the use of special machines and processes. One such approach being explored is laser drilling, which appears

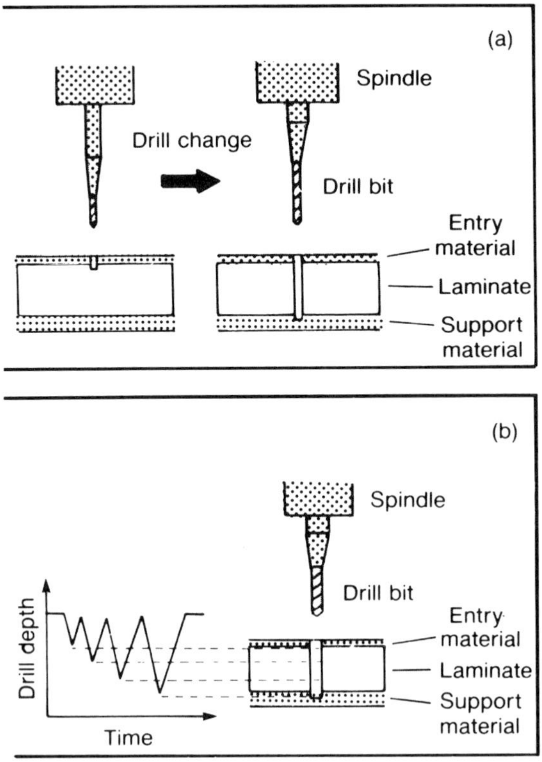

FIGURE 9.4 (a) Two-stage and (b) step drilling.

to be feasible for drilling "blind" through holes in printed boards as small as 0.004 in. in diameter and less.

The use of lasers is receiving much attention for "blind" via-hole drilling. For these applications the use of a CO_2 laser seems to be a good choice, as both epoxy and glass are strong absorbers of radiation at the laser's wavelength, and the high reflectivity of the copper minimizes the coupling of the laser energy into the buried land.

The use of laser drilling and most of the other considerations mentioned here are still under development. There are many more questions than answers. In addition, what works in some cases does not perform as well in others. Therefore, it is clear that much work is yet to be done by the board fabricators and equipment manufacturers in order to develop the optimum process and parameters for specific applications.

9.5 MASS LAMINATION [6]

As defined internationally, "mass lamination" is a technique to manufacture, in large-scale production, multilayer printed boards bearing readily processed conductive patterns on the inner layers and copper cladding on the surfaces intended to be processed by the user to form conductive patterns. (Note: The datum reference will normally be necessary and should be provided.) The process was originally conceived when a base-material manufacturer performed the lamination portion of the multilayer fabrication process. Final processing, such as drilling, through-hole plating, and outer-layer etching, was performed by the purchaser of the mass-laminated product.

Inherent in the mass lamination concept was the use of (1) a copper sheet to form the outer cladding; (2) a large press to do the lamination of large sheets, limited to one or two pre-etched internal layers; and (3) the registration of layers with tooling marks/holes, which were excavated after lamination.

Since the original use of the technique, additional variations of the concept have been employed that are termed "mass lamination." These include:

In-house mass lamination (sometimes called "foil lam"): The complete multilayer board fabrication is performed by a single manufacturer, usually a printed board manufacturer, using the mass lamination techniques of the original concept.

Mini-mass lamination: The mass lamination process using a large press, except that several small panels are laminated at one time in a single press opening by either "vendor" or "in-house" techniques.

Sequential mass lamination [7] (sometimes called "re-lam" or, by maintaining registration of layers using tooling pins, "pinned mass lamination"): The mass lamination process, similar to the original concept, manufactured by either "vendor" or "in-house" techniques, in which there are more than the original three or four layers, with the additional layers being produced by a secondary press cycle.

No matter which type of mass lamination is being used, the main objective is to obtain "economy of scale" (Figure 9.5).

Although some mass laminators are able to provide multilayer boards with six or more layers, the majority of the market consists of boards with three or four layers. However, the greater-than-four-layer segment is going to grow, given such driving forces, for example, as controlled-impedance (Z_0) circuitry.

With "hands-on" raw material expertise, mass laminators can readily alter resin-to-glass ratios and change prepreg thicknesses

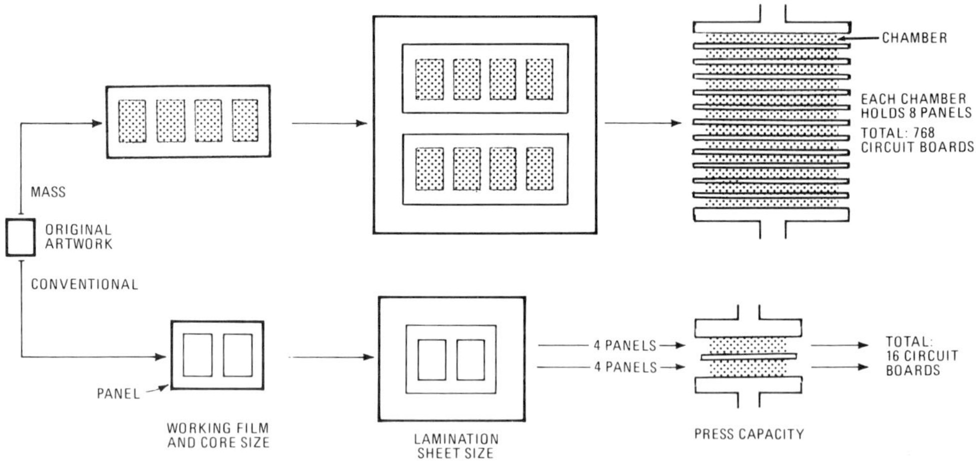

FIGURE 9.5 "Mass lamination" versus conventional lamination (from ref. 7).

to achieve controlled impedance. Also, the growing use of emitter-coupled logic (ECL), which requires additional (and often heavy) power and ground planes, will stimulate the use of mass lamination.

9.6 VACUUM LAMINATION [8]

Vacuum systems, as applied to the multilayer lamination process, are not a new concept. They have been used daily in production environments in some large captive manufacturing shops for several years. However, only recently has there been a marked rediscovery of the advantages of vacuum techniques. The driving force behind this current activity is the new generation of multilayer boards, which have dense geometries and require tight fabrication tolerances.

Basically, the advantages of using vacuum systems include:

Higher yields: This is possible through reduction of the dimensional movements occurring during the lamination process.

Annular ring retention: Designs with small annular rings for improved conductor routing density are possible because of the improved control of registration.

More efficient air removal: Air is driven out of the multilayer layup lamination package very efficiently. Thus, the probability of entrapped air remaining in the package is greatly reduced.

Stress reduction: The lower pressure achieved minimizes fluid mechanical stress in the board for more predictable three-dimensional, true-positional tolerancing.

Improved prepreg thickness control: There is less resin loss because of the use of decreased pressure. Slightly higher thicknesses are thus possible for each prepreg layer, which means that fewer plies are necessary to achieve the desired overall board thickness.

Reduced air and moisture entrapment: Heat and pressure are applied throughout the vacuum-lamination process cycle. This resuts in an inherently more efficient removal of the residual moisture in the prepreg and core material. Consequently, there is a reduction in air and moisture entrapment in the finished board, which results in improved performance characteristics.

On the disadvantage side, it should be noted that low-pressure vacuum lamination is sensitive to tooling setups and prepreg properties. Thus, compared with controlled conventional high-pressure lamination, the vacuum process may not guarantee an equally toleranced multilayer board. In fact, lower pressures result in greater resin retention, a condition further removed from having glass-influenced regions that assist in board thickness uniformity.

9.6.1 Vacuum Hydraulic Lamination

In the vacuum hydraulic lamination (VHL) technique, a vacuum is applied by forming a vacuum assembly, which is placed into each of the openings of a static hydraulic press. In its simplest form, a lamination layup is placed on a carrier plate and a flexible membrane is draped over the package and sealed. A vacuum is drawn on this package and pressure is applied to the assembly.

As the VHL technique evolved, better and more productive fixtures were designed. In one, the flexible membrane is replaced by a piston, which is used for the application of pressure as well as for the transfer of heat. This type of fixture can draw a strong vacuum and still allow flexibility on the height of the lamination layup. In another, a vacuum frame is formed around the lamination package, with flexible seals used to enclose the package. These two VHL concepts have the advantage of allowing traditional tooling and existing static presses to be converted to the newer vacuum technology. Also, very little investment is required to complete the conversion.

A third VHL process encloses the entire platen portion of the hydraulic press. This method has the potential of eliminating the need for specialized fixtures. It is also usable for standard laminating.

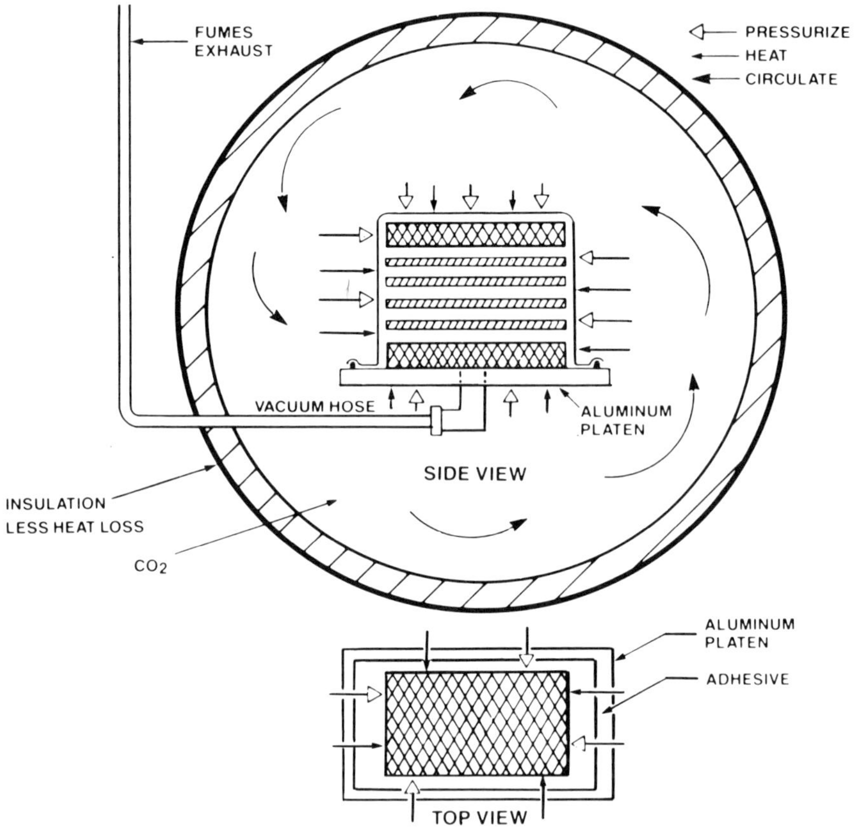

FIGURE 9.6 Vacuum autoclave lamination (from ref. 8).

9.6.2 Vacuum Autoclave Lamination [8]

The vacuum autoclave lamination (VAL) technique uses a pressurized vessel to perform the lamination procedure. In its most basic form, the lamination layup is placed on a carrier tray equipped with vacuum hookups prior to its insertion into the pressure vessel. This forms the vacuum assembly.

The most common VAL method consists of a flexible membrane placed over the layup and sealed at the edges (Figure 9.6). The vacuum assembly is transferred into the pressure vessel and the door is sealed. The vessel is then pressurized with inert gas. After pressurization is completed, enclosed heaters are used to heat the gas to the desired lamination temperature. In order to maintain a uniform temperature, the gas is circulated through the vessel by means of an enclosed fan. At the end of the cycle, the chamber is depressurized and the package is removed.

A unique and significant advantage of the VAL technique is its ability to laminate layups of different area size in the same press cycle. For example, since the applied pressure is not dependent on area, as is the case in a hydraulic process, a 12-in. × 18-in. panel can be laminated adjacent to an 18-in. × 21-in. panel without special adjustments.

9.7 RESIN-COATED COPPER FOIL

Resin-coated copper foil materials are being developed for use as the outer layers of multilayer boards. Using such materials is a way to achieve a "resin-rich" surface on the board. This resin-rich surface has improved smoothness, which is important for fineline circuitry. Also, the increased flexibility of the surface helps to counteract the mismatch of thermal elongation between certain components and the board, especially for surface-mount applications.

The use of resin-coated copper foil should result in fewer problems with poor resin filling (micro-size voids) under the copper foil. If handling characteristics are as good as they should be and if they prevent "weave transfer" to the foil, these new materials should have a significant effect on downstream yield, primarily during imaging. Thus, the use of such foils during the multilayer board lamination process to replace clad-laminate cap layers will probably increase. This will be due to reduced costs and improved reliability.

9.8 COMPONENT CONSIDERATIONS [9]

In addition to providing high-density interconnections, multilayer boards are being developed that also provide capacitance, inductance, and resistance features that replace discrete circuit components. Similar rectangular circuit patterns on adjacent layers of a multilayer board separated by the board's dielectric material form "capacitors." "Inductors" can be simulated by the routing of circuitous conductor patterns.

Planar resistors can be formed on internal layers of a multilayer board using a laminate system composed of an insulating substrate and, rather than a single conductive layer, a two-layer bifunctional cladding. The lower layer, immediately against the base laminate surface, is an electrically resistive thin-film material. The upper layer is copper (see Section 6.6).

The two layers of this bifunctional cladding are in contact with each other over the entire surface area. By means of selective etching, the two layers can be etched differentially, so that separate patterns of conductors and resistors are formed.

The film thickness is fixed when the resistor-conductor laminate is manufactured and is not a board fabrication variable. Thus, since the resistive material is a thin film (from 4 to 6 microinches thick, depending on sheet resistivity), the resulting "resistors" are essentially two-dimensional.

In determining resistor value, the sheet resistivity is the most important design parameter. Presently available materials for multilayer board applications have sheet resistivities of 25 and 100 Ω per square.

Any etched resistor of equal length (L) and width (W)--a "square" resistor—will have a resistance value equal to that of the sheet resistivity value. To get other resistor values, the number of "squares" (L divided by W) must be changed (Figure 9.7). A resistor with a high value would be long and narrow; a lower resistor value would be short and wide. For relatively high values, a long, narrow resistor is usually serpentined to minimize the needed board area used.

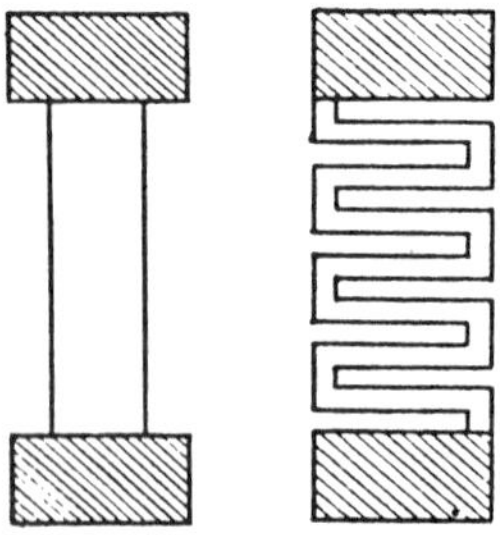

FIGURE 9.7 Planar resistor shapes (from ref. 9).

REFERENCES

1. G. L. Ginsberg, "A Look at Developments in Multilayer Fabrication," *Electronic Packaging and Production*, June 1985, pp. 34-37.

2. H. Winard, "Dense, Fast Circuits Demand Multilayer Boards to Meet Their Needs," *Electronic Design*, August 15, 1985, pp. 51-56.

3. G. Boucher and D. Holland, "The Buried Via Multilayer," *Circuits Manufacturing*, October 1984, pp. 64-70.

4. J. Henshall, "Drilling Small/Deep Holes in PCBs," American Electroplaters Workshop, March 1983.

5. H. Nakahara, "Complex Multilayer Boards Vie for Space in Japanese Computers," *Electronic Packaging and Production*, February 1987, pp. 70-75.

6. R. Keeler, "Mass Lamination Expands the Options of Multilayer Fabricators," *Electronic Packaging and Production*, June 1984, pp. 94-99.

7. P. Marx, R. Forcier, and D. Naseth, "Vacuum Laminating--An Overview," IPC Technical Paper IPC-TP-532, April 1985.

8. R. C. Carter, "Introduction to Vacuum-Press Technology," *PC Fab*, June 1985, pp. 34-42.

9. B. Mahler and P. Schroeder, "Planar Resistor Technology for High Speed Multilayer Applications," October 1985.

10

General Assembly Considerations

Surface-mounted assemblies typically have many solder joints. The solder joints are identical in appearance and spacing. The repetitious nature of making and inspecting such a large number of geometrically identical joints reduces inspection efficiency, so unacceptable joints are likely to be overlooked, even with multiple inspections [1, 2].

Also, defects that are found are often difficult to repair, and repair procedures have the potential of introducing additional problems. In view of this, the most cost-effective assembly procedures (Table 10.1) are those that assure acceptable quality levels through sample-lot inspection. Therefore, acceptable quality must be guaranteed by the proper specification of components and materials and by the use of easily controlled manufacturing processes.

The specifications of interest include (1) tolerances, (2) placement accuracy, (3) solder mask, and (4) solderability.

1. The dimensional tolerances of both the chip carrier and the packaging and interconnecting (P&I) structure: For high soldering yields with 50-mil-center chip carriers, a ±0.002-in. planarity and a ±0.005-in. lateral lead location tolerance is required. Also, surface mounting often requires a 0.5% or better warpage limit.
2. The accuracy of device placement: Studies have shown that leaded chip carriers must be placed within ±0.005-in. of their true position. This value can be somewhat higher for leadless devices, due to the self-centering nature of the soldering mechanisms involved.
3. The registration of the solder mask: To permit acceptable solder fillet formation, the registration of the solder mask must be located within ±0.005-in. or better of its true position.
4. The solderability of the materials to be joined: Solderability specifications for both the chip carrier and the P&I structure should permit acceptable wetting of all joint surfaces, even after inventory storage of the components.

Solderability testing of components is not a new requirement. However, its value to the use of chip carriers and surface mounting is much more pronounced. Since the final solder joint is not always completely visible, it is almost impossible to ensure visually that each terminator has been properly soldered. Therefore, considerable attention must be given to ensuring solderability before components are installed on assemblies. This will probably lead to increased use of automatic solderability testers as such equipment becomes more available and when it can be justified for inspecting components at higher sampling rates than are presently used.

10.1 SOLDER DEPOSITION [2]

Solder must be supplied to the joint sites in a sufficient quantity to close all gaps between the chip carrier leads and the termination land that results from nonplanarity of the leads and/or substrate warpage. In addition, when soldering to gold-plated leads, solder of sufficient volume to keep down the gold concentrations is needed to avoid the possibility of joint failure due to gold embrittlement.

Several methods are potentially available to supply the required amount and location of solder. However, only those with a high degree of control are suitable.

10.1.1 Electroplating

Electroplating is not as well suited as solder depositions, as might first be thought, because of (1) difficulties encountered in controlling alloy thickness where nonuniform geometries exist; (2) long plating times required for thick depositions; and (3) the inability to plate isolated lands of additively produced circuits.

10.1.2 Solder Paste [3]

Plating, the use of preforms, and the use of solder pastes (creams) can provide some or all of the solder needed for surface mounting. When chip carriers are involved, however, the use of solder pastes is preferred in most applications.

Selection of a solder paste is much more complicated than selection of a solder for wave soldering. Therefore, when selecting a solder paste for surface mounting, the following conditions must be considered:

The type of metallization
The amount of oxides and other contaminants on the metallization
Land dimensions
Desired thickness of solder paste deposition
The application method
The amount of slumping that can be tolerated, i.e., the maximum reduction in solder height during reflow
The time between the application of the paste and the attachment of the chip carriers
The time between the application of the paste and the reflowing operation
The desired/actual drying conditions

TABLE 10.1 Assembly Process Flow Comparison

Through board	Surface mount single-sided A	Surface mount single-sided B	Surface mount double-sided	Surface mount through-board mix	Chip on board TAB	Chip on board
Preclean	Preclean	Preclean	Preclean	Preclean	Preclean	Preclean
	Apply solder cream	Apply component adhesive	Apply solder cream side 1	Apply solder cream	Apply die attach adhesive	Apply die attach adhesive
Automatically insert components	Automatically pick/place components	Automatically pick/place components	Apply adhesive side 1	Automatically pick/place components	Automatically place TAB chip	Automatically place chip
Manually insert components	Manually pick/place components		Automatically pick/place components side 1	Manually pick/place components		
	Cure solder cream		Cure solder cream	Cure solder cream	Cure adhesive	Cure adhesive
	Flow melt solder cream		Flow melt solder cream	Flow melt solder cream		

			Clean	Clean		
			Apply second solder cream side 2			
			Automatically pick & place side 2	Automatically insert components		
			Manually pick & place	Manually in-sert components		
Preheat then machine solder		Preheat then machine solder	Flow melt solder cream	Preheat then machine solder	Bond outer leads	Wire bond
Clean	Clean	Clean	Clean	Clean	Clean	Clean
Test	Test	Test	Test	Test	Test	Test
Repair	Repair	Repair	Repair	Repair	Repair	Repair
Retest	Retest	Retest	Retest	Retest	Retest	Retest
					Encapsulate	Protect

The ambient conditions during application of the solder paste and prior to reflow soldering
The reflow method to be used and its temperature profile
The procedure for cleaning the soldered assembly

Defining, let alone satisfying, all of these conditions is not an easy matter. Therefore, choosing the appropriate solder paste for an application involves judicious selection of the components of the solder paste, i.e., the solder powder, rosin/resin system, solvent, activator, and the agents needed to modify rheological characteristics. Needless to say, success can be optimized by having the assembler working closely with the solder paste formulator.

The ideal solder powder has narrow tolerances on the alloy composition and particle size distribution. If the paste is to be applied by screen printing, the solder powder particles should be spherical and from 15 to 55 μm in diameter (Figure 10.1). If the paste is applied by stenciling, slightly larger particles may be preferred. However, a spherical powder with many "fines," i.e., particles less than

FIGURE 10.1 Solder paste with uniformly sized spherical particles.

10 μm in size, will cause an increase in the number of free solder balls. Also, powders that are not spherical can cause clogging of the screen and inconsistent rheological behavior. They are also more susceptible to oxidation due to their greater surface area.

The occurrence of bridging between lands is a major concern. The factors involved are

The degree of spreading of the solder paste during drying/reflow, which may depend on the reflow profile
The downward force on each land, which depends on the weight of the component, the number of lands the component rests on, and the dimensions of the lands
The spacing between the lands

The amount of spreading is most easily controlled by varying the percentage of metal in the solder paste, i.e., higher metal content is clearly better. Land dimensions are also important, as smaller lands are often bridged less often than larger ones. This is probably due to the larger volume of paste that is present in the latter case.

Taking all these considerations into account, a properly designed solder paste has the following characteristics:

10-mil (0.010-in.) line print resolution, or less
Uniform printing
Sufficient tackiness to hold the components in place
Dries and reflows without excessive flow-out
Reflows to form a strong joint free of voids
Contains a sufficiently active flux to produce a virtually solder ball-free surface, yet does not attack the components or the metallization

10.1.3 Screen Printing [4]

Screen printing of solder pastes for surface mounting applications is basically the same technique as is used to deposit thick films on hybrid circuits. The normal screens used are 80, 100, and 200 mesh because of the large particle size. For many applications, the 80-mesh screen is used to deposit about 0.005 to 0.007 in. of wet paste.

The benefits of screen printing (Figure 10.2) are many. For example, it lends itself to high-volume usage in which a uniform deposit of paste is required in intricate patterns. Also, with a good-quality paste with controlled viscosity, it is possible to screen-print 0.006-in. high solder bumps on 0.025-in. square lands, while maintaining resolution for several hours or days.

However, with lands only 0.010-in. apart, for example, resolution and metal retention are critical to control yields and reflowed solder

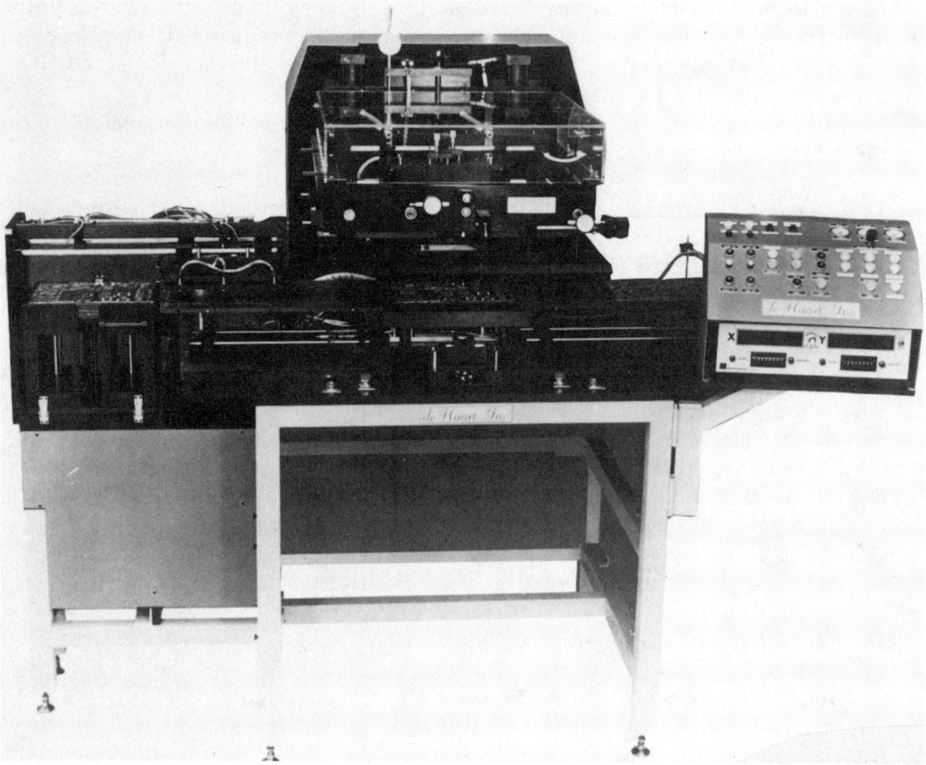

FIGURE 10.2 Automatic solder paste screen printer. (Source: de Haart, Inc.)

heights. Also, the control of reflow solder height is very important, with the use of a quality paste, good printing technique, and controlled reflow practices being essential. Of particular importance is the use of a slower squeegee speed and a greater snap-off distance that is associated with screen printing of other thick-film pastes.

10.1.4 Stenciling [4]

The solder-paste stenciling process is basically the same as screen printing, except that a sheet of metal with appropriate cutouts is used instead of a screen mesh. Thus, variations in the solder thickness deposited depends on the thickness of the sheet metal used.

Another similarity is the solder paste viscosity, with stenciling often using solder pastes at the thicker end of the viscosity range.

Also, stencils usually have a longer life than their screen counterparts.

For some applications the use of screens and stencils is combined. In these instances screen printing can be used for the high-resolution areas and the stencils for thicker depositions. Fortunately, the life of the screen is not adversely affected by this combination.

The use of patterned metal foil stencils enhances both the solder volume deposited and its uniformity. Following stencil printing, the component can be accurately placed and pushed into the paste. Typically, the depth of immersion of the leads into the paste is such that the paste covers nearly half of the lead thickness. Next the assembly is placed in an oven to cure the solder paste, after which the solder is reflowed in a higher-temperature facility.

Stencil printing of solder paste is also ideally suited to high production volumes of similar assemblies with a large number of surface-mounted components. It has been successful with chip carriers with up to 68 leads and for carriers having lead spacings as small as 0.020 in.

10.1.5 Multipoint Dispensing

An alternative method of depositing solder paste is multipoint dispensing. Such dispensing is suitable for applications requiring greater flexibility than stencil printing, such as low production runs with few chip carriers, or assemblies where the design is not fixed and device count or position vary.

Dispensing is also useful for the intermix of technologies, where chip carriers are to be soldered in the midst of previously through-hole soldered components. In addition, the technique may often be required for repair during the replacement of defective components.

10.1.6 Bumped Substrates

"Bumping" the substrate with solder is a method that supplies a sufficiently large and uniform deposit of solder for surface mounting. It consists of a stencil-printed or dispensed deposition of solder paste, followed by both cure and reflow steps before component placement. Thus, smooth solder bumps measuring 0.007 in. in height are formed. The substrate is then cleaned aggressively with solvents and brushed to remove spatter deposits, as required.

A small quantity of screenable tack flux is applied to the peaks of the bumps by stencil-printing before the alignment and placement of the chip-carrier devices. This flux, which can also be cured, holds the device in place during solder reflow.

The bumped-substrate method is particularly well suited to the intermix of through-hole and surface-mounted components, where solder paste is applied before component insertion and wave soldering. Bumping of solder-masked, bare copper circuitry in the manner described will also reveal the solderability of the copper before the addition of components. Also, it should be used where hot-gas solder leveling is not possible.

10.1.7 Solid-Phase Solder Bonding [5]

Large and precise volumes of solder can be added directly to the chip carrier by the use of solid-phase solder bonding. This method encompasses the following steps.

1. Spherical preforms are picked up by a heated vacuum tool in an array that is identical to the land pattern for the chip carrier to which they are to be joined.
2. The tool transfers the preforms to a heated chip carrier and by thermocompression bonding forms a solid-state bond between each preform and its corresponding land on the carrier.

Bonding temperatures are reported to be slightly less than the solder melting point for bonding forces of 2-5 lb per sphere and bonding times of 3-9 sec. Flux must be applied to either the chip carrier or the substrate for the subsequent solder reflow.

10.1.8 Preformed Solder Columns [1]

Leadless chip carriers can be attached with preformed columns made from conventional solders with high-melting reinforcement, i.e., solder columns. The solder columns attach to the termination area on the underside of the chip carrier and therefore preserve the outline of the leadless package (Figure 10.3). This type of direct solder attachment to chip carrier metallization also means that these types of leads are not restricted to perimeter terminations.

The solder columns are available in a holder that restrains them during attachment to the leadless chip carrier. During the attachment reflow step, the solder columns are compressed slightly to achieve planarity of the ends of the columns. The holder may then be dispersed in water and the leaded package attached to the P&I structure.

10.1.9 Hot-Gas Solder Leveling

The hot-gas solder-leveling process has also been considered to supply thick solder on small lands in a uniform manner due to the dominant surface tension of molten solder. However, studies have shown that thick solder, 0.005 in. typically, can be applied with this method, but the thickness standard deviation of 0.001 in. is usually unacceptable. Similarly, inconsistently thick solder can be applied to chip carrier leads by dipping.

10.2 COMPONENT PLACEMENT [6-8]

A wide variety of equipment is being developed or already exists for the handling of surface-mounted components. Most of them are pick-and-place equipment that use a simple vacuum grid mechanism to hold the component on its placement probe. They vary in complexity from basic assembly stations (Figure 10.4) to fully automatic, computer-controlled assembly systems (Figure 10.5). Since the typical vacuum probe holds the component from the top and places it straight down on the substrate without the benefit of side grips, parts placement can be very dense and yet extremely accurate.

Hand placement is not adequate for mounting chip carriers in terms of accuracy and repeatability of the downward force. Any twisting or excessive pressure during placement can force the solder paste off the termination land and onto the substrate material. Therefore, stray solder particles left on the surface may not find their way back onto the land when the solder is reflowed. In the event that this happens, solder balls can result in shorting between conductors/lands.

It is a difficult task to place a chip carrier manually, since two sides of the chip carrier must be viewed simultaneously. This can be virtually overcome by the use of programmable, computer-controlled "pick-and-place" equipment (Figure 10.5). Equipment such as this have placement accuracies to within 0.001 in., which is more than adequate for these applications, at rates of several thousand chip carriers per hour.

Under program control, these assembly systems select a component carrier (Figure 10.6) by pushing it out from the bank of trays, tapes, sticks, etc. (Figure 10.7), and deposits it into the transfer mechanism. This mechanism opens the carrier, centers the component, and locates it for pickup by a vacuum quill in the head assembly.

In the meantime, the machine has positioned the substrate mounted on an X-Y table in the proper location. After the transfer mechanism retracts, the head assembly rotates the component into its correct orientation and moves the component downward to contact the substrate lightly.

At the completion of the alignment cycle, the assembly is ready for the solder reflow operation. Although the equipment shown is

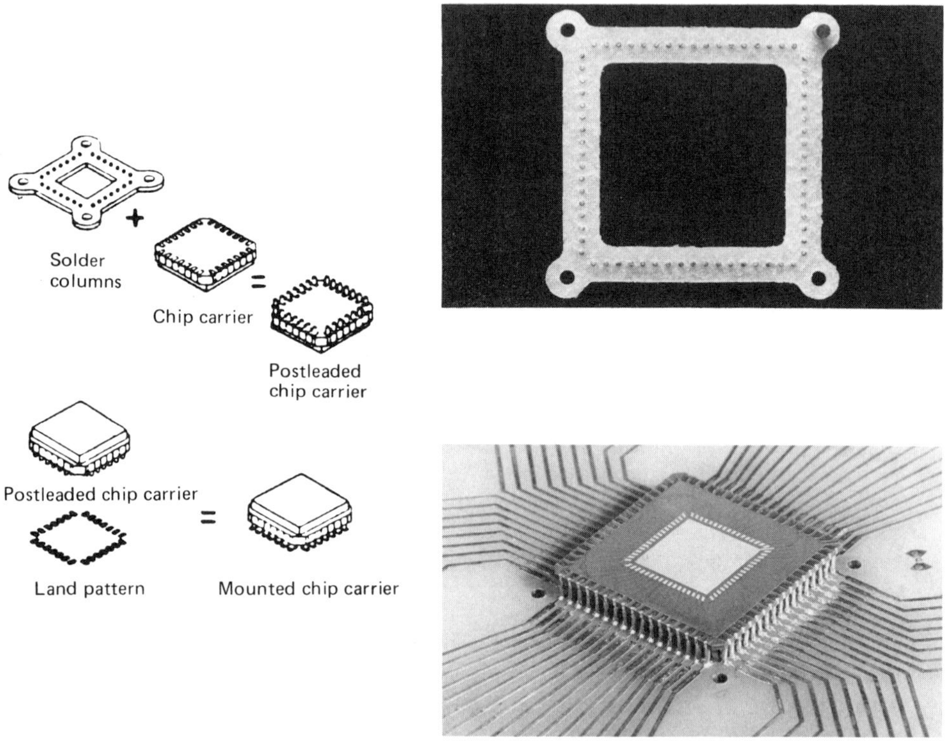

FIGURE 10.3 Solder-column mounting of leadless chip carriers. (Source: Raychem Corp.)

also equipped to do the reflow soldering, it is not uncommon with less sophisticated equipment for the soldering to be done by another piece of equipment.

Because it is not always possible to see the interface between the component and substrate, nonvisual ways are needed to detect mislocated parts. Therefore, as high-production volume demands, human inspection must be supplemented by automated fault detection. (Better yet, prevention of defects must be the watchword when so many defects can be obscured in visual detection.)

Placement accuracy can be determined by the use of automatic, closed-looo, optical pattern-recognition equipment. Such units can be used during the assembly operation to compare actual to preferred component placement. Accuracy can thus be verified and controlled at the placement source.

FIGURE 10.4 Basic surface mount pick-and-place equipment. (Source: Henry Mann, Inc.)

10.3 SOLDERING [2]

Surface mounting of components places a greater emphasis on the quality of the starting materials and on process control than is required for the soldering of through-hole-mounted components. (See Chapter 11 for information concerning the major processes being used for surface mount soldering.) The reasons for this are many.

The higher packaging densities achieved and the reduced solder joint size and spacing dramatically increase the density and total number of solder joints on the printed wiring assembly. But inspection then becomes more difficult and, thus, a more perfect job of soldering is necessary.

The solder joints are almost identical in appearance and spacing. The repetitious task of inspecting large numbers of geometrically identical solder joints reduces inspection efficiency, so unacceptable solder joints are likely to be overlooked even after multiple inspections. To make matters worse, the defects that are found are often difficult to repair, and the repair procedure can introduce additional problems.

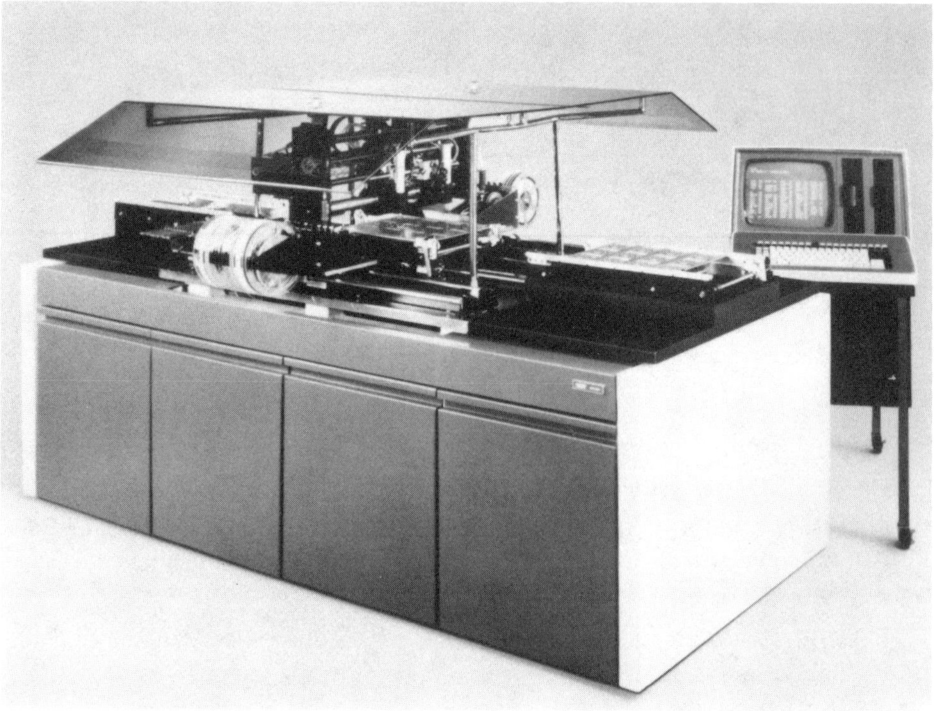

FIGURE 10.5 Surface mount component placement system. (Source: Micro Component Technology, Inc.)

In view of these considerations, the most cost-effective production soldering processes are those that ensure acceptable quality levels through in-process control and sample-lot inspection. Therefore, acceptable quality must be guaranteed by the specification of proper (solderable) components and materials and by using soldering processes that are readily controllable.

10.4 CLEANING

Chip carrier/surface mount assemblies must be cleaned after reflow to ensure the removal of flux and other induced contaminants, as inadequate removal of the flux beneath the chip carriers can cause excessive forces on the solder joints during thermal cycling. (Efficient flux removal is dependent on the height of the chip carrier off the

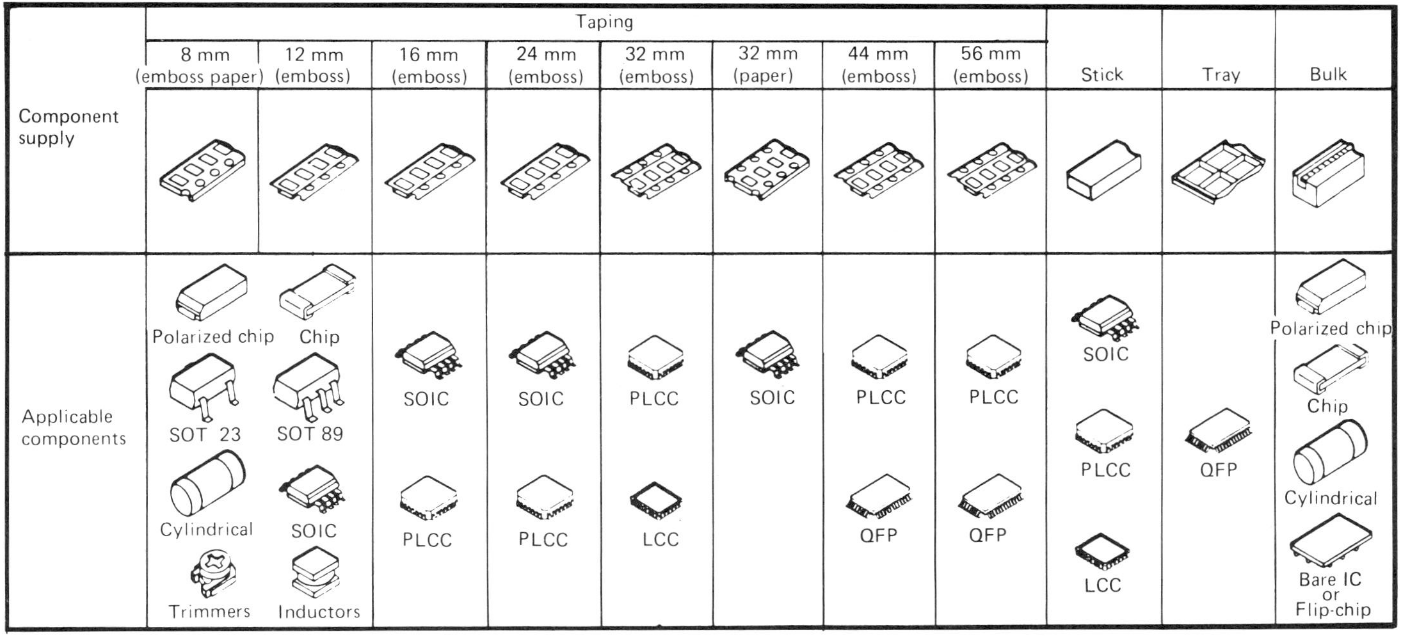

FIGURE 10.6 Surface mount component supply formats. (Source: Universal Instruments Corp.)

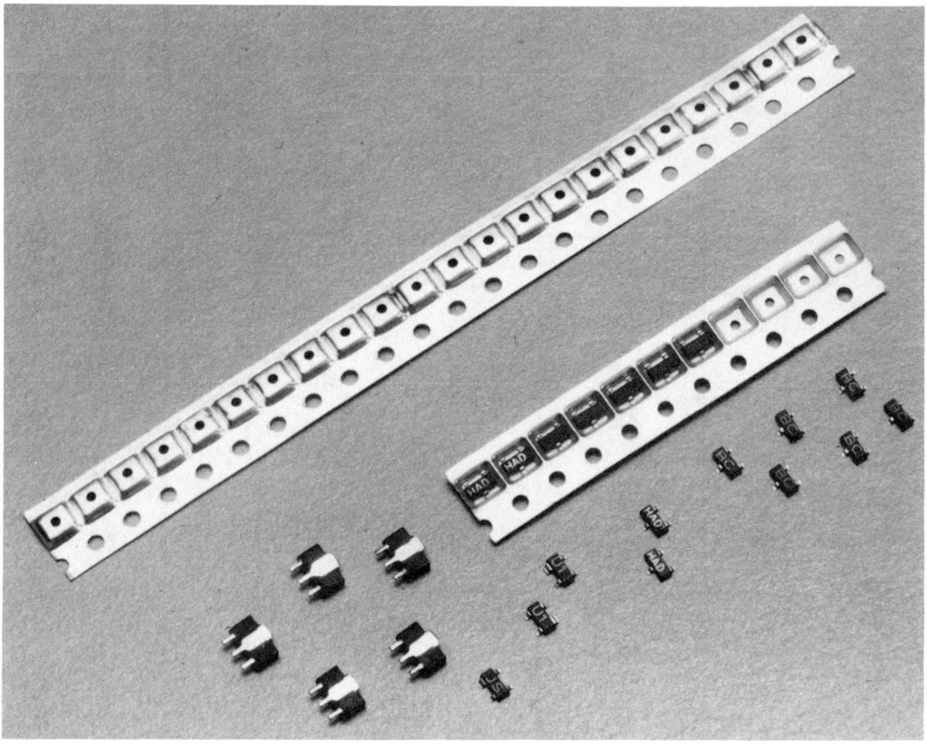

FIGURE 10.7 Embossed tape for small-outline transistors. (Source: Siemens Corp.)

substrate and, therefore, is related directly to the height of the soldering deposition prior to reflow soldering.)

In some instances, an immediate postsoldering vapor degreasing enhances substrate cleanliness. However, it is necessary to perform this operation immediately after soldering, so that the flux residue is not allowed to congeal. If the flux congeals, additional cleaning is required or the possibility exists of leaving flux residue under the components. (Another approach being used successfully is to select mild flux materials, which may, in some instances, be left on the assembly without compromising performance or useful life.)

10.5 INSPECTION [9, 10]

The smaller features sizes and tighter tolerances associated with high-density surface mounting applications require increased inspection and process control. For example, to prevent layer misregistration and drilling inaccuracies, optical pattern-recognition/digital image processing and comparison equipment is emerging that will allow cost-effective inspection of every layer of a high-density multilayer board. Hole sizes, missing or extra holes, conductor widths/spacings, and open/short circuits can also be detected by such systems.

Solder joint quality is often judged on the evaluation of the visible fillet (see Appendix D). However, with surface-mounted chip carriers, the critical solder joint area may be under the component. Also, solder balls beneath the component will not be detected by visual inspection.

Unfortunately, conventional nonvisual testing (Figure 10.8) often reveals only electrical faults, and the use of more sophisticated techniques, such as radiology (Figure 10.9) and sonic scanning, might damage sensitive components. Therefore, new detection methods must be developed and uniform acceptance criteria established.

Because of thermal expansion differences, delamination can occur during the fabrication and assembly of P&I structures. Detection of these delaminations may require the use of new inspection processes and equipment. For example, delaminations will show up as hot spots when power is applied to a circuit because they represent openings in the thermal conduction path of the structure. The use of an infrared microscope could be used to detect these delamination hot spots.

10.5.1 Microsectioning

Surface-mounted assemblies can also be microsectioned after reflow soldering. This is an excellent method of determining the amount of solder at the joint, as well as being useful to verify chip carrier/substrate clearances. By microsectioning across the major axis of the solder joint, the angles and extensions of the solder fillets and the thickness of the solder layer can be determined. The quality of the solder joint—voids, bubbles, cracks, intermetallic layers, etc.—and the adhesion of the substrate metallization may also be inspected. Residues not removed during defluxing and cleaning will also be made visible in microsections that include the body of the component.

FIGURE 10.8 Chip carrier test probe. (Source: AP Products)

10.5.2 Laser/Infrared Analysis [11]

A laser/infrared signature analysis technique will help automatic test equipment detect difficult-to-diagnose solder joints. With this method, the laser injects heat into the joint, while the sensor produces the "signature" to determine how fast the solder joint accepts and gives off heat. Electronically comparing the signature detected from a joint to be tested with each known signature (Figure 10.10), makes it easy to isolate faults that would not only cause immediate failures, but might also create problems after a period of normal operation.

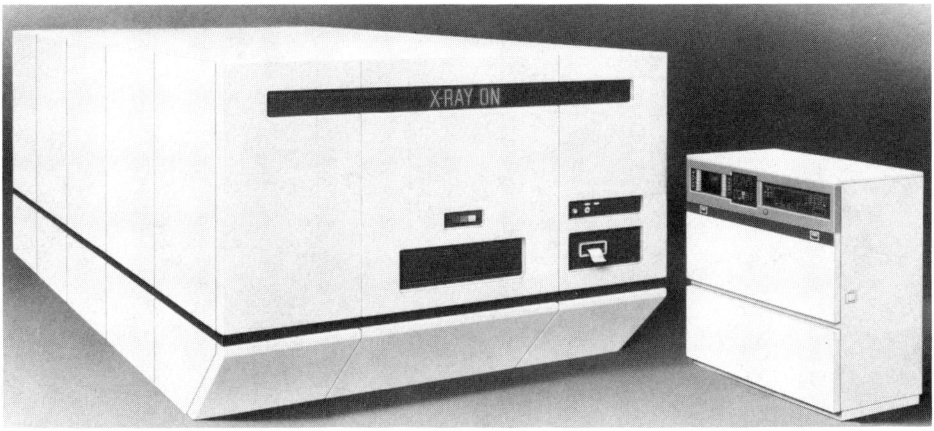

FIGURE 10.9 Automated X-ray solder quality inspection system. (Source: IRT Corp.)

10.6 MODIFICATIONS AND REPAIRS [12, 13]

As assemblies increase in sophistication, their modification of repair becomes more complicated and expensive. Also, the possibility increases that technician errors may cause irreparable damage to the components or substrate. With these considerations in mind, improved on-site repair tools have been developed for surface-mounted components.

Essentially three types of problems exist on surface-mounted chip carrier assemblies: (1) defective components, (2) misaligned components, or (3) incorrect conductor routing. To facilitate correcting each of these problems, the aisle width between components should be wide enough to facilitate inspection/testing and the making of changes.

10.6.1 Component Removal

For defective components it is necessary to remove the component and reflow a new one in its place. For this procedure the original reflow

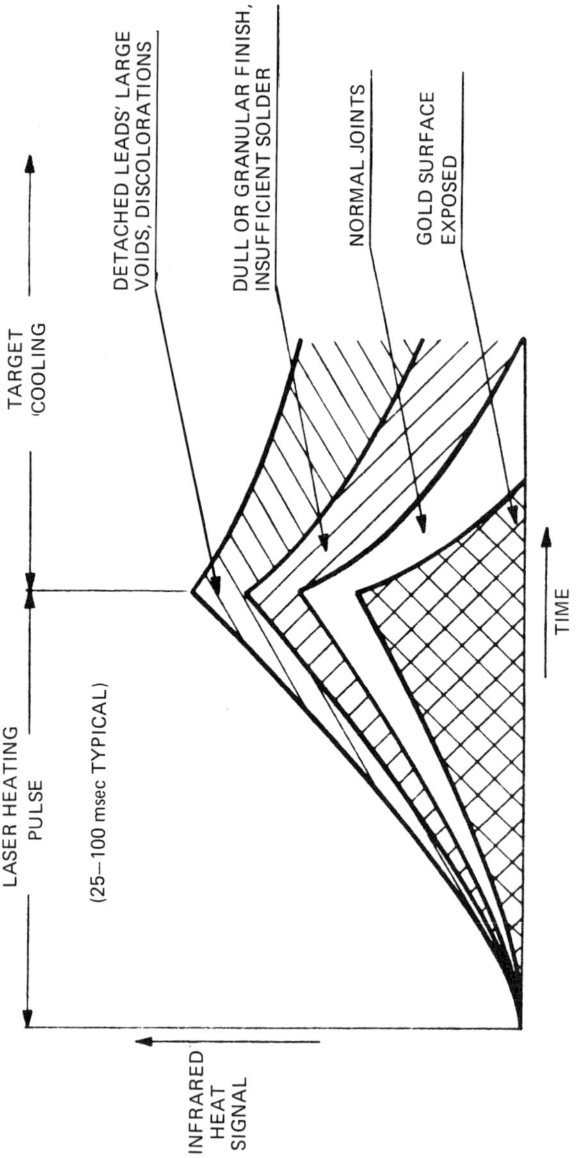

FIGURE 10.10 Typical solder joint infrared signatures. (Source: Vanzetti Infrared Systems, Inc.)

method, or one of the others, is commonly used. However, special equipment, such as the hot-air reflow terminal, can be used to localize the reflow and minimize the disturbance of the good components on the assembly.

Repair tools vary from basic hand desoldering irons (Figure 10.11) to sophisticated repair terminals (Figure 10.12). The latter can typically use directed hot-air exhausts to desolder the components. Temperature-indicating compounds used with the device enable technicians to control heating and reduce field repair damage.

After performing the reflow, adequate cleaning, which often relates directly to the original attachment problem, is critical.

For chip misalignment, it is merely necessary to realign the chip

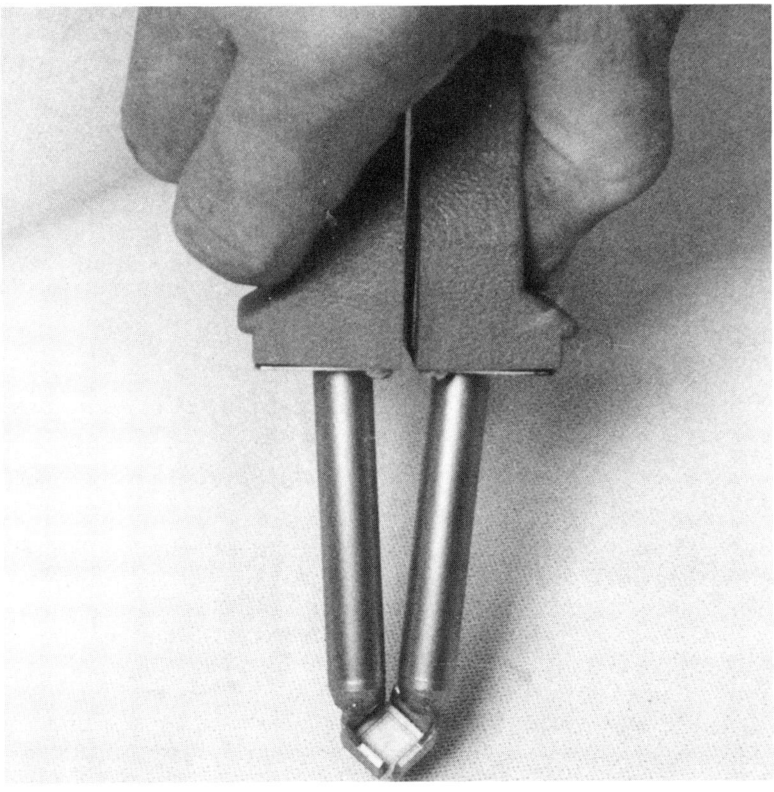

FIGURE 10.11 Hand-held surface mount component replacement tool. (Source: Nu-Concept Systems, Inc.)

FIGURE 10.12 Hot-air surface mount component repair station. (Source: Nu-Concept Systems, Inc.)

carrier during the reflow process. Again, cleaning is a major consideration.

10.6.2 Circuit Modifications

For a conductor routing problem or modification, a means of "white" wire modification is desirable. This often necessitates the electrical isolation of one or more component termination lands for rerouting to a new location. A method to do this involves the use of "cutable links" (Figures 10.13 and 10.14). This involves "fanning out" each termination land to a modification land. To implement a wiring change, one can simply cut the link between the lands and solder the new routing wire to the modification land or its plated-through hole.

The main advantage of the cutable-link method is that modifications can be made easily on the component side of the structure, without disturbing the mounted component. The major disadvantage

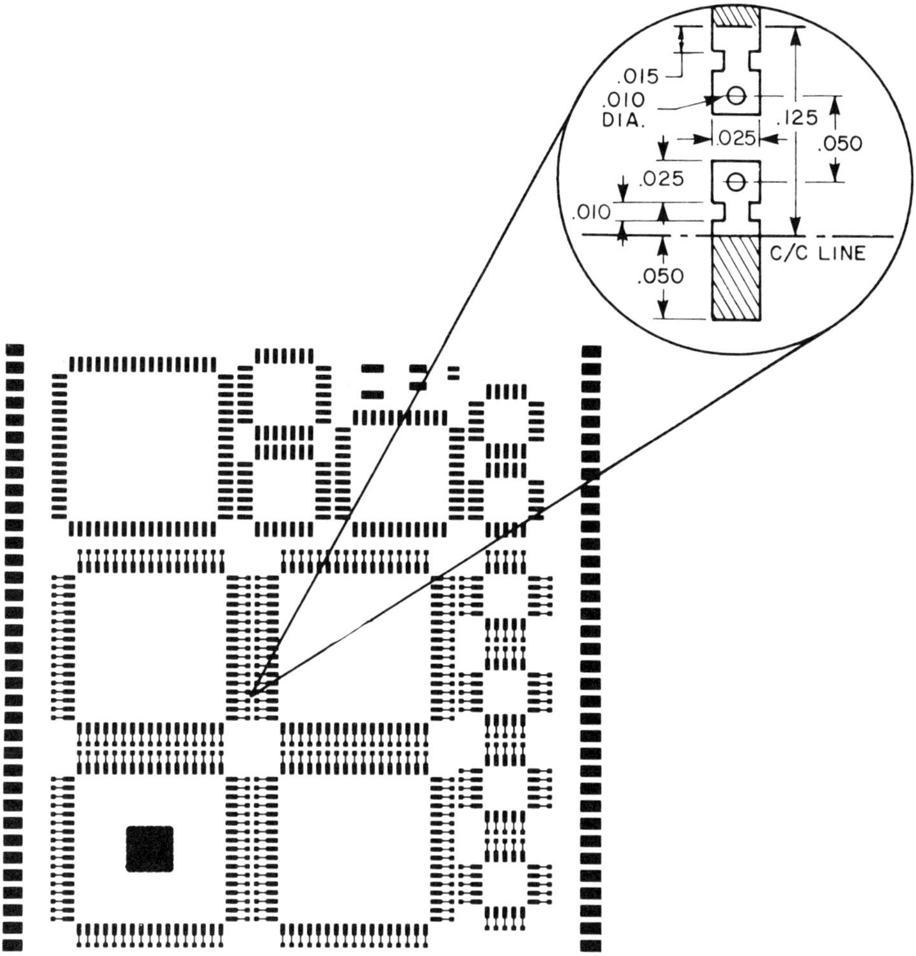

FIGURE 10.13 Single cut-link land pattern.

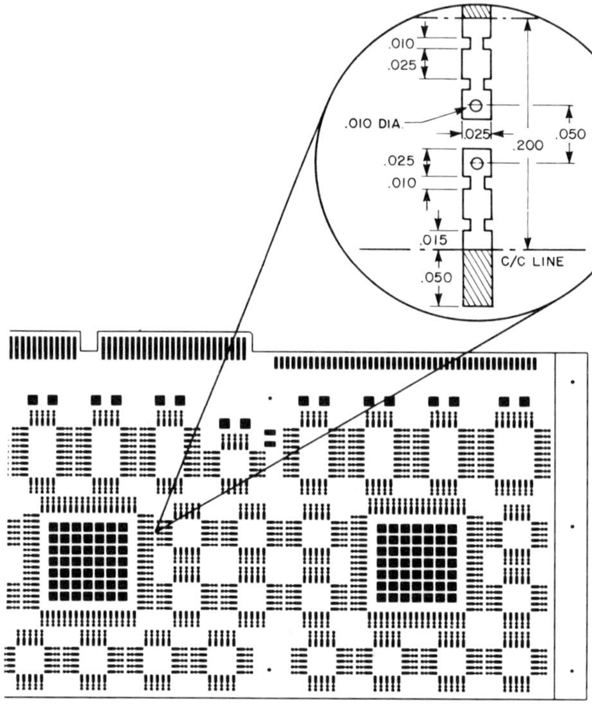

FIGURE 10.14 Double cut-link land pattern.

is that the chip carrier mounting site is enlarged as a result of the extended land pattern.

REFERENCES

1. Guidelines for Component Packaging with Emphasis on Surface Mounting, IPC-SM-780, Institute for Interconnecting and Packaging Electronic Circuits, 1988.
2. D. Schoenthaler, "Soldering Circuit Assemblies in the 1980's," Proceedings, Printed Circuit World Convention II, June 9-12, 1981, pp. 131-140.
3. B. Roos-Kozel, "Designing Solder Paste Materials to Attach Surface Mounted Devices," *Solid State Technology*, October 1983.
4. R. P. Anjar, Sr., "Quality Solder Paste for Use in Microelectronic Applications," *Solid State Technology*, October 1983.

5. R. H. Minetti, "Solid Phase Bonding for Use in the Assembly of Microelectronic Circuits," Proceedings, International Microelectronics Conference, October 20, 1980, pp. 126-129.

6. "Machine Automatically Surface-Mounts Flatpacks and DIPs onto Printed Circuit Modules," *Electronic Packaging and Production*, March 1977.

7. J. Orcutt, "Printed Circuits Get A New Boarder: The Plastic Chip Carrier," *Electronics*, April 7, 1982, pp. 141-145.

8. C. L. Harris and F. J. Cheriff, "Processing Considerations for Using Vapor Phase Soldering," *Electronic Packaging and Production*, May 1981.

9. O. L. Ramsey, "Surface-Mounted Components--The Inspection Challenge," Proceedings, National Electronic Packaging Conference, March 1983.

10. T. S. Wein, "Inspection Challenge of Surface Mounted Device Assemblies," Proceedings, National Electronic Packaging Conference, March 1983.

11. R. Vanzetti and A. C. Traub, "Hidden Solder-Joint Defects Detected by Laser-Infrared System," IPC Technical Paper, IPC-TP-358, April 1981.

12. K. J. Coffin, "Rework of Soldered Subassemblies Containing Hermetic Chip Carriers," Proceedings, National Electronic Packaging Conference, March 1983.

13. C. L. Lassen and M. Motazedi, "High Density Discrete Wiring Offers a Solution to Chip Carrier Design," *Electronic Packaging and Production*, January 1983.

11

Soldering Methods

The actual soldering technique depends to a large extent on the type, materials, size, and shape of both the component and the P&I structure. With each method, consideration should be given to the finish of the component terminals, the finish of the corresponding land pattern, and the possible heating systems to be used to reach the solder process temperature [1, 2].

Many precision reflow soldering techniques have been developed, which vary basically only in the method of heat transfer utilized. For example, radiant, forced hot-gas convection and condensation (vapor-phase) heat transfer methods have been developed for chip carrier/surface mounting applications. Most of these techniques can be used for soldering these assemblies, but each method has unique advantages and disadvantages (Table 11.1), making each more applicable to different types of products.

Three guidelines dominate the considerations for selecting the most appropriate reflow soldering method. First, fixtures and clamps to support and align individual chip carriers should be avoided whereever possible, as they are costly, difficult to inventory and maintain, and their heating capacity usually lengthens the heating time.

Second, heating equipment should not contact the solder joint areas or otherwise become contaminated with flux or its residues. This is important because process control is assisted by maintaining cleanliness and by using a noncontact heating means.

Finally, the selected reflow method should have a large "window" so that reasonable room for error exists in selecting and maintaining the operating parameters without degrading the quality of the end product.

In addition, chip carrier soldering presents other concerns. For example, rapid heating accentuates the solder spatter problem associated with the use of solder paste. This spatter is minimized by properly curing and preheating before reflow. Thus, the rapid heating with condensation soldering potentially causes more spatter than the slower hot-air convection heating method.

Another consideration is that the conveyor devices on most facilities initially used for these applications introduces excessive vibration, which is transmitted to the workpiece so that chip carrier movement and misalignment occur before solder solidification. Therefore, conveyor stability is required for both the horizontal equipment types used with radiant or convection heating and for the vertical types used with condensation soldering.

Each of the following methods can be used for surface-mounted components, as long as the soldering process takes place within the prescribed temperature/time profiles. However, it should be noted that some of the processes are not applicable to all types of surface-mounted components; e.g., some are suited only for use with small-outline components, while others are not so limited.

11.1 WAVE SOLDERING

There are several problems associated with the wave soldering of surface-mounted components using conventional equipment:

TABLE 11.1 Surface Mount Soldering Methods

Technique[a]	Convenient atmosphere control	Minimization of time	Approximate equipment cost ($K)	Process control	Double-sided assembly	Cleaning base	Volume production capability
Vapor phase (belt)	Yes	Excellent	20–30	Excellent	Yes	Excellent	Excellent
Vapor phase (batch)	Yes	Good	10–16	Excellent	Yes	Excellent	Good
Belt, conduction	No	Good	5–10	Good	Difficult	Fair	Fair
Belt, infrared	Yes	Poor	2–10	Fair[b]	Yes	Poor	Good
Belt, convection	Yes	Good	5–10	Fair	Yes	Fair	Good
Wave, conventional	No	Excellent	8–15	Excellent	Yes	Good	Excellent
Wave, dual or bubble	No	Excellent	12–25	Excellent	Yes	Good	Excellent
Hot air/hot gas	No	Poor	1–2	Fair[b]	Difficult	Good	Poor
Soldering iron-style tools	No	Poor	1	Poor	Yes	Fair	Poor
Laser	Possible	Excellent	60–80	Good	Yes	Excellent	Very poor
Hot plate	No	Poor	0.1	Poor	Difficult	Fair	Poor

[a]Soldering iron-style tools and the hot gas machines are used primarily for repair and replacement operations.
[b]Temperature control must be established and maintained.

The densely-packaged components and the nonwettability of the component bodies tends to divert the solder from penetrating to the land area.

The gases evolved from the adhesive and/or flux tend to cause solder pockets.

If a component is positioned so that one or more of its terminals is shielded from the direct flow of the solder, less solder than required will be deposited at that land.

Combining two waves that work in sequence (Figure 11.1) overcomes these problems in most instances. The first wave is a narrow, turbulent jetlike fountain. The function of this wave is to ensure that solder is forced into the component area and that the solder reaches all lands. The narrow contact length permits gas evolved from either the flux or component attachment adhesives to escape. The turbulent jet flow of the wave provides sufficient pressure to overcome the repelling and hydrodynamic turbulence caused by the component bodies, assuring that all lands are wetted with solder.

The second wave embodies the characteristics of a conventional wave soldering machine. Other features of a conventional wave soldering machine are also available, such as preheating, fluxing, and postcleaning.

11.2 VAPOR-PHASE SOLDERING [3, 4]

Vapor-phase (condensation) soldering is a well-known process for surface-mounting chip carriers. It offers several advantages, including (1) rapid heating with precise temperature control, which protects heat-sensitive components from thermal damage; (2) even heating, regardless of the overall geometry; and (3) an inherently clean operation, as only continuously distilled vapor contacts the assembly.

Until recently, a significant limitation was the maximum temperature achievable, i.e., about 215°C. As a result, only a few normal solder compositions could be used, primarily 62Sn/36Pb/2Ag or 60Sn/40Pb. However, other liquids have been developed that can be used up to 250°C and, thus, can be used with other solders, e.g., 95Sn/5Ag. This not only allows for the use of several metallurgies, depending on mechanical factors, but also permits vapor-phase soldering of both sides of an assembly. Typically, high-temperature 95Sn/5Ag could be used to mount components on one side of the assembly and then low-temperature 60Sn/40Pb could be used to solder components on the other side. As liquids with greater temperature capabilities are developed, the utilization of this technique will be greatly expanded.

FIGURE 11.1 Dual wave of surface mount wave soldering machine. (Source: Electrovert, Inc.)

11.2.1 Assembly Details [5]

Although the vapor-phase process offers many advantages over other heating techniques, the condition and preparation of the parts to be soldered are critical factors that must be considered. The following guidelines will help in obtaining the desired results.

The lands on the P&I structure should extend beyond the outline of the chip carrier in order for a good fillet to form and to aid in inspection of the solder joint.

The solderability of all surfaces must meet very high standards.

Pretinning the chip carrier terminals will help to provide the desired amount of solder and will indicate the solderability of the component before assembly.

The solder paste should be of the highest quality and contain the minimum possible amount of oxide.

Curing the solder paste in a low-temperature air oven is recommended. This will evaporate the liquid vehicle and leave the paste in a dry, nontacky form that will hold the chip carrier in position.

No jigs or fixtures are needed for chip carrier positioning. (As long as no overlap has occurred, leadless chip carriers will tend to pull into alignment with their respective P&I structure lands due to the effects of surface tension during the reflow operation.)

As previously mentioned, the transport of the assemblies into and out of the reflow zone should be as smooth as possible to avoid disturbing the molten solder during cooling and solidification.

11.2.2 Equipment

The initial vapor-phase soldering equipment developed were batch-loading units, such as that shown in Figure 11.2. Units like these have proven very useful for low-volume applications. However, their limited suitability for production applications and the complexity of their operation, brought about by their dependence on dual-vapor zones, has necessitated additional developments to overcome these shortcomings. Thus, single-vapor batch systems and inline production systems (Figure 11.3) have recently been developed. The inline unit is fully continuous, with a single, endless conveyor belt moving the product directly through the system. Such equipment can reflow-solder up to 500 assemblies per hour. To illustrate their ease of operation, conveyor speed is the primary operating control, while for a given reflow, vapor peak system temperature remains fixed and constant.

11.2.3 Operational Factors [6]

The operation of a vapor-phase soldering system involves a number of potential problem areas. However, some good results can be expected if the user is aware of these areas and if proper preventive measures are taken by both the user and the equipment supplier.

Under normal operating conditions and in the presence of organic residues, the secondary vapor will undergo some chemical breakdown. Thus, very low concentrations of acids will be formed. These acids could cause corrosion of the cooling coils if allowed to build up to high concentrations. Therefore, a good vapor-phase soldering system will provide for the control of acid formation, usually by using an integral acid stripper.

During soldering, some flux will be washed from the parts and fall into the boiling sump. If allowed to precipitate out after the

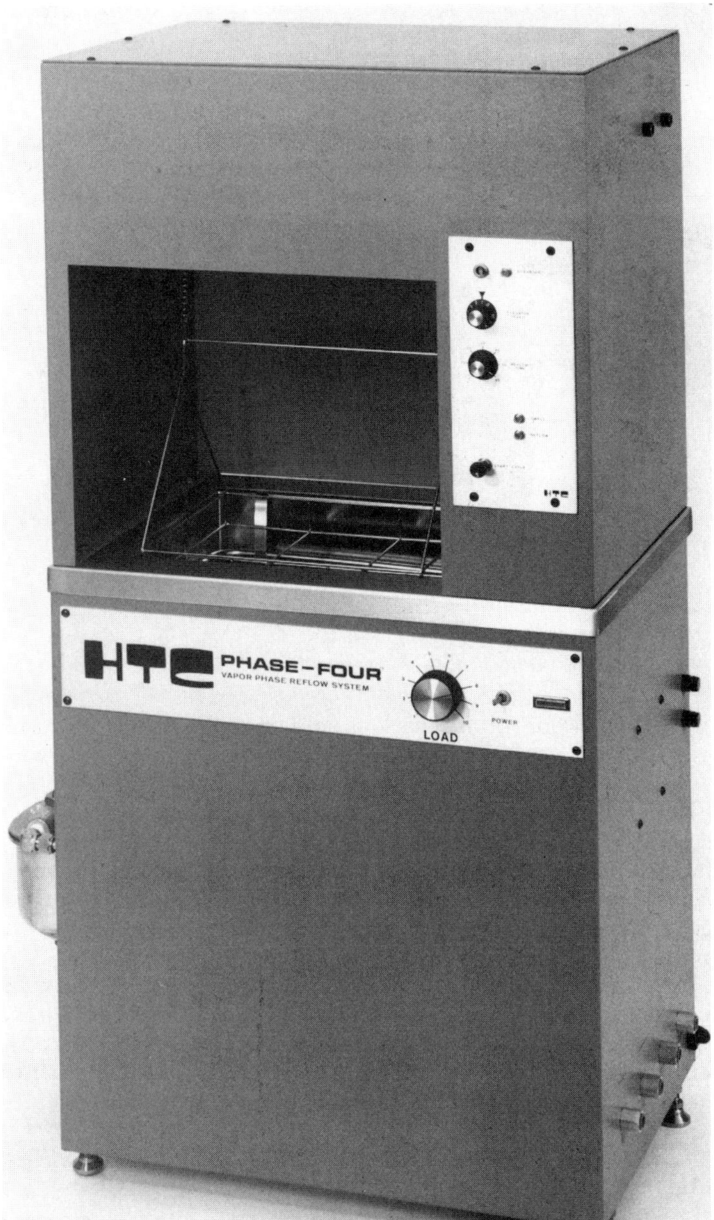

FIGURE 11.2 Batch-type vapor-phase reflow soldering machine. (Source: HTC Corp.)

FIGURE 11.3 Inline vapor-phase reflow soldering machine. (Source: Centech Corp.)

system is shut down, the flux will coat the surface of the immersion heaters, thereby causing inefficient heat transfer. As a preventive measure, a good vapor-phase soldering system must be equipped with a filtration system, preferably one that is activated when the machine is shut down or, under heavy production loads, operated continuously.

The decomposition of the vapor-phase fluids produces toxic gases that could become a health hazard if the user is unaware of this potential danger. Therefore, safety precautions must be a basic aspect of all vapor-phase soldering operations.

11.3 INFRARED SOLDERING [7, 8]

Infrared/radiant reflow soldering can be accomplished in two ways: (1) The radiant energy can be collected using a reflector and focused onto a desired spot on the assembly, or (2) an entire assembly can be soldered by passing it through an inline unit with an infrared oven and a conveyor system (Figure 11.4).

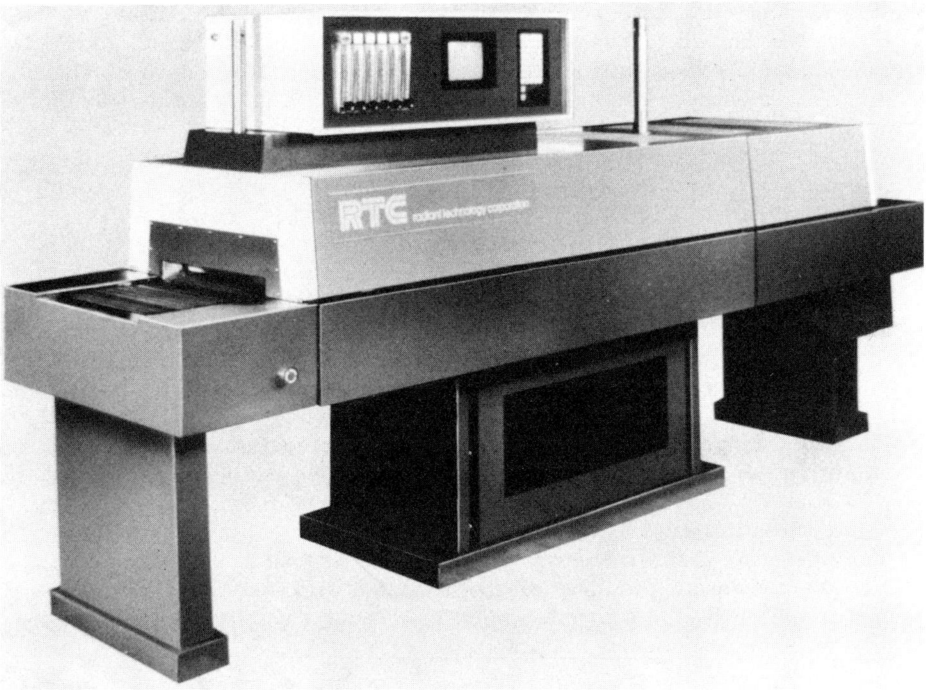

FIGURE 11.4 Inline infrared reflow soldering machine. (Source: Radiant Technology Corp.)

11.3.1 Focused-Energy Heating

Focused-energy infrared heating uses a high-temperature source of radiant energy. The advantage of this approach is that the radiant energy is directed onto a desired spot and the rest of the assembly is not subjected to the heat.

The disadvantages encountered when working with focused infrared energy are that there is a tendency to char the substrate if the radiant energy is not correctly focused onto the soldering area; and a relatively long time is required to position the assembly and make the solder joint.

11.3.2 Oven Heating

When reflow-soldering using an inline conveyorized infrared oven, the assembly is passed through a preheat zone, to avoid thermal shock, and then into a high-intensity zone for from 10 to 45 sec, where the solder reflow occurs. Such systems have advantages over the focused-energy units with respect to being able to more readily handle a variety of surface-mounted components and their suitability to handle production/high-volume requirements.

The following are some disadvantages of using infrared oven heating.

Some substrates and components may be subjected to excessive heat due to the high emissivity of the substrate or component as compared to that of the solder, thereby requiring localized shielding.

Changes in substrate material, P&I structure thickness, color, or conductor configuration could affect the infrared soldering, thus requiring an adjustment in processing variables.

Uneven heating can result in "hot spots" on the assembly, especially when large leadless chip carriers are surface-mounted on thin, printed board laminates. This necessitates reducing the conveyor speed in order to allow for conduction to equalize temperatures across the assembly.

Following the reflow process, lengthy cooling times may be required, depending on the mass of the assembly and associated fixtures. This is more of a problem with high-heat-capacity ceramics than with printed board laminates.

11.3.3 Solderability

There is a marked improvement in solderability of marginally solderable conductors—i.e., platinum/gold, palladium/silver, and copper—with infrared soldering. The principles underlying this process deal with the activation energies of the activators used in fluxes. Two major activators used in electronic-grade soldering fluxes are organic acids and amine hydrohalide salts. Both of these generic classes of compounds absorb energy strongly in the infrared wavelengths (3-4 μm). These energy absorptions are related to the activators' ability to ionize and thereby become active. This gives rise to improved solderability with infrared soldering.

11.4 CONDUCTION-REFLOW BELT SOLDERING [9]

Conduction-reflow belt-soldering equipment (Figure 11.5) conducts the necessary heat required for reflow of the solder through the P&I structure. This is accomplished by a series of heated platens located under a continuously moving belt. The platens and belt are individually controlled so that the reflow profile can be varied for P&I

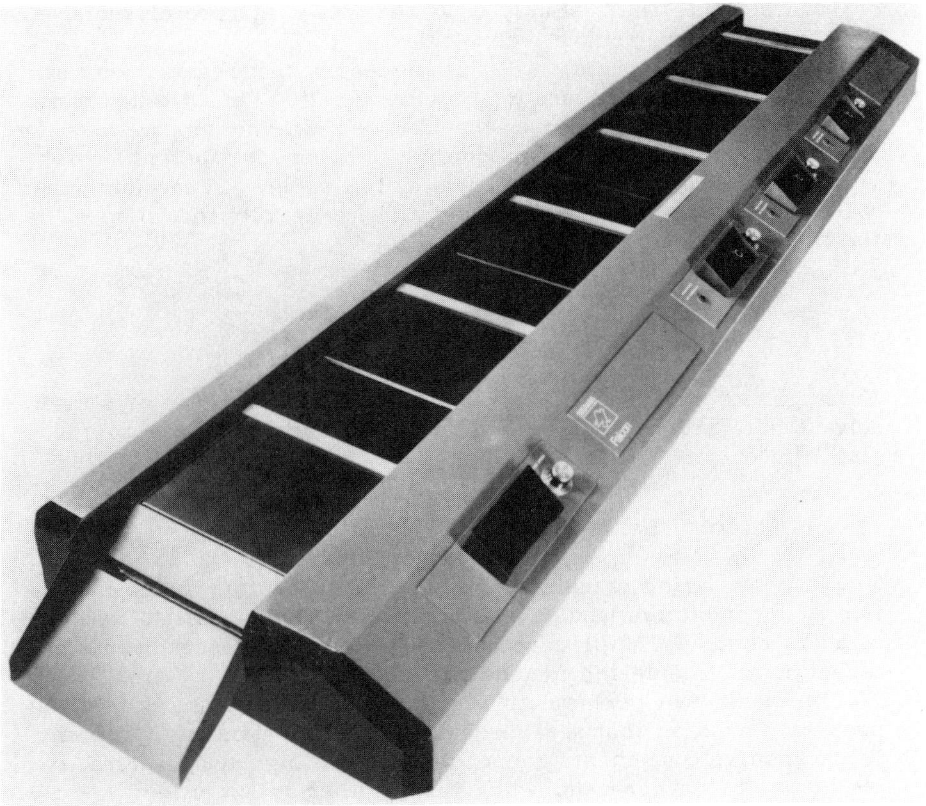

FIGURE 11.5 Inline conduction-belt reflow soldering machine. (Source: Sikama Int'l.)

structures that have significantly different thermal conductivities, are of different mass, or require the use of solders with different melting temperatures. Since heat is conducted through the P&I structure, a thermal gradient can exist across the assembly. Therefore, the cooling rate can also be varied by increasing or decreasing the flow of air over the belt as the assemblies travel off the final platen.

Because the conduction of the heat through the structure depends on the contact of the structure to the belt, a weighted fixture is usually used to apply pressure to multiple points on the assembly. This pressure is applied from the primary (component) side of the structure and is designed not to interfere with the components on that side. However, this is significantly complicated if there are components on both sides of the assembly or if there is a mixture of surface-mounted and through-hole technologies.

Another major limitation of this process is that residual flux can cause components to adhere to the moving belt. Thus, heated metal tracks with automatic index mechanisms (walking beams) are used for production/high-volume operations. Also, there is a potential problem in the effective removal of organic flux vapors. Therefore, the maintenance of these systems is essential, especially with respect to the timely changing of filters.

11.5 LASER SOLDERING [10]

Although not a panacea for all soldering needs, the use of a laser solves many problems associated with terminating surface-mounted chip carriers.

11.5.1 Assembly Details

The laser soldering technique involves the absorption of laser radiation by a conductive land to melt and reflow solder paste to form termination bonds. The primary advantages of using lasers over other techniques for soldering include the following.

The laser can perform its function on precisely the area to be processed, and on that area alone. This is important because many heat-sensitive components cannot tolerate the high-temperature levels to which they are subjected during other reflow soldering processes, such as the vapor-phase and infrared techniques. Also, since laser radiation levels and exposure times can be accurately controlled, only the precise amount of heat need be used, thereby minimizing the heating of surrounding areas.

Unlike techniques in which the application of pressure is required, the use of a laser is a totally noncontact method. Thus, as long as sufficient mechanical contact exists to ensure thermal conductivity between the parts to be soldered, the result of proper component placement, no external force is necessary.

The laser can solder leads as small as 0.001 in. in diameter without the fear of bending, bridging, or other damage.

Because of its rapid processing cycle, laser soldering also minimizes the thickness of the intermetallic layer that can form when working with materials such as copper and gold. Such a reduced intermetallic layer effects a reduction in tin/lead depletion and, thus, minimizes joint brittleness.

Laser soldering lends itself to maximizing board surface usage by being suitable for double-sided soldering.

Fluxless laser soldering to install components that cannot tolerate the corrosive action of soldering flux can be performed after partially completed assemblies have been cleaned.

Finally, laser soldering readily lends itself to automation. When coupled with the appropriate computerized motion control, laser soldering reduces the involvement of an operator in loading and unloading parts.

11.5.2 Equipment

Of the commercially available lasers, the carbon dioxide (CO_2) and neodymium-on-yttrium aluminum garnet (Nd:YAG) lasers have emerged as the most suitable for soldering applications. Both can be operated in a continuous mode and can be interfaced with computer-positioning systems, but that is where the similarities end.

The gas-dynamic CO_2 laser operates at 10.6 μm, while the solid-state Nd:YAG laser operates at 1.06 μm. While both of these wavelengths lie in the infrared zone, the differences between them have a vast effect on the absorption characteristics in materials, most specifically in the type of materials encountered in printed-board laser soldering.

In general, the absorption peak at the CO_2 wavelength corresponds to insulative materials such as plastics, epoxy, and ceramics; while the Nd:YAG output is more readily absorbed by conductive materials, such as metals. Therefore, although both lasers can effectively perform reflow soldering, it is obvious that the Nd:YAG is more efficient for soldering. (Conversely, the CO_2 laser is more effective for board drilling of "etched vias.")

Not quite as obvious, however, are several other important differences. For example, since the Nd:YAG beam is not readily absorbed in insulative materials, it can be scanned across multiple lands without needing to be shut on and off, thereby speeding the operation.

CO_2 laser-soldering systems are limited to spot size of 0.005 to 0.010 in., while comparable Nd:YAG systems can easily produce spot sizes in the 0.001-0.002 in. region.

Finally, the Nd:YAG laser is the ideal choice for fluxless soldering applications. Unlike the CO_2 laser, whose energy absorption is through the flux, the Nd:YAG output is effectively coupled directly into the metallic land. However, with this method, clean oxide-free parts are required, as the laser energy does not remove any of the oxides that inhibit solder wetting and flow.

11.5.3 Operational Factors

The CO_2 laser beam is initially reflected from most metals, and this reflected beam can cause serious problems. Therefore, when soldering at off-normal angles, on curved lands, or on densely populated circuits, the reflected energy can be of sufficient magnitude to damage plastic component cases, connector bodies, or the P&I structure itself. Fortunately, the Nd:YAG laser's radiation, although not 100% absorbed, generally tends to have minimal reflected energies and this, coupled with its generally shorter exposure times, minimizes reflected beam damage.

If laser soldering achieves more widespread acceptance, sophisticated fully automated systems might evolve. Such a system could couple an automatic pick-and-place system directly to the laser, which would solder the components at the same point as they are placed on the substrate. Using this technique, the assembly process would be greatly accelerated, with rates of from 20 to 50 solder joints per second being realistic.

11.6 COLLET-TOOL SOLDERING [11]

Leaded chip carriers can be individually surface-mounted with a collet tool. With the use of this device, the collet containing the chip carrier is lowered to the substrate and heat is applied. The solder melts and a mechanism pushes the carrier against the molten lands. The collet retracts, but it holds the carrier in place until the solder solidifies, after which it is removed.

11.7 HOT-AIR SOLDERING

With hot-air soldering, the prepared substrate with the positioned component is put on a metal conveyor that passes under two heat nozzles. Hot air passes through the two tandem-mounted nozzles. The temperature of the air coming from the first nozzle preheats the devices to +110°C. The air from the second nozzle heats the devices to the soldering temperature of +260°C. Special attention has to be paid to the air pressure, because too much air pressure can blow a part out of position or upset the prepositioned solder.

REFERENCES

1. J. Resendes, "Production Techniques with S.O. Packages," *Electronic Packaging and Production*, March 1982, pp. 115-128.

2. D. Schoenthaler, "Soldering Circuit Assemblies in the 1980's," Proceedings, Printed Circuit World Convention II, June 9-12, 1981, pp. 131-140.

3. R. P. Anjar, Sr., "Quality Solder Paste for Use in Microelectronic Applications," *Solid State Technology*, October 1983.

4. C. L. Harris and F. J. Cheriff, "Processing Considerations for Using Vapor Phase Soldering," *Electronic Packaging and Production*, May 1981.

5. "Mounting of Chip Carriers to PCB's and Ceramic Substrates Using Vapor Phase Soldering," *Insulation/Circuits*, March 1981.

6. H. W. Markstein, "Vapor Phase Soldering," *Electronic Packaging and Production*, July 1983.

7. C. A. Capillo, "The Assembly of Leadless Chip Carriers to Printed Wiring Boards," IPC Technical Paper, IPC-TP-425, October 1982.

8. S. Stach, "A Comparison of Infrared and Vapor Phase in Surface Mount Automation," Proceedings, National Electronic Packaging Conference, March 1983.

9. J. K. Hollomon, Jr., "Manufacturing Equipment for Surface Mount Technology," Proceedings, National Electronic Packaging Conference, March 1983.

10. O. L. Ramsey, "Surface-Mounted Components—The Inspection Challenge," Proceedings, National Electronic Packaging Conference, March 1983.

11. J. Orcutt, "Printed Circuits Get a New Boarder: The Plastic Chip Carrier," *Electronics*, April 7, 1982, pp. 141-145.

12

Chip-on-Board Technology

There are several different ways to package and assemble integrated circuits (ICs) on printed wiring boards (Figure 12.1). Yet, except in hybrid circuits, IC chips have been used almost exclusively in packages supplied by the IC manufacturer. However, during the past several years, bare IC chips have begun to be used, both singly and in multiples, on printed wiring boards in what is referred to as chip-on-board (COB) technology [1, 2]. The factors that lie behind the increasing interest in COB include size, cost, semiconductor technology, encapsulation, and automation [3].

Size [4]

The benefits of using COB all stem from the absence of the IC package. A wire-bonded chip takes up about a quarter of the area of a dual-inline package (DIP) and is also more space-efficient than leadless chip carrier (LCC) packages. Also, with a lower profile, COB can be used in applications not possible with the other packaging methods (such as in "smart" credit cards). As an example of both size and cost benefits that can be obtained with COB technology, Figure 12.2 shows a memory board's progressive reduction in size (and percentage of overall system cost) as it is changed from conventional through-hole (DIP) technology to surface-mounting (LCC) technology and, finally, to COB technology.

Cost [5]

As just mentioned, lower overall system cost is another factor behind the selection of COB technology. A similar consideration relates to the fact that the cost of making circuit interconnections is much

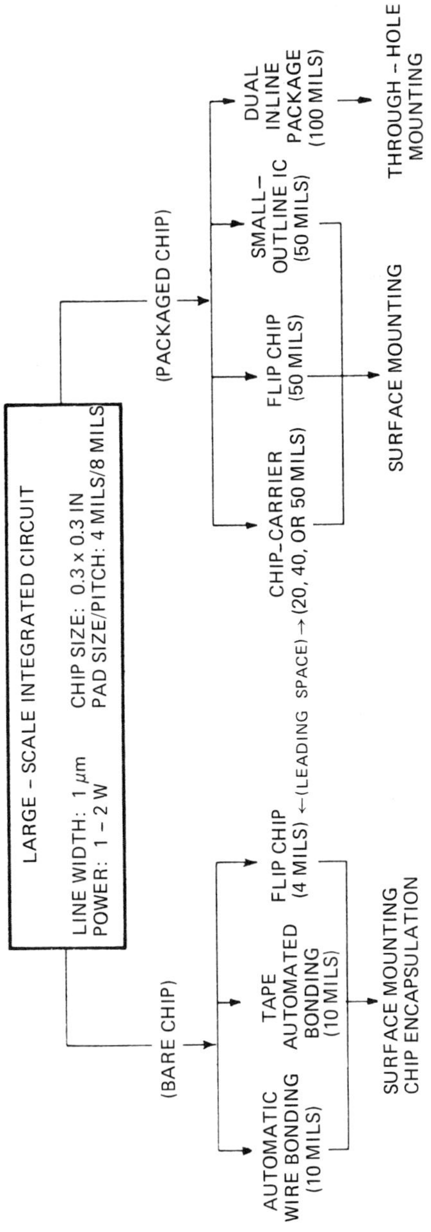

FIGURE 12.1 Integrated circuit packaging and assembly options.

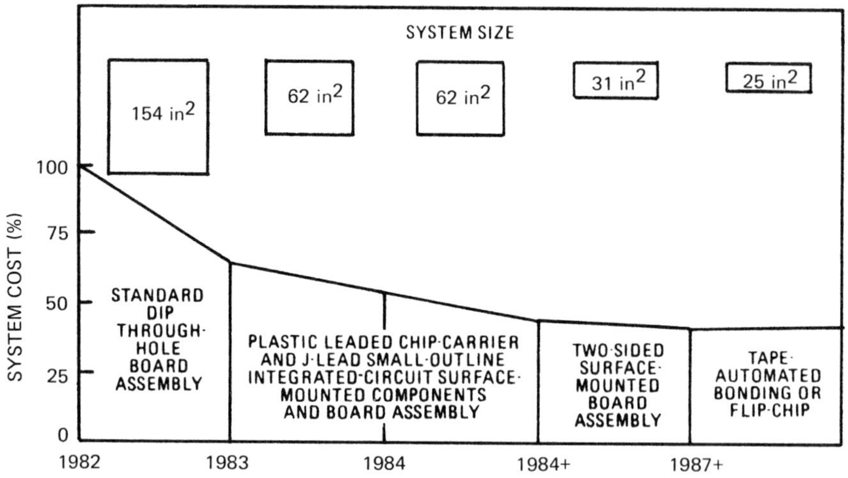

FIGURE 12.2 Packaging technology comparison.

less when they are made directly on the IC chip, as opposed to making the interconnections on a higher-level packaging scheme, as shown in Figure 12.3. Therefore, the trend has been toward having more sophisticated IC chips. This, unfortunately, has been associated with the need for more input/output terminations and more sophisticated packages for the integrated circuit, such as LCCs and pin-grid array (PGA) packages. However, since COB technology eliminates the need for the chip package, it also avoids the cost of the expensive package and custom tooling that is often required.

Semiconductor Technology

Trends in semiconductor technology have also paved the way toward COB technology. Low-power CMOS devices, already taking a major share of the IC chip market, are ideal for the power-limited COB technology. And the trend toward custom and semicustom ICs should mean additional importance for COB.

Encapsulation

Reliability has improved with the development of encapsulating materials that more closely match the thermal expansion properties of conventional printed-wiring board materials. In the past, the expansion mismatch between the encapsulant, usually a "glob" of epoxy, and the board would excessively stress the joint between the chip

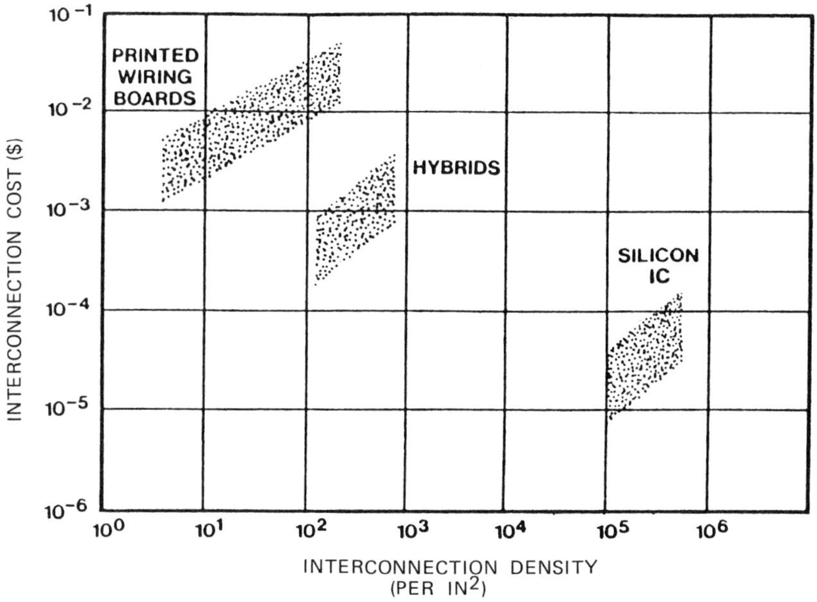

FIGURE 12.3 Interconnection cost versus density.

and the board. However, recent developments with improved sealants have reduced the significance of this problem, making COB attractive for use in more applications.

Automation

Companies that have had limited results competing in labor-intensive assembly markets may pay more attention to COB, as many, if not all, of the processes involved can be automated.

12.1 SEMICONDUCTOR DICE (CHIP) TYPES [6]

12.1.1 Back-Bonded Chips

Back-bonded chips comprise components that are traditionally installed in discrete packages. In transistors, for example, the bottom of the chip serves as the collector and metal pads serve as the base and emitter. The chip's collector is bonded directly to the mounting substrate, and discrete connections are made to the base and emitter. For COB applications, chip-and-wire technology is ideally suited for use with back-bonded chips.

The use of back-bonded chips has several advantages:

Availability: Nearly every IC available in discrete-package form also comes in chip form.
Size: The small size of back-bonded IC chips allows for very high packaging densities.
Heat transfer: The bottom of the chip dissipates most of its heat. Since in COB technology this surface is in intimate contact with the board, metallization on the board can be used as a heat sink.
Cost: Because back-bonded chips require no additional processing by the IC manufacturer, they cost less than face-bonded chips.

The use of back-bonded chips also has some disadvantages:

Handling and testing: The chip's small size makes it very difficult to handle. A few chips are usually lost during the chip-to-board transfer. Also, the tiny dice are difficult to test and attach.
Fragility: Exposed bonds make the completed interconnection susceptible to handling damage.

12.1.2 Face-Bonded Chips

Face-bonded chips are dice with supplementary features that provide access to the input/output pads. For COB applications, common forms of these components include beam-tape (TAB) devices and flip chips.

The advantages of face-bonded chips include the following:

Mounting: Bonding circuit connections to metallization accomplishes both mechanical mounting and electrical interconnection in one step. It also accommodates semiautomated bonding operations.
Ruggedness: Because the chips are somewhat enclosed, the resulting assemblies are more reliable than those using back-bonded components.
Inspection: With some of these devices, it is readily possible to check for proper bonding.
Pretest and burn-in: Face-bonded chips can be readily pretested for critical circuit parameters. They can also be subjected to preassembly burn-in.

However, the use of face-bonded chips does have some disadvantages:

Availability: Relatively few semiconductor components come in face-bonded chip form.
Inspection: Some face-bonded components, such as flip chips, do not readily facilitate the inspection of the bonded interconnections.

Cost: Face-bonded components cost more than back-bonded chips because they require extra processing by the semiconductor manufacturer.

Heat transfer: Face-bonded components dissipate heat poorly to the board, as the only connection to the substrate is through the chip leads.

12.2 PRINTED-BOARD CONSIDERATIONS

12.2.1 Layout Guidelines [7]

The layout of a board for COB applications, as with conventional designs, begins with the positioning of components and the routing of conductors. A prime factor that should be kept in mind is that exact placement of components is not always possible. Therefore, a good layout for volume production should be relatively immune to variations in part placement. Ideally, discrete parts should be able to be mislocated by up to 0.010 in. and up to 10° in rotation.

Figure 12.4 and Table 12.1 show layout features and limits beyond which cost and layout penalties will be incurred. (*Note*: Although the following information pertains specifically to chip-and-wire technology, similar considerations apply to the other COB technologies.)

Automatic wire/lead bonders require special layout considerations to optimize their operational efficiency, including:

Uniform wire/lead length.

Preferred wire/lead lengths of from 0.060 to 0.080 in.

A narrow range of chip-to-board "step down." (*Note*: The values in Table 12.1 assume a maximum step-down of 0.020 in.)

A uniform and homogeneous bonding surface. Since an automatic machine is often not selective, it will not hunt for the best bonding spot on the substrate metallization.

Conservative layout practices for the effective utilization of automatic equipment. (The manual wire/lead bonding machine operator can often inspect and compensate while bonding. The automatic machine cannot; it is precisely repetitious. Also, its bonding defects, though fewer, are more difficult to detect.)

When wire/lead bonds cross uncommon conductor metallization, the conductor metal should be covered with protective dielectric, such as solder mask, to prevent sagging wires/leads from causing shorts on the board. (*Note*: The thermosonic wire bonding process is particularly sensitive to wire sagging.)

The design guidelines in Table 12.2 should be used when die attachment, wire/lead bonding, and encapsulation are required.

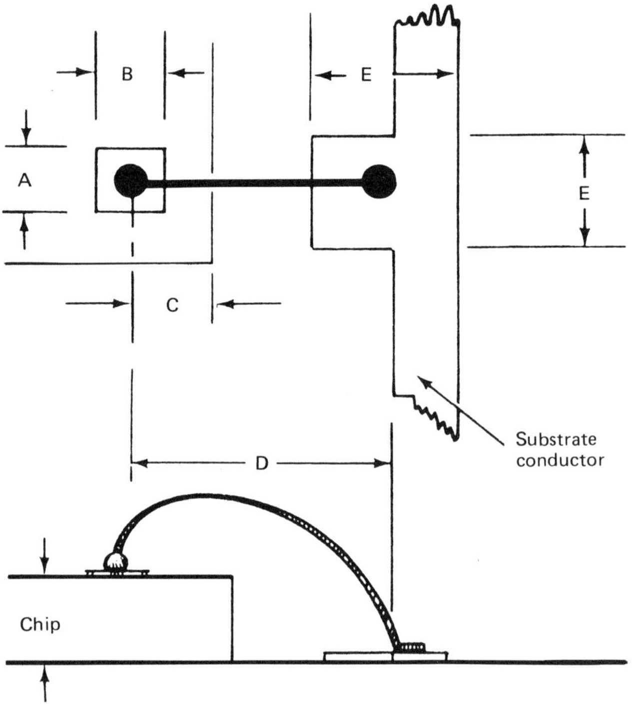

FIGURE 12.4 Wire bonding layout features (see Table 12.1).

TABLE 12.1 Wire Bonding Feature Limits (0.001 in.)[a]

Bonding process	A (min)	B (min)	C (max)	D (min)	E (min)	D (max)
Gold wire bonding						
Thermocompression	4	4	30	25	10	100
Thermosonic	4	4	15	40	10	100
Aluminum wire bonding						
Ultrasonic	2	4	30	25	10	100

[a]See Figure 12.4.

TABLE 12.2 General Chip-on-Board Design Guidelines

Guidelines	Reasons
Die-to-Board Attachment	
Chip attachment land should be at least 0.020 in. larger than the die size on all four sides	To allow sufficient tolerances for the chip attachment adhesive and die placement during assembly
Multichip-to-Board Attachment	
The location of multiple chips on the board should be equally spaced and on the same axis	To simplify die attachment, wire/lead bonding, and encapsulant/cover placement automation
Lead Bonding	
The bonding land on the board should be at least 0.020 in. from the chip attachment land	To avoid bridging of the chip attachment adhesive
The width of the wire/lead bonding land on the board should be at least 0.010 in. The bonding area should be at least 0.010 in. × 0.030 in.	To allow sufficient area for bond placement and rework
Bonding Wire/Lead	
The length of the bonding wire/lead between the chip and board lands should not be greater than 0.100 in.	To minimize wire/lead sagging
Spacing between adjacent wires/leads should be a minimum of 0.010 in.	To avoid shorts
The tip of the bonding wire/lead should preferably have a square configuration	To enable easier judgment of the reference points taken during automatic wire/lead bonding
Solder Mask	
The solder mask opening should be at least 0.050 in. from the edge of the bonding lead	To avoid bond placement on the solder mask

12.2.2 Board Substrate Materials [9]

The selection of an appropriate board substrate material depends a great deal on the cost trade-offs associated with the end-product application and the type of COB technology being employed. For most low-cost applications, where thermocompression wire bonding is not used, conventional printed-wiring board materials are sufficient.

Where thermocompression wire bonding is employed, consideration must be given to the high temperatures associated with this process. When cost is an overriding consideration, localized bond-site charring and delamination are often tolerated when conventional board materials are used. However, when this is not allowed, special high-temperature epoxy/glass and polyimide/glass board materials are often used.

For special COB applications, the use of flexible printed-wiring, conductive-polymer, and molded-thermoplastic board materials has also been considered.

12.3 CHIP ATTACHMENT

12.3.1 Chip Handlers

The chip-on-board assembly industry has now begun to closely follow similar manufacturing methods and procedures of those already well established in the semiconductor and printed circuit industries. However, due to the relatively small size of this market and the associated geometric complexities, which necessitate extremely accurate placement of the chips, there has been a slower rate of progress in automating COB chip-handling operations.

Low-Volume Equipment [10]

With this in mind, various approaches have been taken to solve COB manufacturing problems with cost-effective solutions for the particular applications involved. A typical approach for low-volume applications (Figure 12.5) combines proven technical components that have been used for computer graphics with system components that have been developed for front-end semiconductor manufacturing.

The low-volume COB assembly equipment shown in Figure 12.5 is a computer-controlled, pick-and-place machine for selecting devices from chip tray carriers, vibratory feeders, and tape input. The system orients and aligns devices on an automatic centering nest, then accurately assembles them on the boards at preprogrammed locations. Such a system can also be equipped with provisions for die-attachment epoxy dispensing, wafer frame or die tray loader elevators, pattern recognition for orienting devices, and vibratory feeders.

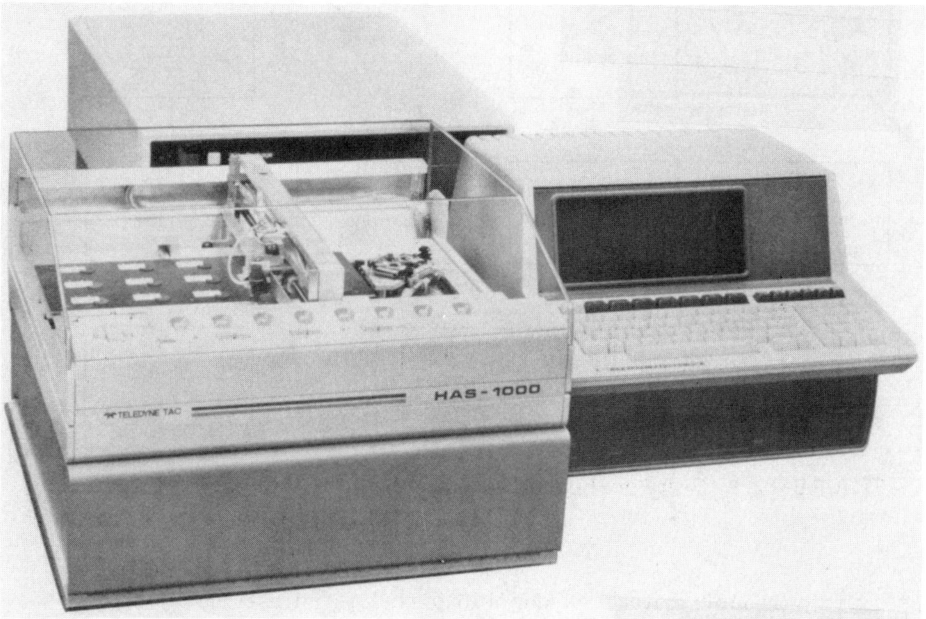

FIGURE 12.5 Typical low-volume chip-on-board chip-handling system. (Source: Teledyne Corp.)

The basic system (Figure 12.6) consists of a transfer module and a controller. The system controller provides the interface and control electronics; the transfer module is where the assembly actually takes place. This particular system has been adapted from a flatbed plotter with an 11-in. × 16-in. work area that can be used with many combinations of input and output fixturing.

Virtually any type of semiconductor device can be handled by equipment such as this, with chip dimensions ranging from 0.020 in. to 0.300 in. Several collet stalls are used to hold collets of varying size, so that several combinations of chip sizes can be programmed in a single assembly operation. A vacuum force is typically used to pick up the chips, thus eliminating distortions or scratches from mechanical fingers. A vacuum-sensing circuit is also often used to detect if a device was picked up and transported correctly. If no device is sensed, the operator has the option to continue, retry, stop, or skip to the next chip.

Once the part has been picked up by the collet, it is placed on the centering nest and squared. The motorized centering nest has

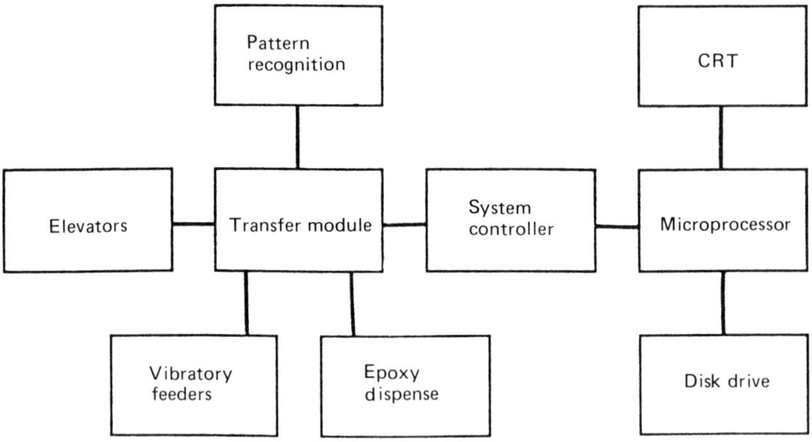

FIGURE 12.6 Chip-on-board chip-handling system block diagram.

a programmable rotational capability that is adjustable in 1.8° increments.

The more sophisticated units have a pattern-recognition feature that is performed by a camera system mounted directly over the centering nest. The orientation is then electronically compared with stored images, and the device is turned to the correct orientation. The chip can now be turned to the proper angular alignment, picked off the nest, and placed onto the board. The motion is also programmable to control chip placement pressure. With a system such as this, throughputs of up to 1200 chips/hr can be realized.

Fully Integrated Systems [11]

High-volume, fully integrated systems are also being developed for multichip applications. A typical system (Figure 12.7) processes up to 12-in. × 12-in. boards in 50-unit magazine-to-magazine equipment.

A fully integrated system could be configured to include equipment for adhesive silk screening, pick-and-place die attachment, wire/lead bonding, and chip sealing. Such a system would operate in the following manner. First, bare boards located in 50-unit magazines are presented to the input feed magazine of the silkscreener, where adhesive is deposited. Then the boards with adhesive are automatically transferred to the pick-and-place machine, where the chips are placed on the board over the adhesive. The populated boards are then put back in the magazine for further processing.

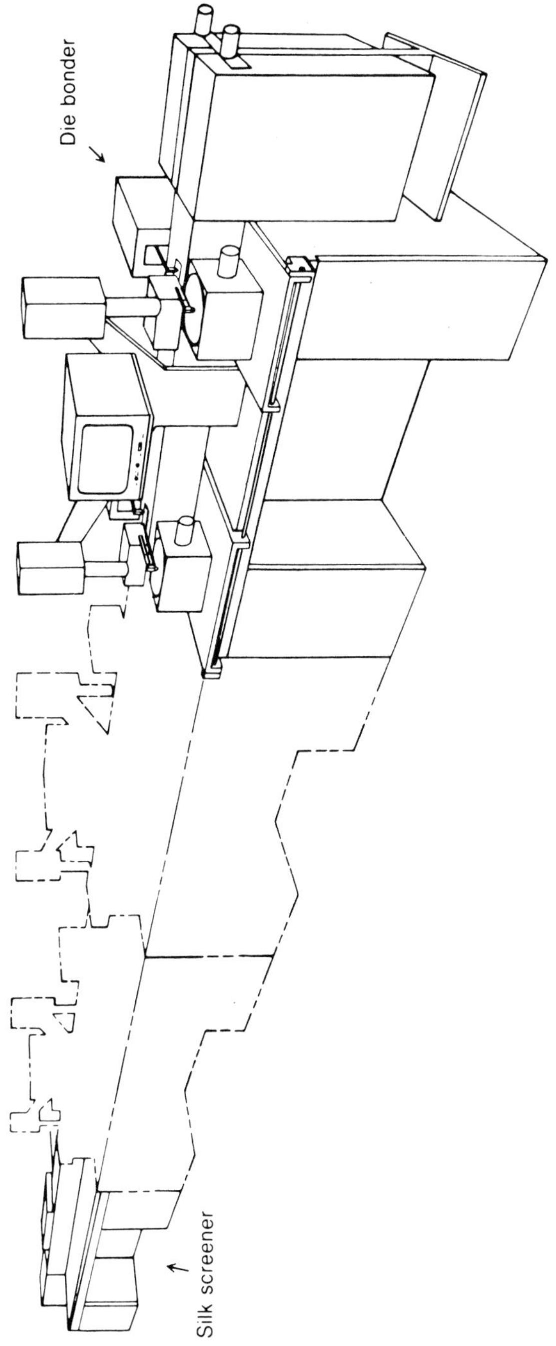

FIGURE 12.7 Fully integrated chip-on-board assembly system.

The magazines are removed from the output of the chip attachment unit manually and put into an oven to cure the adhesive. The boards, after curing in the magazines, are put into the wire/lead bonder's magazine feed. The chips are then bonded to the boards and automatically put back into the magazines. These magazines are placed manually into another wire/lead bonder for inspection and repair.

The inspected magazines are then taken to the chip-sealing unit. The boards are environmentally sealed and returned automatically to the magazines. These sealed units are then taken to another oven for curing of the sealing compound.

If more sophistication is desired, a control console can be added to oversee the operation of the system, consisting of a closed-circuit television monitor, a keyboard, and a tracking assembly that allows an operator to intervene by remote in the operation of the equipment.

12.3.2 Chip Bonding Adhesives [6, 7, 12]

Depending on the COB wire/lead attachment method being used, it is often necessary to attach the chip to the board with an adhesive. The better-known general types of adhesives are listed in Table 12.3.

Epoxy

Because of their superior properties, conductive and nonconductive epoxy bonding methods are usually best for attaching chips to boards. (They also rival passive chip solder bonding.)

In condutive epoxy bonding, metal particles (usually silver or gold) are added to the basic epoxy resin. When the resin cures, the metal particles remain in contact in suspension and thus provide electrical connectivity and thermal conduction. Nonconductive epoxies, on the other hand, contain a basic resin or incorporate additives to promote heat dissipation.

Epoxies designed for use with bare chips do not contain volatile solvents. Also, because they do not significantly outgas, even at high temperatures in a low vacuum, they can be used to bond chips to boards without forming voids between them.

Epoxies have several other advantages, including:

Low-temperature cure: Most epoxies cure near 150°C. This low temperature does not degrade the electrical properties or the reliability of the chips, as do some other bonding methods.

Ease of processing: Epoxy bonding equipment is straightforward to use. In many instances it can be deposited on all designated board areas simultaneously with the same silk screening equipment as is used for processing the board itself.

TABLE 12.3 Chip Bonding Adhesive Types (ISHM-1402/IPC-H-855)

Type	Advantages	Limitations
Phenolics	Very high bond strength	Used mostly for structural applications, possibly corrosive, difficult to process at low temperatures
Polyurethanes	Easy to rework	Not suitable for temperatures above 120°C, relatively high outgassing, some decomposition
Polyamides	Easy to rework	High moisture absorption, high outgassing, variations in electrical insulation properties, especially when exposed to high humidity
Polyimides	Very high-temperature stability	High cure temperatures, require solvents as vehicles
Silicones	High-temperature stability, easy to rework, high purity, low outgassing	Moderate to poor bond strength, high coefficient of thermal expansion
Epoxies	Some are easy to rework by thermomechanical means, some are low outgassers, easy to process, can be filled to 60-70% with a variety of conductive or nonconductive fillers	Depending on type of curing agent used and degrees of cure: outgassing, catalyst leaching, corrosivity
Cyanoacrylates	Very rapid setting ($\approx$10 sec) give very high initial bond strengths	Bond strengths often degrade under moist or elevated temperature ($\leq$150°C) conditions

Ease of handling: It is possible to store boards containing wet epoxy for a long time before mounting the chips.

High yield: Devices bonded with epoxy exhibit yields approaching 100%.

Ease of repair: After the epoxy heats to its softening temperature, it is easy to remove a faulty chip and replace it with a good one.

Ruggedness: An epoxy bond's mechanical strength is sufficient to withstand most rough handling and most shock and vibration environments.

As with other chip mounting processes, epoxy bonding involves trade-offs. In particular, its use dictates that certain precautions be taken, including:

Heat control: When thermocompression wire bonding is used, a bonder with a heated probe tip must be used, as the heat associated with some types of thermocompression wire bonding will soften the epoxy. (However, this precaution does not apply when thermosonic wire bonding is used.)

Precision tools: If a vacuum-hold chuck is used to mount the chips, each component should be checked to be sure that a positive seal forms against the probe tip. Otherwise, the epoxy might be drawn to the top of the chip.

Chip processing: Sinter semiconductor chips have gold backing in order to alloy the gold with the silicon. If the gold is evaporated, the epoxy can pull it away from the chip and form a high-resistance contact.

Polyimide

The use of polyimides and silver-filled glass adhesives is increasing due to their lower cost and ability to withstand higher temperatures than the epoxies. The polyimides used are thixotropic pastes containing approximately 70% silver powder in a polyimide resin that has been dissolved in a high-boiling solvent.

To minimize the quantities of solvents and other vapors released during cure, the polyimides used in chip attachment adhesives are low-molecular-weight resins that cure by an addition reaction. They are supplied and used in much the same way as the epoxies.

The main objection to their use is that they are organic polymers and it is difficult to remove all of the organics. If all the organics are not removed, they may outgas and cause chip degradation.

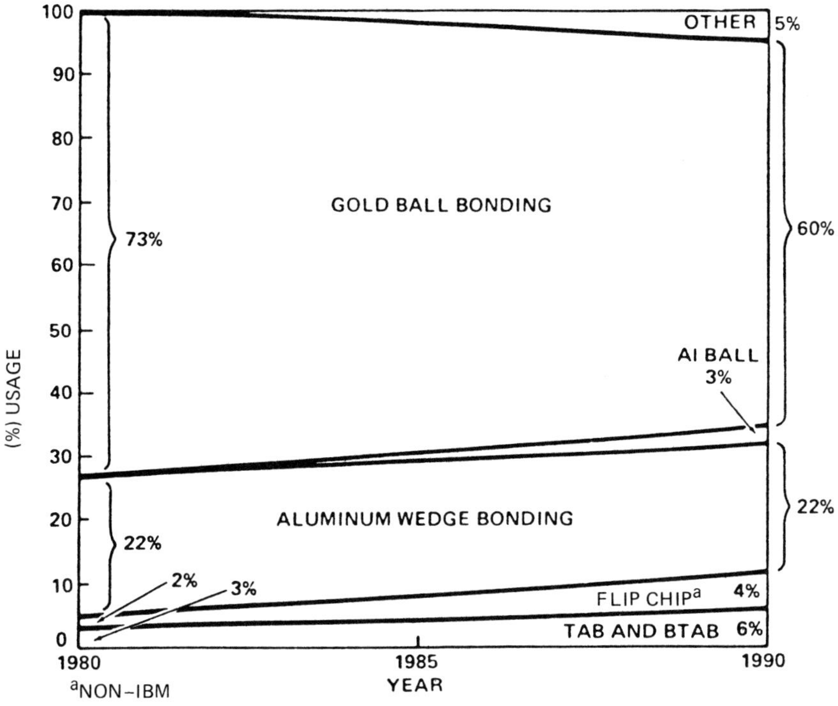

FIGURE 12.8 Chip termination techniques.

12.4 CHIP TERMINATION TECHNIQUES, GENERAL [14, 15]

The three basic chip attachment and termination techniques, or derivatives thereof (Figure 12.8)—wire bonding (chip-and-wire), tape automated bonding (TAB), and controlled-collapse (flip-chip) bonding—have been used in some cases for over 20 years in semiconductor packaging and in hybrid assemblies. All of these techniques have been refined to the point where they are suitable for COB applications (Table 12.4).

12.4.1 Wire Bonding

Automatic wire bonding rates, for example, have been increased by a factor of 15 over manual rates. The rate for automatic wire bonders

TABLE 12.4 Chip-on-Board Process Comparison

Property/requirement	Wire bond	TAB	Flip-chip[a]
Circuit density	Medium	High	High
Special circuit process	No	Yes	No
Circuit cost	Low	High	Medium
Bonding process	Simple	Complex	Moderate
Profile	High	Low	Low-medium
Easy pretest	No	No	No
Special chips	No	Yes/no[b]	Yes
Development status	Mature	Emerging	Mature
Density capability	Med. high	Very high	Medium
High speed	No	Yes	Yes
Reliability	High	High	High
Aluminum pads OK	Yes	No	No
Gang bonding	No	Yes	Yes
TCE stress	No	Sometimes	Yes
Thermal characteristics	Good	Good	Poor
Substrate choices	Wide	Limited	Limited

[a]Controlled-collapse chip connection.
[b]Yes for bumped chips—alternatively, circuit is bumped.

exceeds 10,000 bonds per hour in some applications. Also, wire-bonding labor content has decreased by an even larger factor, since one person can operate more than one bonder.

12.4.2 Tape Automated Bonding

Tape automated bonding (TAB) rates have always been high for inner-lead bonding within a semiconductor package, such as a DIP. (One thousand chips per hour is a commonly quoted rate.) Also, chip placement rates for TAB chips on multichip assemblies can be relatively high with appropriate mechanization.

12.4.3 Flip-Chip (Controlled-Collapse) Bonding

The use of flip chips with controlled-collapse bonding has potential for very high assembly rates. For example, with only the use of simple optomechanical assists, manual chip placement rates as high as 200 per hour have been achieved. This results in a very high assembly/bonding rate if a chip has many input/output terminations.

12.5 CHIP-AND-WIRE TECHNOLOGY [6, 16]

Chip-and-wire technology is commonly used in COB applications to make the majority of interconnections between the IC chip and the board. Unlike some of the other COB technologies, this technology requires the COB assembler to make two bonds per chip interconnection (one at the chip metallization and one at the board land).

There are three basic types of wire bonding (Table 12.5) [17]:

1. *Thermocompression*: Thermocompression wire bonding uses a combination of heat (approximately 270°C) and compression to connect the wire to the IC chip and to the board. The basic method actually encompasses three different bonding processes: ball, stitch, and wedge.
2. *Ultrasonic*: Ultrasonic wire bonding relies on a wire-to-conductor land interface's mechanical resonance to absorb the energy needed to weld the wire to the land. An ultrasonic oscillator generates the resonant frequency.
3. *Thermosonic*: Thermosonic wire bonding heats the substrate to 150°C and uses ultrasonic energy to provide the remainder of the energy required.

By the time a chip reaches the interconnection stage of assembly, a considerable investment has been made in materials and labor. Therefore, high yields in these late process stages can result in greater cost savings than can be realized at earlier stages of the process. For this reason, COB manufacturers carefully follow precisely defined processing procedures, and place stringent demands on the materials and equipment used in wire bonding.

12.5.1 Bonding Wires [7]

Both gold and aluminum wire are used for COB applications. The gold wire are typically alloyed with small amounts of beryllium copper to control grain growth during bonding. Aluminum wires are typically alloyed with 1% silicon.

TABLE 12.5 Wire Bonding Process Comparison (ISHM-1402/IPC-H-855)

Parameter	Thermocompression (gold wire)	Automatic thermosonic (gold wire)	Ultrasonic (gold and aluminum wire)
Develop and control	May be difficult to control mert to tip	Easiest to control	Must control acoustic energy and force
Speed	∼6 wire bonds per minute	∼2000 wire bonds per minute	∼30 wire bonds per minute
Current carry (0.001 in.)	∼0.55 A	∼0.55 A	<0.40 A
Heat required	Pulse capillary heated work stage optional	150°C work stage (max)	No heat required (typical)
Acoustic required	None	Controllable	Controllable
Force required	Most force required	Less than ultrasonic and thermocompression	Less than thermocompression

Direction	Omnidirectional (360°)	Omnidirectional (360°)	Straight line bond ±7.5° off true line)
Looping	Some control—operator dependent	Best control	Some control—operator dependent
Pad size (0.001 in. diam.)	Ball size dependent 2X to 5X diameter	Ball size dependent 2X to 5X wire diameter	Wedge size dependent 1.2X wire size
Bond head clearance	Largest head size max clear required	Smallest head size min clear required	Large head size—no deep access
Sensitivity to contamination	Most sensitive	Less sensitive	Least sensitive
Purple plague	Most prominent	Increases with reduced hermeticity and increased temperature	Increases with reduced hermeticity and increased temperature

The information in this table is for general application. Some parameters, i.e., heat, force, may be adjusted for specific applications.

The following factors influence the choice of bonding wire and equipment:

Maximum processing temperature, including rework (Table 12.6)
Size of the bond sites on the IC chip (0.001-in.-diameter wire for thermosonic ball bonding requires a land size greater than 0.0025 in.^{2})
Board substrate materials
Final operating and nonoperating environment
Geometry of the board assembly, which may restrict bonding tool access.

The fusing current of the wire is dependent on the material type, its diameter, its length, and the wire termination method being employed. The termination method is particularly important, as it determines the degree of heat sinking required at the chip or board substrate.

Table 12.7 lists wire sizes and typical resistance values for aluminum and gold wires. (For most applications, the current should be limited to less than 0.5 A.)

Both gold and aluminum fine wire, in sizes from 0.0005 in. to 0.002 in. in diameter, may be purchased in increments of 0.0001 in. Larger sizes of wire are available in diameters of 0.003, 0.005, 0.008, 0.010, and 0.015 in.

The procurement of both gold and aluminum wire should be in accordance with ASTM F-72 and ASTM F-487, respectively, except that fine aluminum wires with 1% magnesium should be specified per ASTM F-638. All of these references may be found in Part 43 of the ASTM Annual Book of Standards.

12.5.2 Board Metallization [17]

Gold Wire Bonding

Gold plating is usually used on the wire bonding lands of the board. A major criterion for evaluating its suitability is bond yield. A good gold wirebond should have a pull strength of from 8 to 10 g for 0.001-in.-diameter wire, with failure occurring in the wire and not between the wire and bonding land (bond lift).

For gold (thermosonic) wire bonding yields approaching 99.5%, the board plating should be a minimum of 0.000040 in. (40 microinches) of 99.99% pure soft gold over 0.000150 to 0.000200 in. (150 to 200 microinches) of conventional nickel. Also, the materials must be formulated specifically to meet the stringent demands of automatic wire bonding if bonding rates of up to 4 wires per second are to be achieved. For these applications, it is particularly important that

TABLE 12.6 Bonding Wires for Chip-on-Board Applications

Wire diameter (in.)	Attachment method	Substrate temperature (°C)	Comments
(Gold) 0.0007–0.001	Thermocompression (hot substrate)	>300	Limited repair ability
(Gold) 0.0007–0.002	Thermosonic ball or wedge	25	Quality of bonds improved if the substrate is raised to 125°C. Can bond chips without raising substrate temperature
(Aluminum) 0.0007–0.010	Ultrasonic wedge	25	Low-temperature process, restricted due to tool size

TABLE 12.7 Aluminum and Gold Wire Sizes and Ratings

Wire minimum diameter (in.)	Standard wires		Resistance (Ω/ft)		
	Aluminum	Gold	Aluminum[a] 1% Si (hard)	Aluminum[a] 1% Si (annealed)	Gold[b]
0.0007		X	36.06–39.86	34.62–38.273	27.71–29.43
0.001	X	X	17.67–19.53	16.0 –18.75	13.58–14.42
0.002		X	4-41–4.88	4.23–4.68	3.40–3.61

[a]ASTM F487.
[b]ASTM F72.

the plating be optically dense and consistently provide smooth, flat bonding surfaces, which are necessary to increase yields and quality.

Aluminum Wire Bonding

Aluminum (ultrasonic) wire bonding can be done reliably with board land plating that is similar to that for gold wire bonding. However, it is reasonable to expect to achieve bond strengths of only 3 g minimum and automatic wire bonding rates of up to 2 wires per second.

Conductive Polymer Circuitry [18]

Printed boards produced with polymer conductive materials on low-temperature substrates are now being made that are wire-bondable. The application of a plated metal surface over the printed thick-film polymer conductor pattern makes this possible. The printed polymers currently being used for this type of board construction are thermal-set, epoxy-silver conductors, and non-noble nickel polymer.

12.5.3 Thermocompression Wire Bonding [19, 20]

Thermocompression wire bonding is one of the most frequently used bonding processes. Its principle is to join two metals using heat and pressure, but without melting. The elevated temperature maintains the metals in an annealing state, as they join in a molecular metallurgical bond. The process is quite involved, but, generally speaking, the softer the metals, the more readily they bond together.

A major disadvantage associated with thermal bonding, however, is that many materials cannot withstand the high temperatures required. Aluminum, for example, is difficult to bond by thermal techniques, because a hard oxide forms at temperatures of 250-300°C, thus inhibiting bondability.

Thermocompression Wedge Bonding

The oldest form of thermocompression bonding is known as wedge bonding. This technique requires two separate alignments, X and Y, and is rather slow. However, it is very useful with small-diameter wires. It is also quite simple to produce two bonds with the same wire on the same land, thereby improving joint reliability.

Thermocompression Ball Bonding [7]

A technique more suitable for higher bonding rates is known as ball bonding. With this method (Figure 12.9), the wire to be bonded

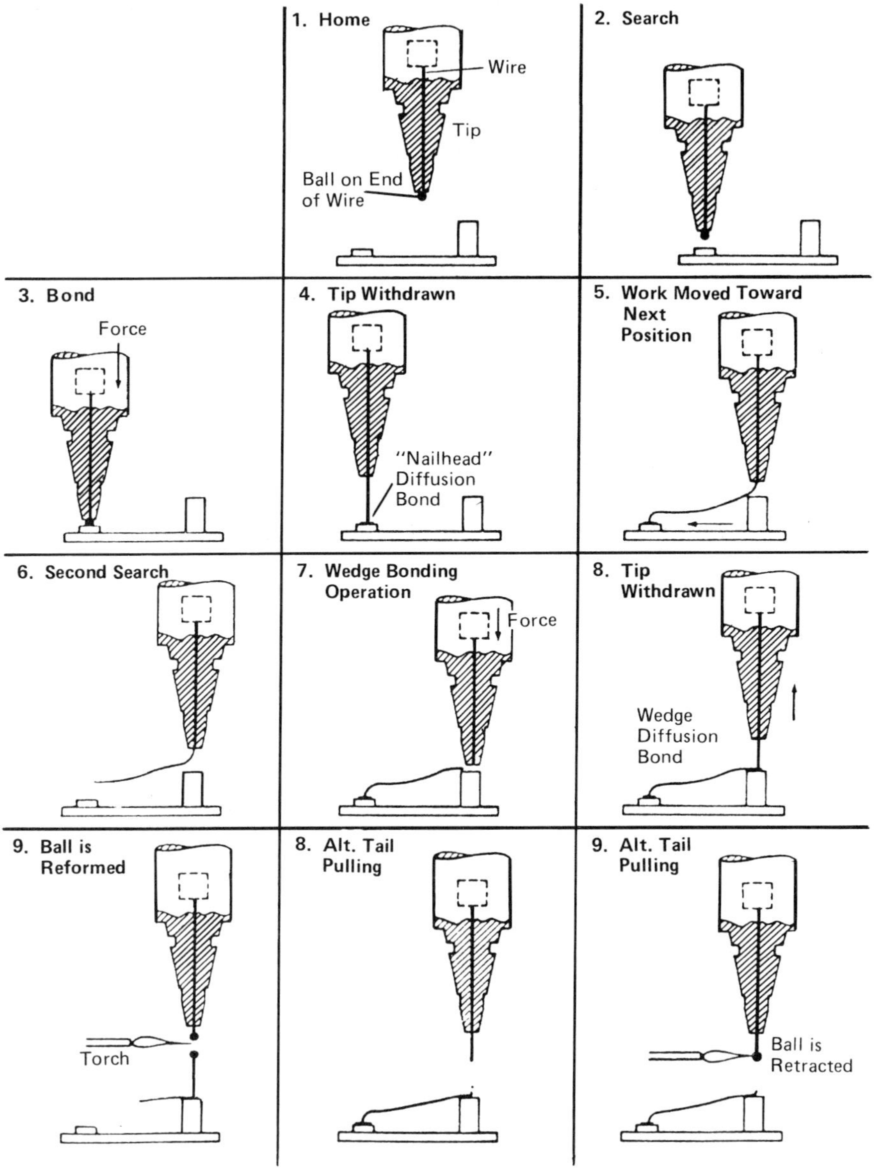

FIGURE 12.9 Mechanics of thermocompression ball wire bonding.

is fed through a capillary, which is usually heated to between 300 and 400°C. An open flame (or spark discharge) is then used to melt the end of the wire to form a small ball. Only one alignment is necessary, thus allowing for faster bonding rates. The bonding tool then forces the ball onto the bonding land (Figure 12.10) and the bond is formed. Wire cutoff is achieved by a flame-off operation, which also produces the ball used at the next wire termination.

The advantages of using thermocompression ball bonding include:

It is omnidirectional; i.e., after the first bond is made, the bonding head may move in any direction to make the second bond.
No unique fixturing is required during bonding.
It is the easiest process to develop and control.

Its disadvantages include:

It requires a high working temperature.
It requires relatively large bonding lands and wire bonding clearances for high-yield/high-rate production (see Figure 12.10).
It is the wire bonding process that is most sensitive to surface contamination.
It has relatively slow bond times (i.e., the time during which the bonding force is applied).
Aluminum wire is not usable, since it will not ball properly during the flame-off operation.

Thermocompression Stitch Wire Bonding

A technique frequently used as a compromise between wedge and ball wire bonding is stitch bonding (Figure 12.11). In this technique, a cutoff arrangement is used in place of the flame-off. This allows for the use of gold and aluminum wire and smaller bonding areas. The cutoff process also forms the wire for the next stitch-bond operation.

12.5.4 Ultrasonic Wire Bonding [7, 19, 20]

Ultrasonic bonding (Figure 12.12) is a different concept of bonding that employs a rapid scrubbing or wiping motion in addition to pressure as a means of achieving the molecular bond. The scrubbing action removes any oxide films that might be present. A slightly larger area of contact is necessary because of the scrubbing action. Also, extreme care must be taken so as not to damage the chip during the ultrasonic wire bonding operation.

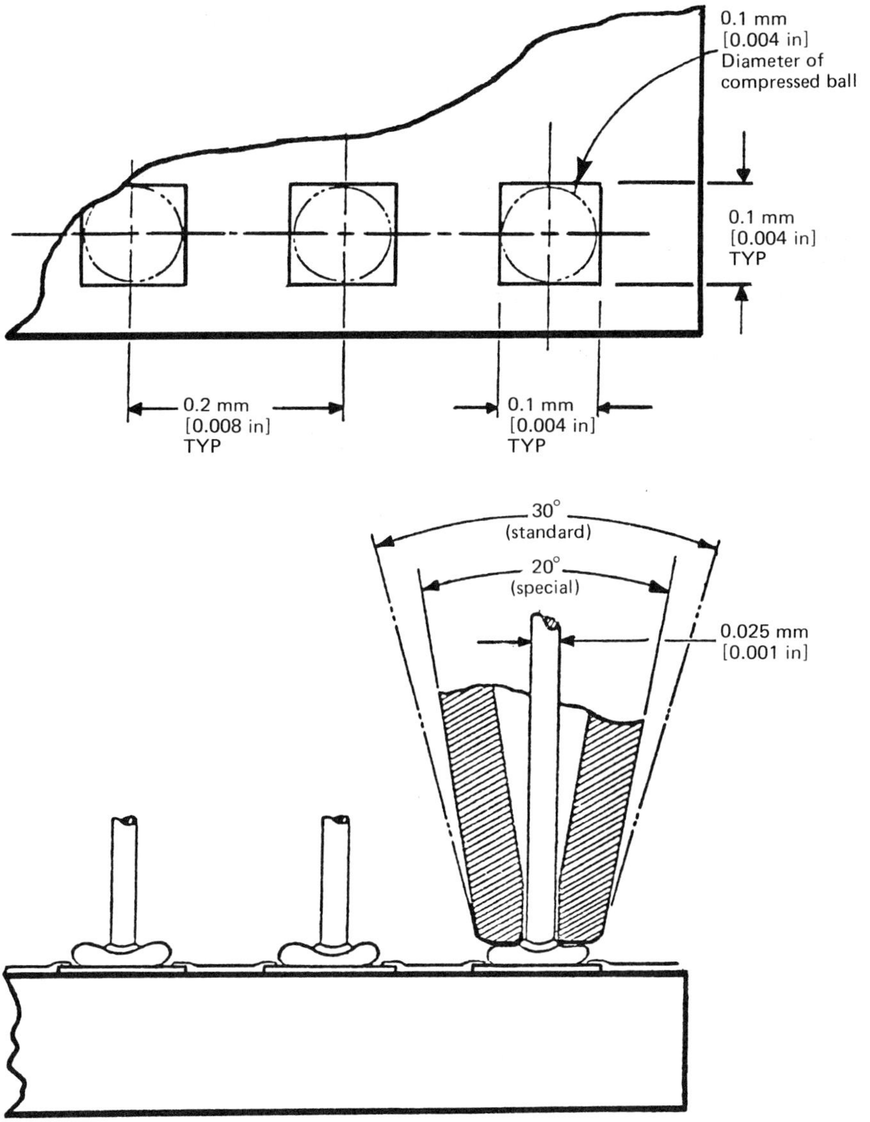

FIGURE 12.10 Thermocompression ball wire bonding parameters (IPC-CM-770).

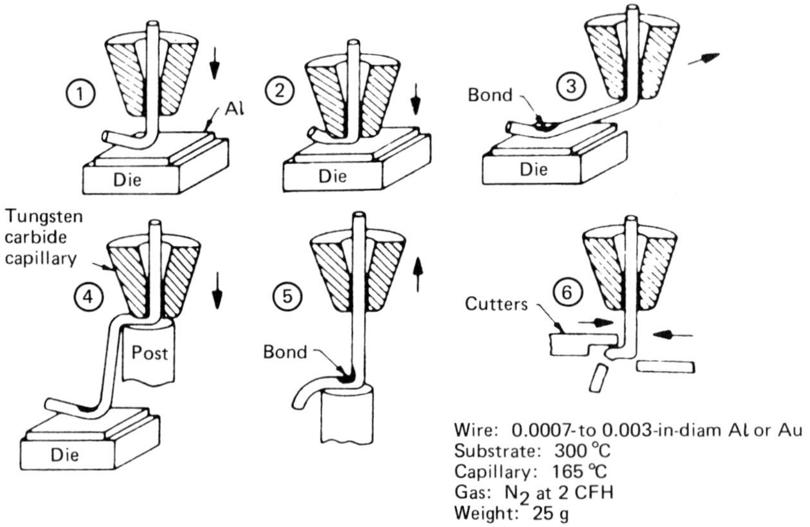

FIGURE 12.11 Thermocompression stitch wire bonding.

Because ultrasonic bonding can create bonds between a wide variety of dissimilar materials, it is an extremely flexible process. Thus, both gold and aluminum wire are compatible with this technique.

Void-free junctions are produced by ultrasonic wire bonding with relatively few foreign material inclusions, making it a desirable means of creating high-quality, low-resistance electrical junctions. Also, bonding dissimilar metals at low temperatures eliminates, or greatly reduces, the formation of intermetallic compounds and allows bonds to be made in the immediate vicinity of temperature-sensitive components without adverse effects.

The advantages of ultrasonic bonding include:

It is a room-temperature process.

It is the wire bonding process that is the least sensitive to surface contamination.

It is workable with smaller lands than thermocompression ball bonding (e.g., for a 0.001-in.-diameter wire, a land size of 0.0015 in. × 0.003 in. is practical).

Disadvantages include:

It is unidirectional (i.e., after the first bond is made, the bonding head can move in only one direction toward the second bond site).

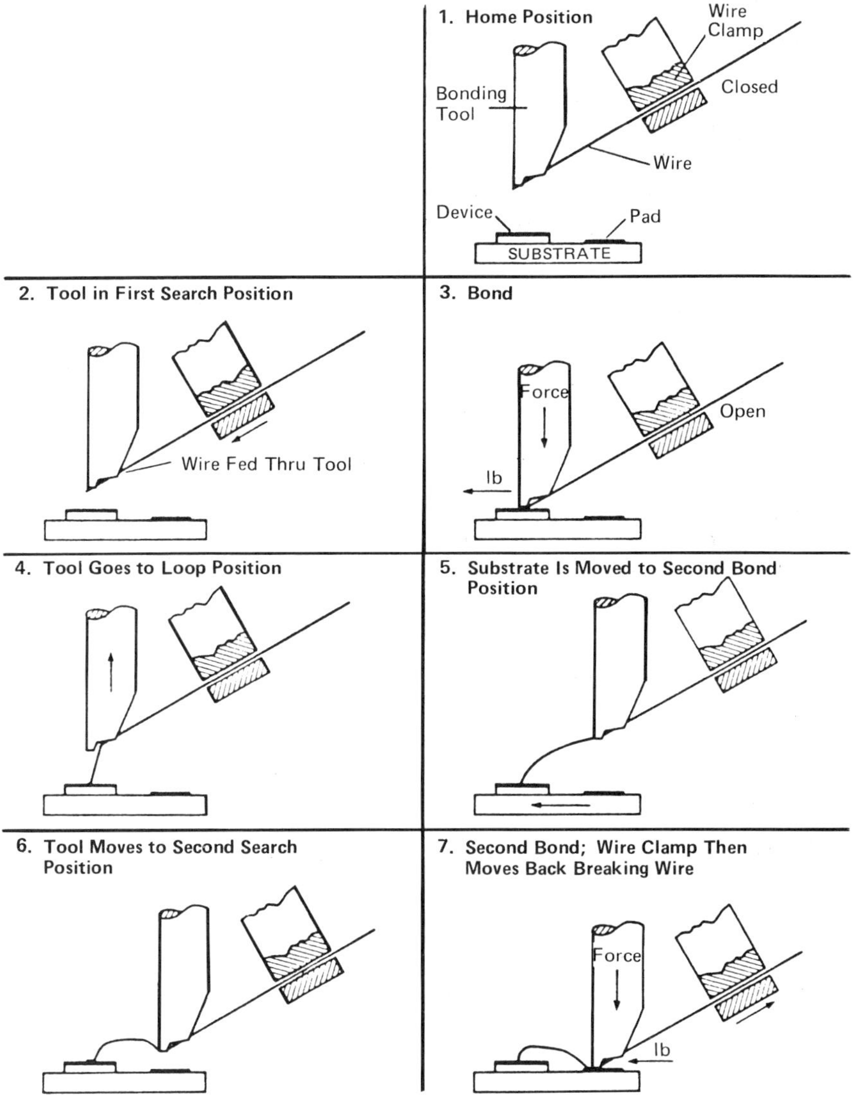

FIGURE 12.12 Mechanics of ultrasonic wire bonding.

Component or package-wall proximity must allow for bonding-tool and wire-clamp clearances.

Acoustical energy measurement problems make the process difficult to characterize and control.

12.5.5 Thermosonic Wire Bonding [7, 19, 20]

Ultrasonic energy is commonly used together with thermocompression wire bonding methods, resulting in a technique known as thermosonic wire bonding. This technique (Figure 12.13) relies on the vibrations created by the ultrasonic action to scrub the bond area to remove any oxide layers and also to create the heat for bonding. This, combined with the pressure of the tool forcing the wire into the bonding land, provides the final bond. Since this can take place at temperatures around 120 to 150°C, thermosonic wire bonding can be done with low-temperature materials. It can also be used with aluminum wire.

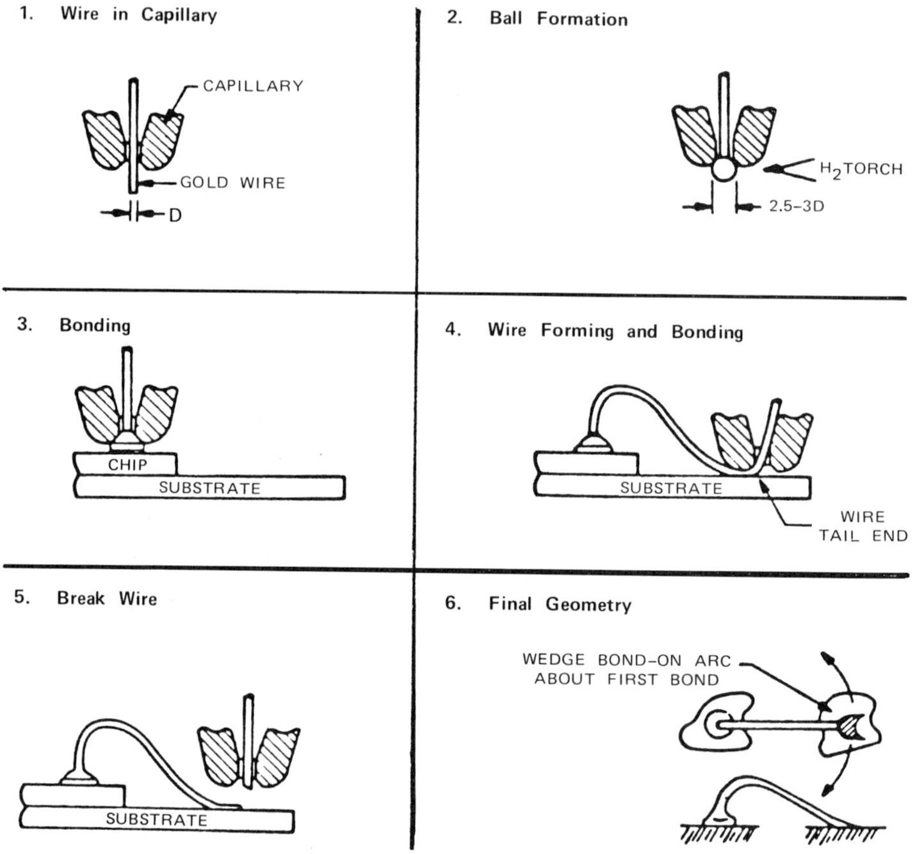

FIGURE 12.13 Mechanics of thermosonic wire bonding.

The advantages of using thermosonic wire bonding include:

It uses moderate bonding temperatures.
It is omnidirectional.
It is less sensitive to surface contamination than is thermocompression wire bonding.

Disadvantages include:

It requires a minimum land size of 0.004 in. × 0.004 in.
It requires control of acoustical energy.
It requires special circuit fixturing.
It has some wire looping limitations.

12.5.6 Applications [21-23]

Chip-and-wire technology has found uses in COB applications with a wide variation in size and complexity. For example, video-game cartridge packaging is entering a second generation of evolution. The first-generation cartridges consisted of plastic DIPs wave-soldered onto printed wiring boards. The current generation is being built with direct COB construction, which eliminates the first level of packaging. (Further generations are expected to incorporate even more efficient and cost-effective approaches, such as TAB or flip-chip mounting.)

Slightly more sophisticated COB packaging will incorporate bubble memories, in the manner shown in Figure 12.14. In these applications, the bubble substrate (chip) will be mounted on a small, but complex, multilayer board, which will interconnect the chip and the I/O leads of the coils wrapped around it. The entire COB subassembly will be enclosed with magnets in a shield to complete the assembly.

Perhaps the most sophisticated chip-and-wire COB application to date is the multichip assembly shown in Figure 12.15. This COB assembly contains 22 VLSI (very-large-scale-integration) IC chips, interconnected by the multilayer board. Lids placed over each of the wire-bonded chips complete the assembly.

12.6 TAPE AUTOMATED BONDING [20, 24-27]

A relatively recent concept that appears to have a bright future for COB applications is known as tape automated bonding (TAB). This technique utilizes photo-imaging/etching processes to produce fabricated conductors on a dielectric tape in movie-film format.

The most visible aspect of the TAB system is the carrier tape, which is stored on reels similar to movie film, in widths from 8 to

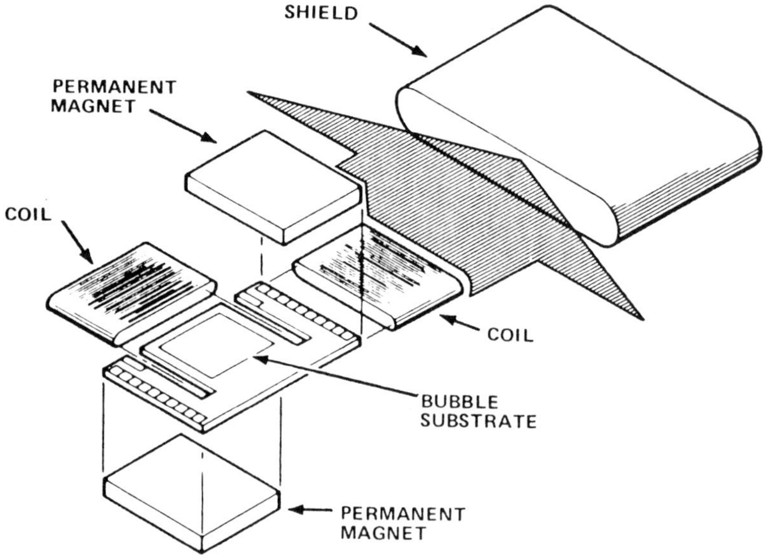

FIGURE 12.14 Bubble memory chip-on-board assembly.

70 mm. Windows are punched at specific locations in the tape and a thin, often 0.0014-in.-thick, conductive foil is bonded to the tape, which is usually either mylar or polyimide. A conductive pattern is then etched in the foil to give the desired interconnection circuitry with "beam-type" leads extending over the windows in the tape.

In subsequent processing (Figure 12.16), the beams are simultaneously bonded to the chip, which is located precisely under the window. The exact location of the chip with respect to the sprocket holes on the tape carrier permits automated tape-handling equipment to be used that accurately positions the individual chips for the subsequent processing operations. For COB applications these include testing, burn-in, and mounting onto the board.

TAB provides a means of mounting bare IC chips to printed boards with the following advantages:

The bonding areas on the chip can be hermetically sealed when a gold bump is constructed on the bonding land.

Less gold is required for TAB than is needed for wire bonding.

FIGURE 12.15 Multiple-chip chip-on-board assembly. (Source: Hewlett-Packard Corp.)

TAB provides a means for pretesting and "burn-in" of the chips prior to their final mounting on the board.
Bonding lands can be as small as 0.002-in.^{2} on 0.004-in. spacing.
Input/output counts can be as high as 300 or more leads.
Very-low (0.030-in.)-profile COB assemblies can be made.
TAB provides a method for automated (and robotic) simultaneous bonding of all the leads to the chip and to the board.
The TAB process chips require only a fraction of the board surface area required for mounting packaged chips (see Figure 12.17).
The rectangular TAB lead provides a lower conductance than does a round wire, enhancing its use in high-speed applications.

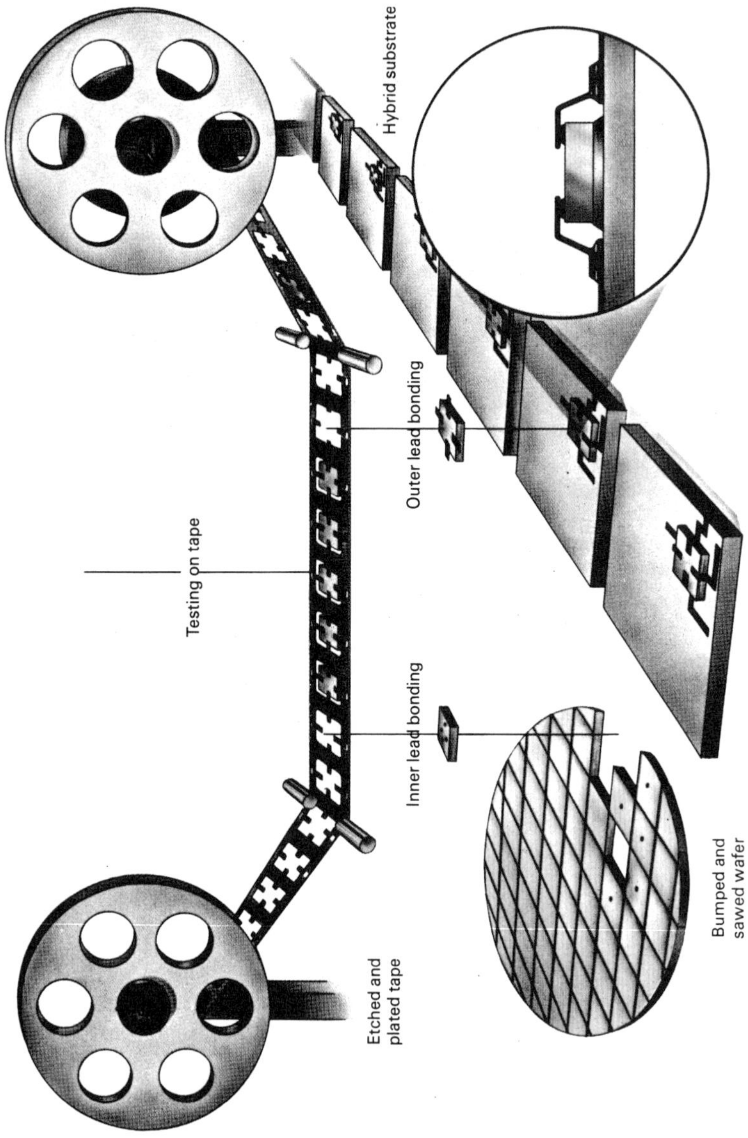

FIGURE 12.16 Tape automated bonding (TAB) process steps. (Source: Farco s.a.)

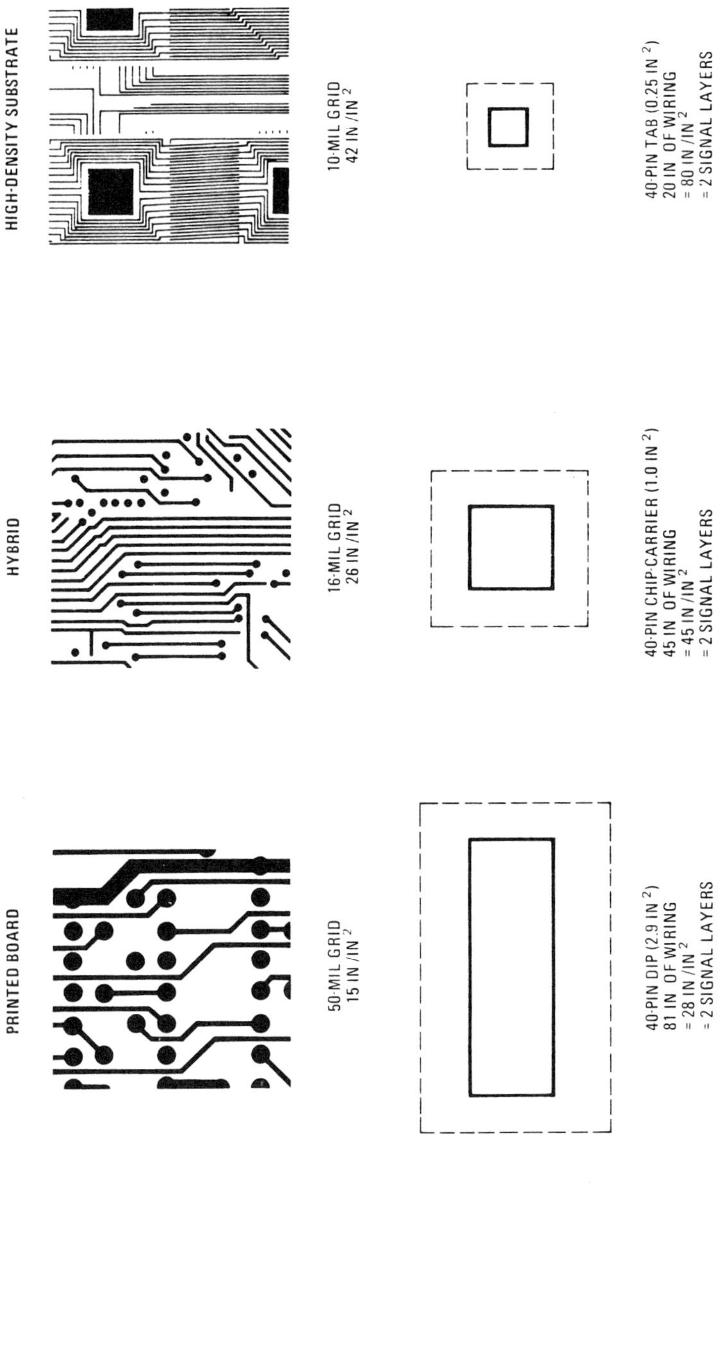

FIGURE 12.17 Chip packaging technology density comparison.

The disadvantages associated with using TAB process include:

It requires specially designed equipment to match each application, at both the chip and the board interfaces.
In general, TAB chips cannot be purchased economically in small quantities.
At present, there is a lack of commercially available "bumped" wafers or chips.
TAB bonding and bumping equipment is rather sophisticated.

12.6.1 Types of Tape Automated Bonding

The initial stage of TAB wafer preparation is much the same as that for wire-bonded wafers. A pinhole-free silicon nitride passivation layer or, in some cases, silicon dioxide or polyimide, is deposited at low temperatures. The passivation is selectively removed, leaving a good portion of the aluminum land exposed. The chip is now at a state where it can be presented either for wire bonding or for additional TAB fabrication.

Basic (Bumped-Chip) TAB

At this point in the preparation of a chip slated for TAB, it can be processed in one of two ways. In the basic TAB approach (Figure 12.18), a barrier metal, such as titanium-tungsten, is deposited over both the exposed aluminum and over the passivation on the periphery of the land. This, plus the addition of 0.001-in.-high gold bumps, helps to ensure the reliability of the TAB connections to the tape during the inner-lead-bonding (ILB) process.

The bumps on the chip are electroplated onto the barrier metal at each land position. Copper bumps can also be used, and both copper and gold can be tin-plated. This completely seals the chip and, thus, further enhances TAB's reliability as a COB assembly process.

Bumped-Tape TAB (BTAB)

The other processing approach (Figure 12.19) puts the bump on the tape rather than on the chip. This approach is known as BTAB. In either case, the bumps are necessary to elevate the etched tape conductors above the chip to prevent shorting of the leads.

Area TAB

One of the most recent interconnection technologies is known as area TAB. In this approach, circuit patterns are fabricated on the tape. This allows the chip designer to put the I/O lands at any position on the chip, as opposed to being on the periphery of the chip.

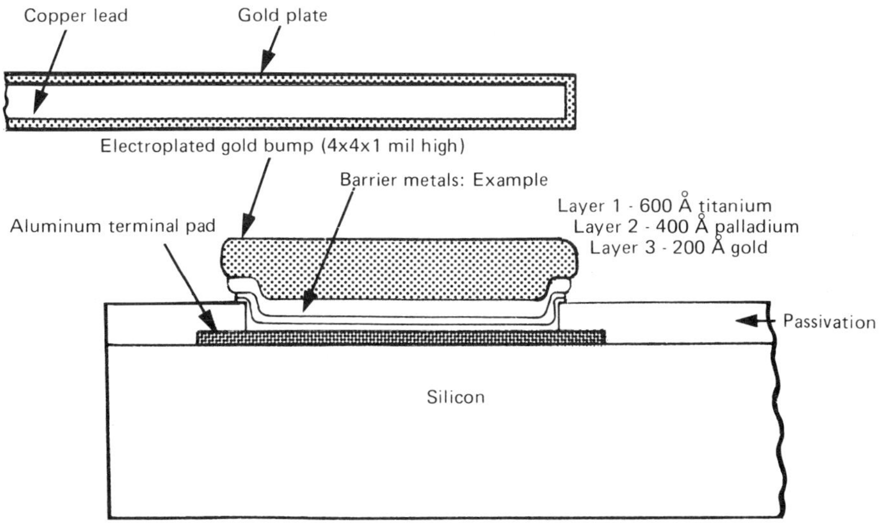

FIGURE 12.18 Bumped-chip TAB interconnection geometry.

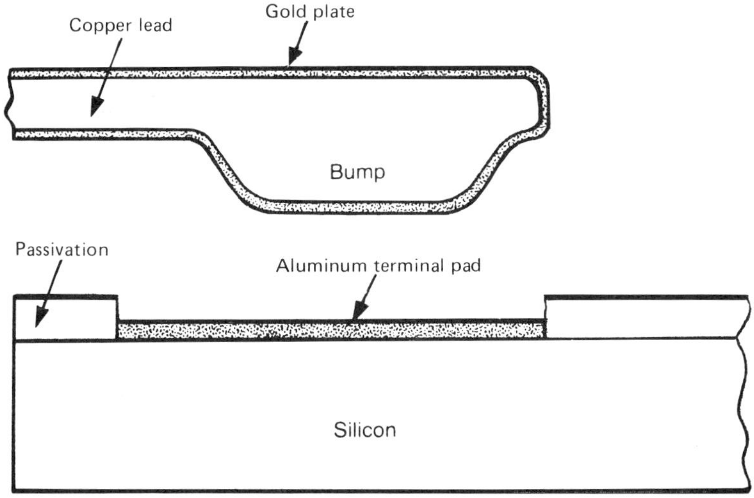

FIGURE 12.19 Bumped-tape TAB (BTAB) interconnection geometry.

This reduces the signal path lengths required and therefore reduces signal delay.

12.6.2 Lead Bonding

Inner-Lead Bonding (ILB)

After the tape is fully fabricated, the next operation is to make the inner-lead bonds between the chip and the etched cantilevered conductors on the tape. The positioning for this operation is controlled by the holes on the tape. Monitoring the chip position is done using a television monitor or microscope.

A mass bonding operation is then performed by thermocompression bonding of all interconnections at one time. Once the chips are bonded to the tape, the points of connection between frames are punched out to allow the individual chips to be electrically tested and, if required, burned in.

Outer-Lead Bonding (OLB)

The outer-lead bonding operation transfers the chip, with its leads, to the board. Outer-lead bonding is a two-step process that is usually done sequentially on the same piece of equipment. As previously mentioned, OLB equipment is predominantly custom because of the wide variety of chips and boards involved. However, for a given combination of chip and board types, only the excising tool, thermode, lead-forming tool (if required), and work fixtures need be customized.

The first step in the OLB process is to excise the leaded chip from the carrier tape. The leaded chip is then transferred to the board in a way that precisely maintains the original relationship of the chip with respect to its tape sprocket holes. Thus, the chip will be placed on the board at a position that is accurately known.

Depending on the TAB mounting configuration (Figure 12.20), a secondary lead-forming operation may be included as a part of the process. When required, care must be taken to assure that the leads are shaped down to the board level without shorting them against the electrically active edge of the chip.

The OLB process can then be completed in one of several mass bonding ways. The connection between chip and board can be done using thermocompression bonding, reflow soldering, or by using conductive adhesives.

12.6.3 TAB Applications

COB applications for TAB vary widely, depending on which of its main attributes are being taken advantage of. For example, where

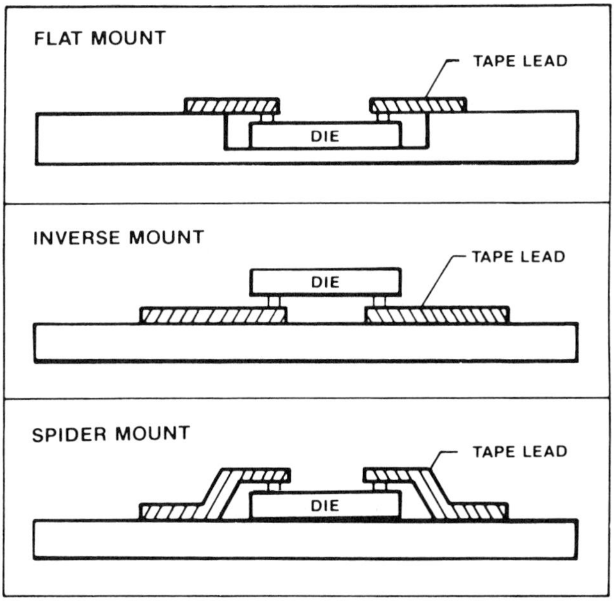

FIGURE 12.20 Tape automated bonding mounting techniques.

low cost is the driving force in single-chip applications, assemblies such as camera flash-firing units have been made in very large quantities. Also, where extremely low assembly profile is important, TAB is used to assemble "smart" credit cards that are only 0.030-in. thick. Such cards are expected to find wide use in banking and telephone applications (Figure 12.21).

The need for small-size and relatively low-cost "dogtags" for military personnel is also being satisfied using TAB. In addition, the direct replacement of conventional through-hole and surface-mount printed-wiring board assemblies with several TAB chips is being used for applications such as the display driverboard assembly shown in Figure 12.22, which uses twelve 80-lead TAB integrated circuits (Figure 12.23).

12.7 SPECIAL CHIP CONFIGURATIONS [20]

Special chip configurations, such as flip chips and beam leads, require different bonding technologies. Flip-chip, or face-down, and beam-lead bonding utilize techniques such as ultrasonic bonding, welding, and soldering.

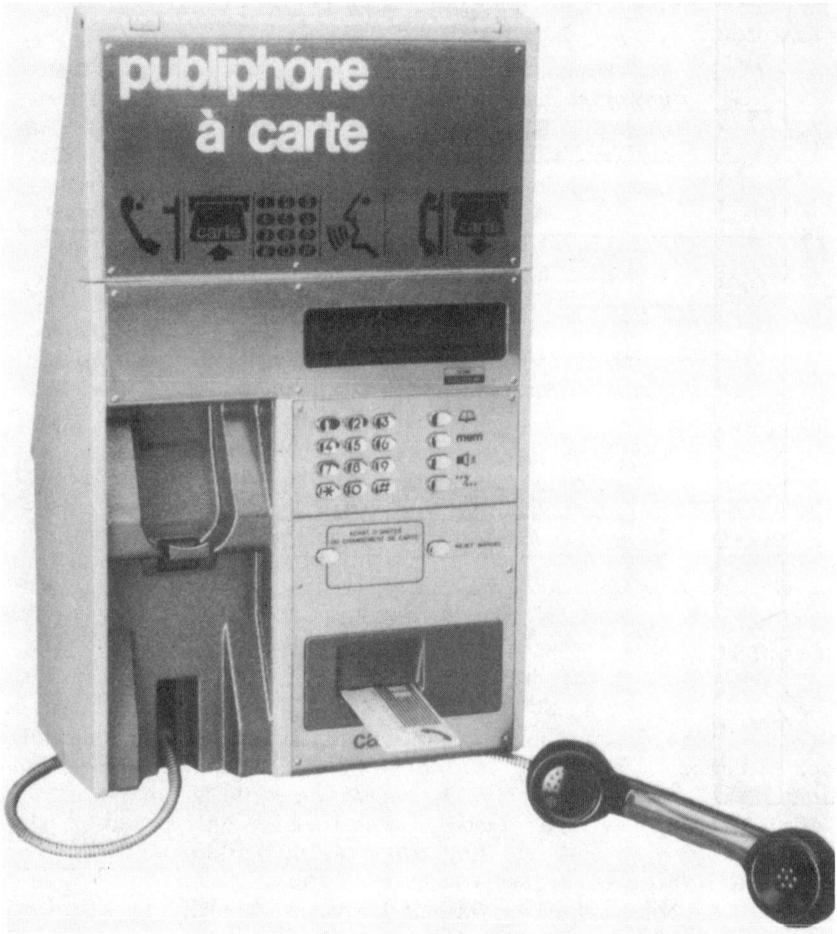

FIGURE 12.21 TAB smart-card telephone.

12.7.1 Flip-Chip Technology

The original flip-chip concept employed small, solder-coated copper balls sandwiched between the chip termination lands and the appropriate lands on the interconnecting substrate. The resultant solder joints were made when the unit was exposed to an elevated temperature. However, the handling and placement of the small-diameter balls was extremely difficult, and the operation was costly.

In a more advanced technique, a raised metallic bump or lump, usually solder, is provided on the chip termination land. This is

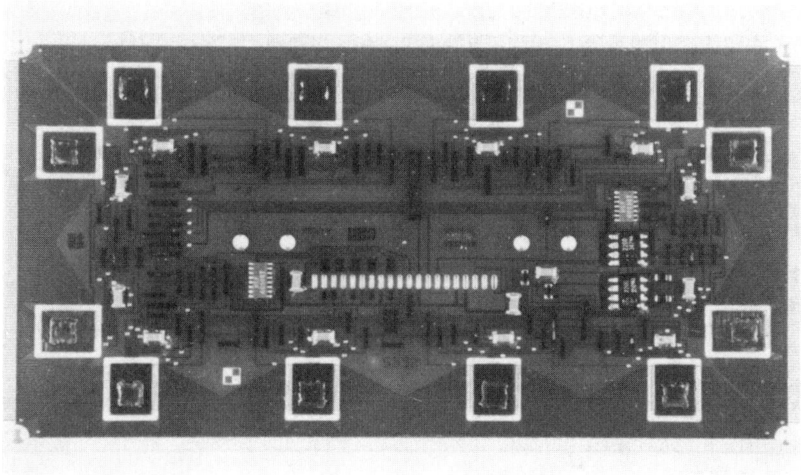

FIGURE 12.22 Display driver board. (Source: International Micro Industries)

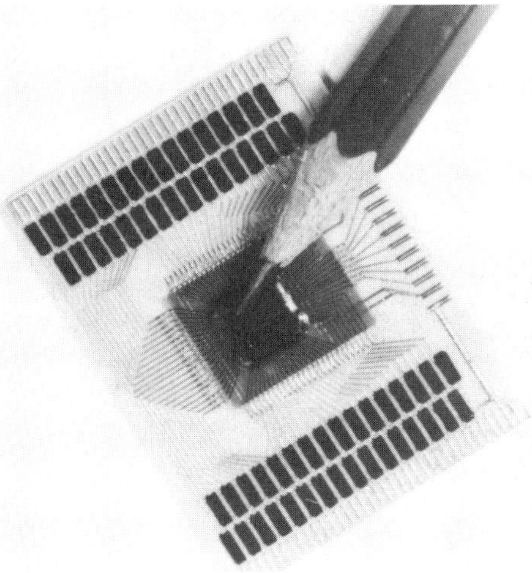

FIGURE 12.23 Display driver 80-lead TAB integrated circuit. (Source: International Micro Industries)

normally done on all lands of all of the chips while they are still in large-wafer form. The individual chip is then aligned to the appropriate circuitry on the substrate and bonded in place using reflow soldering techniques. In this way, the interconnection bonds between the chip and the substrate are made simultaneously, thus reducing fabrication costs.

The primary advantages of using flip-chip packaging are [7]

Fast throughput times
Efficient use of board area

The major disadvantages include [7]:

The inability to inspect the assembled chip visually
The limited availability of bumped chips
Difficult flux removal
Thermal transfer complications

A solution to the availability problem is found in using the approach by which the bump is produced as part of the circuitry on the interconnecting substrate, rather than as part of the chip. Thus, the COB assembler can use any of the devices that are available in bare-chip form.

12.7.2 Beam-Lead Technology [30]

The other major special chip configuration is beam lead (Figure 12.24). In this technology, the lead is produced with plating techniques during the chip fabrication processing on the undiced wafer.

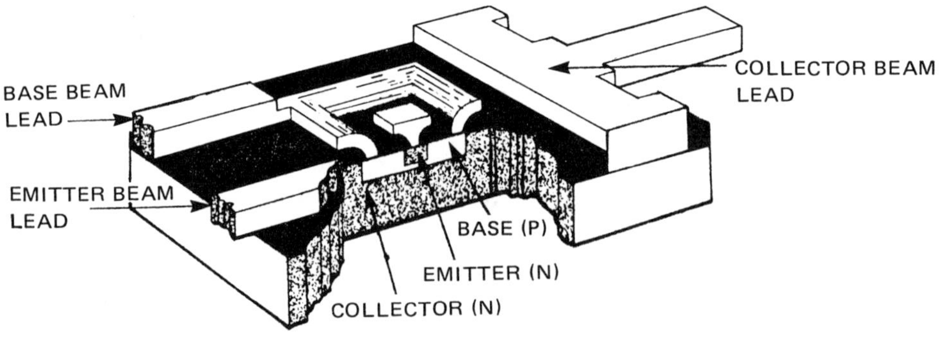

FIGURE 12.24 Beam-lead transistor.

The advantages of using beam-lead devices include:

Ease of alignment of the chip to the interconnecting substrate
Its compatibility with a number of lead-bonding techniques, including soldering, thermocompression, and thermosonic bonding
When compared to wire bonding, the reduction in the number of bonds required to interconnect a chip to the substrate
The inherent reliability of the beam-lead structure due to its superior electrical, mechanical, and thermal characteristics

The major drawbacks to using beam-lead technology include:

The limited availability of beam-lead chips
The larger area required during chip processing, which means that fewer devices can be produced per wafer

12.7.3 Special Chip Applications [31]

The use of flip-chip and beam-leaded devices has been limited to very special packaging applications. Whether or not these can be considered to be COB applications depends on your definition of "board," as most of the applications pertain to the use of ceramic substrates. In general, such applications have been referred to as "hybrids." These techniques have not been applied to conventional printed wiring boards to a significant degree.

Perhaps the most publicized application of these devices is shown in Figure 12.25. This very sophisticated assembly can contain up to 133 flip chips. As many as 20,000 logic circuits or 30,000 memory bits have been packaged on a single substrate using this technology.

12.8 CHIP PROTECTON [32]

Increasingly, many applications are requiring that the COB assembly protect the chips from the atmosphere—i.e., that the chips be sealed, hermetically when practical. The means for doing this fall basically into two categories, "glob-top" coatings and lids.

In selecting the technique to be used for the seal, several important properties must be considered:

Sealing temperature (and time): The technique chosen must be able to form a seal at a low enough temperature and in a short enough time so as to minimize the heating effects on the chip and board components.
Thermal expansion: The thermal expansion of the sealing material/device should also closely match that of the board in order to

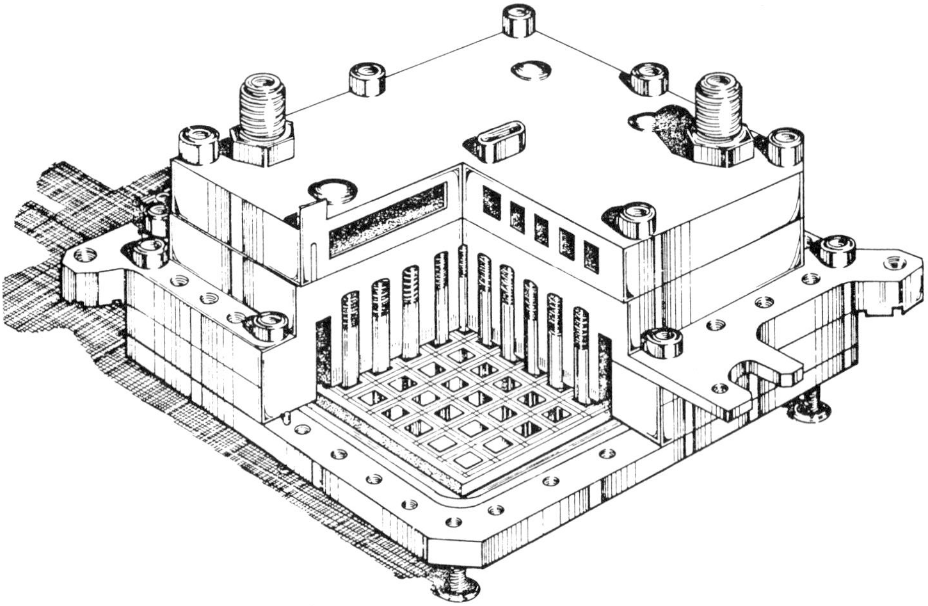

FIGURE 12.25 High-density flip-chip assembly. (Source: IBM Corporation.)

minimize thermal stresses and thus maintain the integrity of the seal.

Hermeticity: The seal must provide the degree of hermeticity required and maintain this level when exposed to the equipment operating and storage environment.

Cost: The sealing technique must be cost-effective, not only with respect to material cost, but also with respect to application and replacement costs, when applicable.

Repairability: If it is necessary to be able to replace a chip or repair a wire/lead termination after the chip has been sealed, it is important that the seal can be readily "broken" and replaced.

Stability: When coatings are used, they should be sufficiently stable that they do not tend to excessively stress the wire/lead bonds or the die attachment bond when exposed to the equipment operating and storage environments.

12.8.1 Protective Coatings

RTV Dispersion Coating [33]

Typical RTV dispersion coatings are one-component, room-temperature-vulcanizing (RTV) silicone rubber coatings supplied as a xylene

TABLE 12.8 Characteristics of Chip-on-Board and Surface Mount Components

Characteristics	Chip-and-wire	Tape automated bonding	Small-outline	Chip carrier
Pretest	Not practical	Possible	Easy	Easiest
Change and repair ease	Difficult	Difficult	Easy	Easy
Chip availability	Excellent	Limited	Good	Very good
Multiple sourcing	Very good	Poor	Good	Very good
Footprint commonality	Poor	Poor	Very good	Very good
Packaging density	High	High	Moderate	Very good
Standardization	Good	Limited	Limited	Good
Thermal performance	Good	Good	Fair	Good
Substrate choices	Limited	Limited	Broad	Broad
Assembly investment	High	High	Moderate	Moderate
Support investment	High	High	Moderate	Moderate
Outside assembly services	Many	Few	Some	Some
Maintenance skills	High	High	Moderate	Moderate
Change risk	High	Moderate	Low	Low
Assembly test	Complex	Complex	Easy	Easy
Documentation	Variable	Variable	Basic	Basic
Logistics support	Factory only	Factory only	Field change.	Field change
Inspection (circuit)	Lot acceptance	100% testable	100% testable	100% testable
Component burn-in	Not practical	Possible	Easy	Easy

dispersion. (No mixing is required.) The curing process uses a cross-linking mechanism that generates methanol during cure. Once applied and exposed, the material vulcanizes by reaction with moisture from the air to form a soft, resilient elastomeric coating, which will withstand long-term exposure to temperatures as high as 250°C (482°F).

Silicone Dielectric Gel [33]

Special controlled technology can yield a silicone gel that affords the nonflowable permanence of a solid, but also gives the freedom from large mechanical and thermal stresses of a fluid. Chemically, a typical silicone gel is very similar to a silicone fluid, but with just enough cross linking to prevent separation of the individual polymer chains and give nonflow thermal-set properties. The fully cured dielectric gel is a soft, jellylike material that exhibits tenacious pressure-sensitive adhesion to virtually any substrate.

Epoxy Coatings [28]

Epoxy coatings are also available for COB self-crowning or "glob-top" applications (Figure 12.23). Typical materials are two-component, liquid epoxy/anhydride systems that have been formulated for their superior thermal shock performance, substrate adhesion, moisture resistance, and glass-transition temperatures in the range of 165-180°C.

12.8.2 Sealing Lids

Metal lids, such as those shown in Figure 12.15, can be used to seal individual chip sites. For COB applications, soldering is generally used to attach the lid to the board.

The use of preforms is often the most convenient method for applying the solder, in addition to the solder coating on the board. The heat required to reflow the solder can be applied using one of the soldering processes associated with conventional surface-mounting technology.

12.9 SUMMARY AND CONCLUSIONS [34]

There is no one best solution to the choice of an IC packaging technique. The most suitable packaging approach must be determined by a system analysis that considers the application, overall costs, manufacturing capabilities required, and the degree of maintainability (modification and repair) needed. The risks to be taken to achieve the desired results are also major considerations.

Table 12.8 provides a summary of some of the characteristics of a candidate technology to be compared when considering the use of COB, as opposed to the use of small-outline (SO) component and chip-carrier surface-mount packaging alternates. Many of these trade-offs are quite difficult to make. Therefore, each potential user should make his or her own comparisons based on the end goal being sought. The user should also include other considerations, such as those relating to design, procurement, and management acceptance, when appropriate. It is then expected that in several instances the use of chip-on-board technology will be the best choice.

REFERENCES

1. G. L. Ginsberg, "Chip and Wire Technology: The Ultimate in Surface Mounting," *Electronic Packaging and Production*, August 1985, pp. 78-83.
2. J. Tuck, "Chip-on-Board Technology," *Circuits Manufacturing*, March 1984, pp. 78-83.
3. Mah El Refaie, "Chip-Package Substrate Cushions Dense, High-Speed Circuitries," *Electronics*, July 14, 1982, pp. 135-141.
4. J. Lyman, "Surface Mounting Alters the PC-Board Scene," *Electronics*, February 9, 1984, pp. 113-116.
5. F. Dance, "Part 1: Can PWBs Meet the Surface Mount Challenge?," *Electri-Onics*, May 1983, pp. 25-27.
6. J. Sergent, "Part 1: Designers Guide to Thick Film Hybrid Circuits," *EDN*, October 14, 1981, pp. 341-345.
7. Hybrid Microcircuit Design Guide, ISHM-1402/IPC-H-855, October 1982.
8. PCI, Inc., Santa Clara, Calif.
9. J. Lyman et al, "VLSI Spurs Progress in Both Packaging and Interconnection...," *Electronics*, October 6, 1983, pp. 165-171.
10. D. Bauks, "Automated Hybrid Assembly Systems for the Electronic Factory of the Future," *International Journal for Hybrid Microelectronics*, October 1983, pp. 40-42.
11. Mech-El Industries, Inc., Woburn, Mass.
12. L. F. Miller, "A Survey of Chip Joining Techniques," *Solid State Technology*, September 1969, pp. 119-122.
13. D. M. Dowdle, "Various Methods of Direct Chip Attachment to Flexible Circuitry," IPC Technical Paper IPC-TP-640, March 1987.

14. A. H. Mones and R. K. Spielberger, "Interconnecting and Packaging VLSI Chips," *Solid State Technology*, January 1984, pp. 119-122.

15. G. Dehaine, K. Kurzwell, and P. Lewandowski, "Crossing the 200 Wire Barrier in VLSI Bonding," Proceedings, International Symposium on Microelectronics, ISHM 1984, pp. 353-357.

16. J. Henderson, "Thick Film Processing and Higher Yields, Part 4: Wire Bonding," *Electri-Onics*, September 1984, pp. 43-46.

17. Guidelines for Chip on Board Technology Implementation, IPC-SM-784, Institute for Interconnecting and Packaging Electronic Circuits, 1989.

18. G. Keitel, "Solderable and Wire Bondable Conductive Polymer Circuitry," *International Journal of Hybrid Microelectronics*, November 1982, pp. 357-361.

19. P. Singer, "Hybrid Wire Bonding Advances," *Semiconductor International*, July 1984, pp. 66-73.

20. W. Staley, "Hybrid Microelectronic Interconnections," *Electronic Packaging and Production*, October 1982, pp. 78-84.

21. D. Rose, "Packaging and Assembly: The 1983 Semiconductor Technology Forecast," *Semiconductor International*, January 1983, pp. 56-60.

22. S. Fundakowski, "Bubbles Bursting with Military Potential," *Defense Electronics*, July 1984, pp. 83-88.

23. S. Leishman, "Finstrate: A New Concept in IC Substrates," *IPC Technical Review*, April/May 1984, pp. 12-17.

24. G. L. Ginsberg, "Chip-on-Board Profits from TAB and Flip-Chip Technology," *Electronic Packaging and Production*, September 1985, pp. 140-143.

25. "What Is Tape Automated Bonding (TAB)?," International Micro Industries, Cherry Hill, N.J.

26. D. Meyer, "Tape-Automated Bonding Suits Surface Mount," *Electronic Products*, October 1, 1984, pp. 71-75.

27. C. Lassen, "Wanted: A New Interconnection Technology," *Electronics*, September 27, 1979, pp. 113-121.

28. W. Pastone, "Tape-Carrier Packaging Boasts Almost Unlimited Potential," *EDN*, October 20, 1974.

29. S. Weinstein, "Smart Credit Cards: The Answer to Cashless Shopping," *IEE Spectrum*, February 1984, pp. 43-49.

30. H. E. Culver, R. L. Michals, and D. R. Robillard, "Integral Spiders (I/S)," *Proceedings, International Journal of Microelectronics*, October 27-29, 1975, pp. 302-303.

31. L. S. Goldman and P. A. Totta, "Area Array Interconnections for VLSI," *Solid State Technology*, July 1983, pp. 91-97.

32. J. B. Blum, "The Potential Use of Sol-Gel Technology in Microelectronics Packaging," Proceedings, Electronic Components Conference, March 1984, pp. 407-410.

33. D. Sierawski and R. Niemi, "Silicone Protection on Microcircuits," Proceedings, International Electronic Packaging Society Conference, November 9-10, 1981, pp. 430-433.

34. D. Amey, "Surface Mounted Components," Annual ISHM Surface Mounting Workshop, November 15, 1982.

13

Microminiature Multichip Modules

Traditionally, the differences among printed-board assemblies, hybrid circuits, and integrated circuits have been related to the base materials that are used, i.e., organics, ceramics, and silicons, respectively. However, using the base material as a determining factor no longer seems to be appropriate, due to the wide range of material combinations now being used in the electronics packaging and interconnection industry [1].

In general, the characteristics of all electronic circuit assemblies can be viewed as being application-specific. Thus, if their base materials are not considered, it is possible to define new categories of electronic circuit assemblies. In this view, the hierarchy of electronic circuit packaging elements is as follows:

1. Printed-board assemblies (application-specific electronic assemblies)
   a. Usually larger than 10 in.^{2} in base size
   b. Usually provide interconnection for packaged components, both through-hole and surface-mounted
   c. Usually use copper foil/plating as the primary interconnection wiring
   d. Use various base materials, including nonorganic or ceramic materials with or without reinforcements, constraining cores, or supporting planes.
   e. May also provide mounting for unpackaged semiconductors, which are subsequently protected by an individual coating and/or cover
2. Multichip modules (application-specific electronic subassemblies)
   a. Usually less than 10 in.^{2} in base size
   b. Usually provide interconnection for surface-mounted unpackaged semiconductor chips, which are subsequently protected by an overall coating and/or enclosure
   c. Usually use thick- and/or thin-film passive components and interconnection wiring
   d. Usually interconnected and mounted on a printed-board assembly
   e. Usually use either ceramic or silicon base materials
   f. Contain more than one discrete active (integrated circuit) device.
3. Integrated circuits (application-specific electronic components)
   a. Usually 1 in.^{2} or less in size
   b. Consist of several active circuit devices and interconnection wiring, usually made simultaneously by bipolar or metal-oxide semiconductor (transistor) fabrication processes
   c. Base material usually silicon glass (or gallium arsenide), sometimes enhanced by other materials
   d. Usually mounted and interconnected in a printed-board assembly (packaged) or multichip module (unpackaged)

e. When packaged, usually protected by a permanent inseparable coating or encapsulant

Although these descriptions are not perfect, they do provide a basis for defining the variations in using these products in different applications.

13.1 INTRODUCTION [2-4]

Future electronic packaging needs will increasingly emphasize the speed and density of interconnections. However, conventional printed-board assemblies and integrated circuit packages rely heavily on interconnections of dimensions that are much larger than those encountered on the integrated circuit chip. A great deal of performance is therefore sacrificed in these external signal paths.

External paths are not the only cause of loss of performance. Since the percentage of the semiconductor substrate area in conventional board-level packaging can be as low as 5%, the combined effect of relatively large external connections and low active-component density is substantial.

In addition, if one examines the paths typically found in multiple plug-in printed-board assemblies, another cause for loss of performance is identified. Wires must be routed with some length of connection on the backplane. Although the volume density for such multiple plug-in board systems is sometimes fairly high, the use of connectors imposes limitations on the signal paths that prevent the achievement of maximum practical density.

Also, much of current technology is motivated by mechanical requirements for assembling and repairing multiple integrated circuit assemblies. Therefore, future performance improvements can be obtained only by adapting this circuit partitioning and repair strategy to finer-pitch dense wiring or by modifying (abandoning) the repairability requirements altogether in the interest of speed.

Microminiature multichip modules (Table 13.1) are attractive for these system applications where interchip delays are critical to performance. The use of microminiature multichip modules is also attractive for those applications where a functional unit can be clearly defined.

Fortunately, the level of development activity directed toward microminiature multichip module packaging is increasing. As will be shown, new substrate materials are being investigated, including silicon wafers. Both organic and inorganic materials are being evaluated as dielectric layers. Fine-line lithography is being used with thin-film metallization, either aluminum or copper, to fabricate circuits with 10-μm (0.0004-in.) feature sizes.

TABLE 13.1 General Microminiature Multichip Module Parameters

Technology	Number of chips	Number of I/Os	Number of transistors[a] ($\times 10^3$)	Area (mm^2)	Chip area fraction[b] (mm)
Mitsubishi HTCM	9	624	108	66 × 66	0.13
Hitachi RAM	6	108	?	27.4 × 27.4	0.06
Honeywell SLIC	110	240	?	80 × 80	?
NEC SX	36	2177	144	125 × 125	0.15
IBM 4381	36	882	252	64 × 64	0.19
IBM 3090	100	1800	600	150 × 150	0.10
NTT	25	900	?	85 × 105	0.18

[a]Based on an average of four transistors/gate.
[b]Chip-to-module area ratio.

As shown in Figure 13.1, however, microminiature multichip modules are more expensive per unit substrate area than conventional packaging technologies. Thus, the packaging technologies described in this chapter place great emphasis on minimizing substrate size, while extending electronic circuit assemblies to new heights of cost-effective performance that are not otherwise achievable.

13.2 DATA PROCESSING APPLICATIONS [6]

Faster machine cycle time has always been a major objective in designing computers. From the standpoint of hardware implementation, it is therefore important to place as many logic gates and memory devices in as close proximity as possible. Thus, the challenge in designing and manufacturing general-purpose computers is to achieve these goals while offering a good cost/performance ratio and flexibility for various applications.

The various packaging approaches taken by some of the major large-scale, general-purpose computer manufacturers (Table 13.1) reflect the abilities of each manufacturer at each electronic circuit

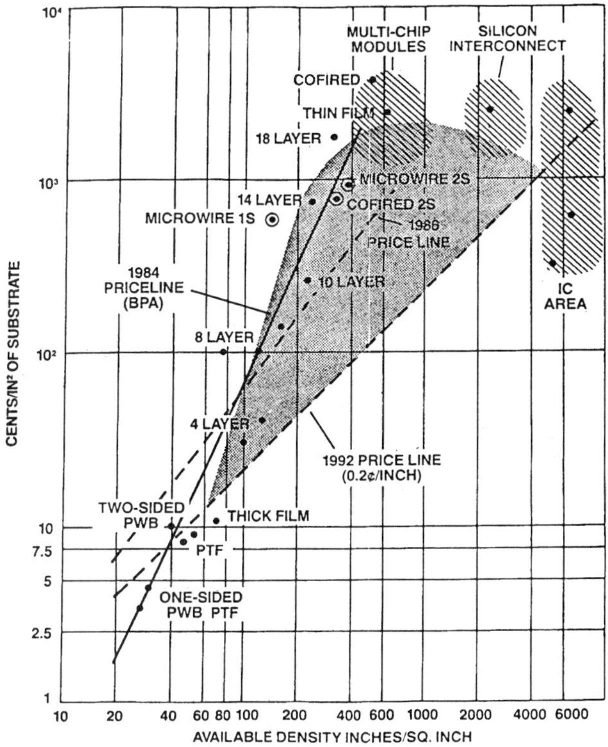

FIGURE 13.1 Packaging technology price-density relationships.

packaging level. As shown, IBM, Hitachi, and NEC take the traditional three-package level approach, although there is a difference in where middle-level printed-board assemblies and microminiature mlutichip modules are used. Fujitsu, on the other hand, eliminates the middle packaging level altogether in its system.

Common to all the computer systems in Table 13.1 is the use of high-pin-count emitter-coupled logic (ECL) devices, which operate with subnanosecond switching time. When in full operation, these ECL devices emit an enormous amount of heat. So, from the very beginning, thermal management considerations, as will be shown, are a very important aspect of hardware design.

The bottom line seems to be where to concentrate interconnections and support hardware. The question is, then, whether a selected configuration can be manufactured, and if so, whether it is

FIGURE 13.2 Solid logic technology. (Source: IBM Corp.)

cost-effective. The following descriptions of the technologies used by each company should shed more light on this subject.

13.2.1 IBM Multichip Module Technology [7, 8]

In 1964 the IBM Corporation System 360 introduced a new kind of circuit package called solid logic technology (SLT). Each SLT package, or multichip module, consisted of up to four single transistor chips and screen-printed thick-film resistors mounted on a 0.5-in.^{2} alumina ceramic substrate (Figure 13.2). The 0.015-in.^{2} transistor chips were attached flip-chip fashion (face down) with three 0.005-in.-diameter copper balls, which also provided a path for heat dissipation. These balls were soldered to thick-film conductors that fanned out to the module pin connections. Twelve to sixteen module pins were available for connection to the next higher level of packaging, forming the beginning of the pin-grid array (PGA) package.

In 1969 IBM introduced a newer packaging technology called the monolithic systems technology (MST). The chips had become larger and now averaged six to seven circuits per chip, requiring 16 terminations, though MST chips with up to 25 circuits and 19 terminals were also used.

The increased chip size created problems with respect to solder joint failure (due to coefficient of thermal expansion mismatch) and reduced productivity (due to the complexity of placing more copper balls). IBM's solution to these problems was called the controlled-collapse chip connection, or C-4. The C-4 technology replaced the copper balls with a lead-alloy pad. Surface tension of the molten solder joint aligned the chip on the substrate and created a strong interconnection.

Another benefit of the C-4 process included a reduction in space under the substrate for chip termination by making the connections beneath the chip, leaving more room for I/O pins. This in turn resulted in shortened chip-to-pin interconnection distances, which translated into higher signal speeds and ultimately lower power levels, since the resistance of the interconnections was reduced.

The next evolution of the C-4 technology was developed for the intermediate-performance IBM 4300 processor introduced in January 1979. The circuit packaging element was now reffered to as a multichip module, or MCM. The MCM was a circuit module having a 35- or 50-mm^2 multilayer ceramic substrate for mounting up to nine large-scale integration (LSI) chips, each having up to 704 logic circuits (Figure 13.3). The substrates had from 18 to 23 metallized ceramic layers and 5 m of wiring.

The 121 C-4 chip-to-substrate interconnections were arranged in an 11 × 11 array on 0.25-mm centers. The back side of the 50-mm^2 MCM was populated with 361 I/O pins in a 19 × 19 array on 0.25-mm centers; the 35-mm^2 MCM had 196 I/O pins in a 14 × 14 array. These large pin-grid arrays were then mounted and interconnected by multilayer printed-wiring boards.

A typical 50-mm^2 MCM contains six LSI logic chips utilizing a total of 4000 circuits. This corresponds to the circuit count of 18 fully populated printed-board assemblies in an IBM System/370 Model 148 computer. Thus, in a grand leap from its predecessors, essentially simple space savers, the MCM provided most of the signal and power distribution that was conventionally accomplished on printed wiring boards.

The culmination of C-4 packaging technology evolution came about in the early 1980s with the development of the thermal conduction module, or TCM (Figure 13.4), which is the heart of the high-performance, liquid-cooled IBM 3081/3090 processors (Table 13.2). (The IBM 4381 mid-range processor is an air-cooled adaptation of the basic TCM concept.) The TCM holds up to 133 high-performance bipolar LSI chips and dissipates up to 300 W, permitting the circuits to operate at full potential by removing certain limits on signal and power distribution and thermal dissipation.

The microminiature multichip module subassembly within the TCM (Figure 13.5) is keyed to the use of a 90-mm^2 multilayer ceramic substrate. The cofired multilayer ceramic substrate (Figure 13.6) has

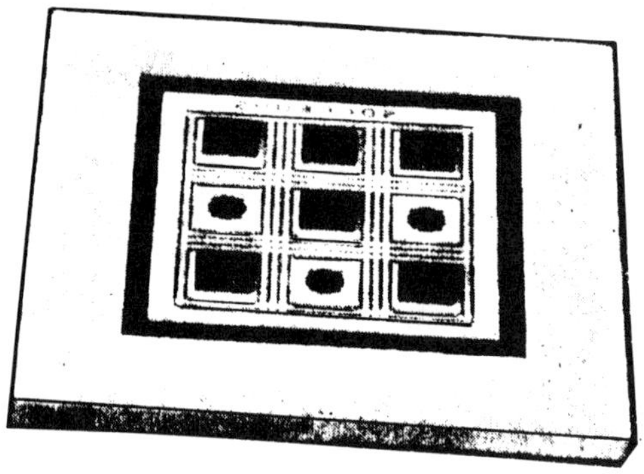

(a)

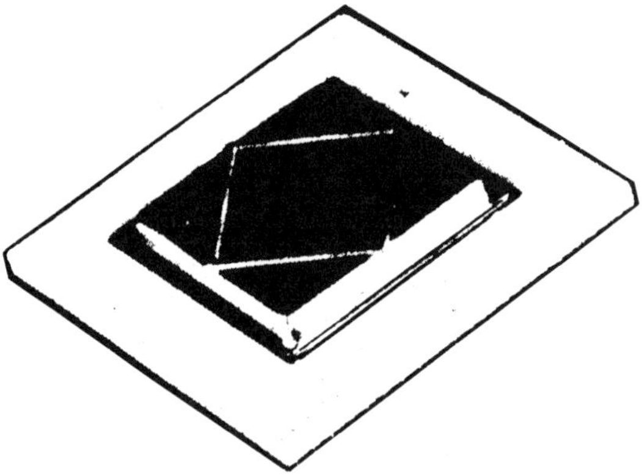

(b)

FIGURE 13.3 Multichip modules: (a) uncapped MCM; (b) capped MCM. (Source: IBM Corp.)

from 28 to 33 interconnection layers, typically 130 m of wiring, and more than 350,000 metallized vias for interlayer connections. Over 10,000 C-4 flip-chip solder joints connect the integrated circuit chips to the substrate (Figure 13.7).

The bottom of the substrate contains a pin-grid array consisting of 1800 I/O pins arranged in 60 rows of 30 pins each. The pins are positioned on a 1.25-mm × 2.5-mm grid. Nine of these microminiature multichip modules then plug into a multilayer board to complete the assembly of the electronic circuits.

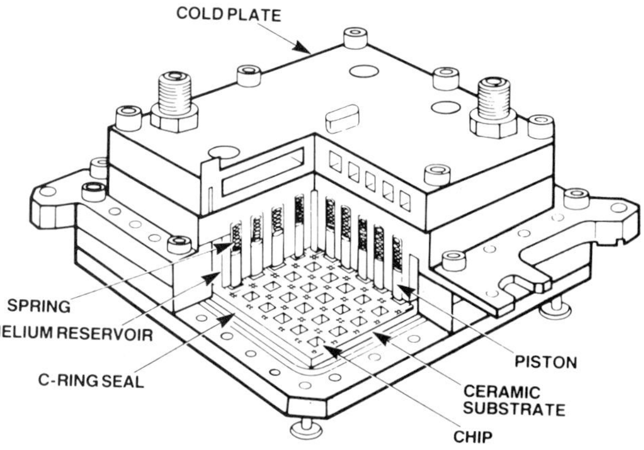

FIGURE 13.4 Thermal conduction module. (Source: IBM Corp.)

13.2.2 Fujitsu M-780 Packaging Technology [9]

Before proceeding with descriptions of the microminiature multichip-module packaging technologies being used by other electronics companies, let us look at an alternative approach. For example, the Fujitsu, Ltd. (Kawasaki, Japan) M-780 central processing unit (CPU) (Table 13.2) is built entirely with surface-mounted, packaged integrated circuits.

The CPU is a double-sided multilayer printed board assembly that contains 168 gate arrays in identical-sized gullwing packages, (Figure 13.8). The 540-mm × 460-mm multilayer board has 42 polyimide layers with 60-μm (2.5-mil) etched-copper interconnection wiring. Each of the 22.5-mm^2 integrated circuit packages on both sides of the CPU substrate has 180 leads on 0.4-mm centers.

The M-780's main storage unit includes a horizontal motherboard with up to 32 vertically stacked plug-in daughterboards, each of which can contain up to 128 megabytes of memory (Figure 13.9). Other main modules, the memory control unit and channel processor, are packaged as single multilayer printed-board assemblies.

13.2.3 Hitachi Packaging Technology [10, 11]

Another example of high-density, high-speed surface mount technology is used in the Hitachi, Ltd. (Kanagawa-Ken, Japan) M-68X computer (Table 13.2). The logic section of the M-68X consists of a printed-board assembly that mounts and interconnects 72 LSI flatpacks, either bipolar gate arrays or memory modules. Its 419-mm × 280-mm polyimide-glass multilayer board has a total of 20 layers of

TABLE 13.2 Packaging Technologies of Large-Scale General-Purpose

	IBM 3090	Fujitsu M-780
Date announced	2-12-1985	11-14-1985
Logic LSI Chip		
Logic type	ECL	ECL
No. gates	612[a]	3,000/10,000
Speed, ps	425[b]	180
No. I/O pins	176	180
Lead pitch		20 mils
Max. pwr/chip, W	6	8.5/9.5
Remarks	Flipped chip	ML ceramic
Module		
Material	Ceramic	
Dim., mm	90 × 90 × 5	
No. layers	36	
Cond. width, μm	100	
No. chip/mod	100	
No. I/O pins	1,800	
No. gates/mod	61,200	
Wiring density	400 in./in.^{2}	
Power	40 W/in.^{2}	
Card		
Material		
Dim., mm		
Dim., in.		
No. layers		
Cond. W, mm (mil)		
Channels/grid		
PTH diam., mm (mil)		
IVH diam., mm (mil)		
Aspect ratio		
No. I/O pins		
Comp. mounting		
No. packages		
Backboard		
Material	Epoxy glass	Polyimide G
Dim., mm	600 × 700 × 4.6	540 × 488 × 7.3
Dim., in.	24 × 28 × 0.18	21.2 × 19.2 × 0.29
No. layers	20	42

Computers

Hitachi M-68X	NEC ACOS-1500	NEC SX-AP
3-11-1985	2-28-1985	—
ECL	ECL	ECL
2,000/5,000	300/2,000	1,000
200/250	250/350	250
160	52/132	176
6	2.2/5.0	5.7
SiC substrate	TAB	Flipped TAB
Al wire bonded		
	Ceramic	Polyimide/ceramic
	80 × 80	100 × 100 × 2.75
	4	5 + 5
	50	25
	56	36
	420	2,177
	16,800	36,000
		270 in./in.^{2}
		16 W/in.^{2}
Polyimide G	Polyimide G	
420 × 280 ×	432 × 394 × 3.8	
16.5 × 11.0 ×	17 × 15.5 × 0.15	
20	16	
0.1 (4)	0.1 (4)	
2/1.27 mm	3/2.54 mm	
	0.6 (24)	
	0.3 (12)	
8.5	6.3	
1,776	1,344	
100% SMT	Pin	
	9 modules	
Polyimide G	Polyimide G	Polyimide G
570 × 420 ×	531 × 455 × 4.8	545 × 460 × 4.9
22.4 × 16.5 ×	20.9 × 17.9 × 0.19	21.5 × 18.1 × 0.19
22	18	16

(continued)

TABLE 13.2 (Continued)

	IBM 3090	Fujitsu M-780
[Backboard]		
Cond. W, mm (mil)	0.076 (3)	0.06 (2.36)
Channels/grid	4/2.5 mm	2/1.2 mm
PTH diam., mm (mil)	0.44 (17)	0.35 (14)
IVH diam., mm (mil)	0.08 (3.1)	0.12 (4.7)
Aspect ratio	10.4	20.8
No. packages	9 TMCs	336 LSIs
No. I/O pins	1,800	1,400
Remarks	-41,644 PTHs -7,880 IVHs -ZIF connected -Full additive plating for conductor patterns and PTH -IVH laser drilled -Cond. liquid cool -60 in./in.^{2}	-80,000 PTHs -196 PTHs/in.^{2} -150,000 IVHs -PTH electroplated -Stepped drilling -Diagonal wire -Cond. liquid cool -290 in./in.^{2} -2-sided SMT

[a] 612 circuits.
[b] Loaded speed, equivalent of 300 psec unloaded.

interconnections: 2 surface layers, 8 signal layers, and 10 power and ground planes.

The 2000- and 5000-gate array packages are 22 mm^2 in size and have 160 I/O leads, 40 on each side, with a lead pitch of only 0.5 mm (20 mils). Cache-memory modules are in 29-mm × 39-mm ceramic flatpacks with two rows of 48 leads on 0.75-mm (30-mil) centers. Proprietary laser-reflow soldering is used to make the nearly 20,000 connections between the flatpacks and the multilayer board.

Hitachi also uses microminiature multichip modules in its S-810 array processor for large-scale scientific applications. In this computer the memory chips are packaged in 31-mm × 29-mm flatpacks that have a silicon carbide (SiC) ceramic base. The 16-mm-high RAM multichip modules (Figure 13.10) have 108 I/O gullwing leads.

A relatively large silicon "mother chip" is bonded to the module's SiC base. Then up to eight 1-W ECL 1.9-mm × 4.0-mm chips are

Hitachi M-68X	NEC ACOS-1500	NEC SX-AP
0.1 (4)	0.1 (4)	0.1 (4)
2/1.27 mm	3/2.54 mm	3/2.54 mm
	0.8 (32)	0.6 (24)
	0.25 (10)	0.2 (8)
9.5	6	8.2
21 cards	13 cards	12 TMCs
21,840	1,344	—
-PTH plated with full additive process with tent and etch -7 kW -Air cool	-Partly additive for PTH -1.2 kW -Air cool	-Partly additive for PTH -ZIF connector -Pwr plane 300 μm Cu -Max. 3 kW -Max. 432,000 gates -Cond. liquid cool

"solder-bump/flip-chip" reflow-soldered to the mother chip to interconnect the memory circuits.

A silicone gel is used to encapsulate the bare chips within the module, to protect them from humidity and alpha particles, before a final protective cover is attached. A finned aluminum heat sink is attached to the reverse side of the SiC base with a 50-μm (2-mil)-thick conductive silicone rubber adhesive to complete the modular assembly.

In addition, Hitachi has developed a microminiature multichip pin-grid array (PGA) module. This 45-mm × 45-mm module (Figure 13.11) has 208 I/O pins. Unlike the RAM module just described, the PGA module uses a more conventional alumina ceramic base. However, it does use a 20-mm^2 mother chip to interconnect CMOS gate arrays in a "silicon-on-silicon" arrangement. Flip-chip reflow soldering and a silicone protective gel are also used.

FIGURE 13.5 Microminiature 133-chip module. (Source: IBM Corp.)

13.2.4 NEC Multichip Module Technology [12]

Packaging technologies for the NEC Corporation (Tokyo, Japan) systems (Table 13.2) have been developed to achieve the highest performance, i.e., 1300 megaflops (floating-point operations) per second, as compared to the 630 megaflops achieved by the Hitachi S-810 array processor. [In terms of millions of instructions per second (Mips), NEC's ACOS-1540 supercomputer is rated at 108 Mips, the Fujitsu M-780-40 exceeds 100 Mips (perhaps as high as 140 Mips), and the IBM 3090-400 is rated at 50 Mips.]

The packaging technology used by NEC to achieve these extremely high data processing capabilities features three levels of packaging: (1) flipped-TAB carriers, (2) microminiature multichip modules, and (3) a high-density multilayer printed-board assembly interconnection system with zero-insertion-force connectors and high-speed coaxial cabling.

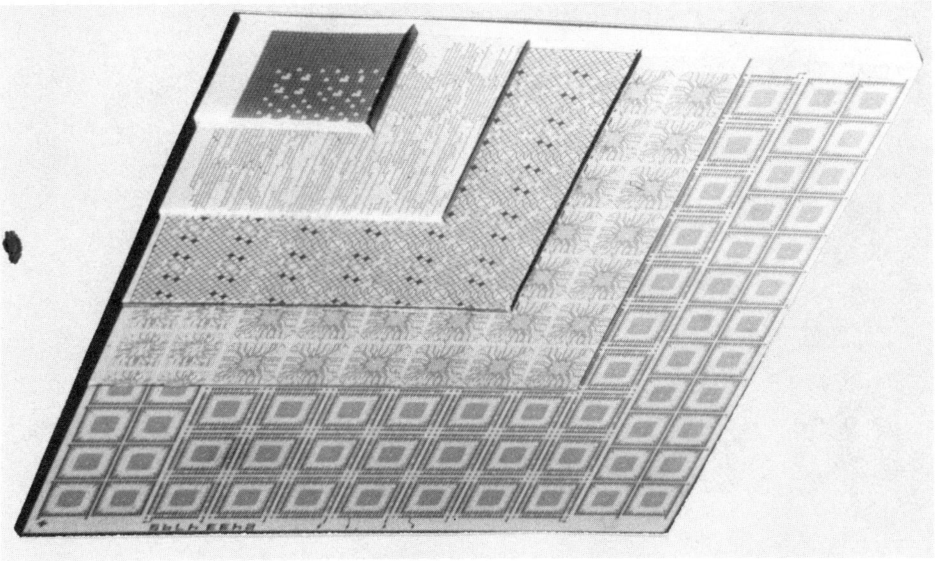

FIGURE 13.6 Cofired multilayer ceramic substrate, cutaway. (Source: IBM Corp.)

The logic flipped-TAB carrier consists of a 1000-gate LSI integrated circuit with 176 TAB (tape automated bonding) leads packaged in a 12-mm^2 chip carrier called the FTC. The materials used, such as copper-tungsten alloys, have low thermal resistances and coefficients of thermal expansion (CTE) that are well matched with the silicon integrated circuit chip.

The multilayer ceramic substrate of the FTC has 169 I/O bumps arranged in a 13 × 13 matrix on its bottom surface and 176 TAB outer-lead bonding pads on its top surface. The LSI chip is fixed in a face-down position and the TAB leads are then simultaneously gang-bonded to the FTC substrate.

Four 1-kbit bipolar RAM integrated circuits are packaged in a multichip, 14-mm^2 RAM FTC. The profile of the RAM FTC is almost the same as the logic FTC without the multichip structure (Figure 13.12). That is, four bipolar RAM chips are fixed individually in a face-down configuration and 52 TAB leads of each chip are gang-bonded on the surface of the substrate. The RAM FTC also has 169 I/O bumps.

FIGURE 13.7 Closeup of TCM substrate with reflowed VLSI chips. (Source: IBM Corp.)

The high-density microminiature multichip packages (MCP), which mount and interconnect the FTCs, consist of a multilayer substrate (MLS) with fine lines, higher-speed connections, and a high number of I/O terminals. Figure 13.12 shows a cross-sectional view of the MCP.

The basic substrate of the MLS is a 100-mm^2/2.75-mm-thick alumina ceramic substrate with tungsten-metallized inner layers for power and ground and 2177 I/O pins brazed on its botton surface on a 2.54-mm staggered grid (Figure 13.13). Its signal layers consist of four

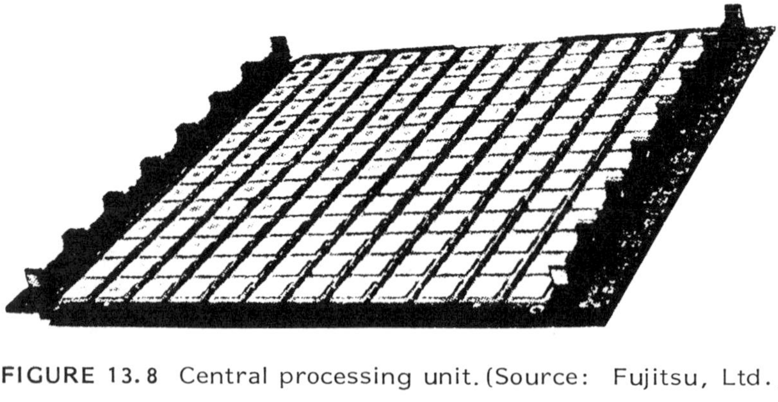

FIGURE 13.8 Central processing unit. (Source: Fujitsu, Ltd., M-780)

FIGURE 13.9 Main memory storage unit. (Source: Fujitsu, Ltd., M-780)

polyimide insulative layers and five thin-film conductor layers, including two ground layers with two signal layers sandwiched between them, and a top layer with FTC attachment pads.

By far one of the most innovative new developments is the use of polyimide insulative layers on the multichip package substrate in

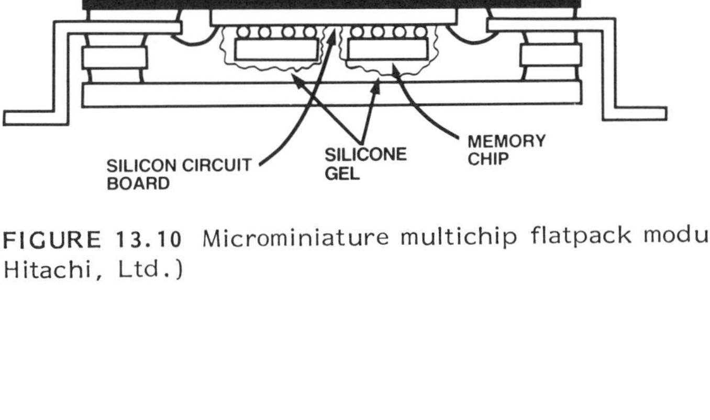

FIGURE 13.10 Microminiature multichip flatpack module. (Source: Hitachi, Ltd.)

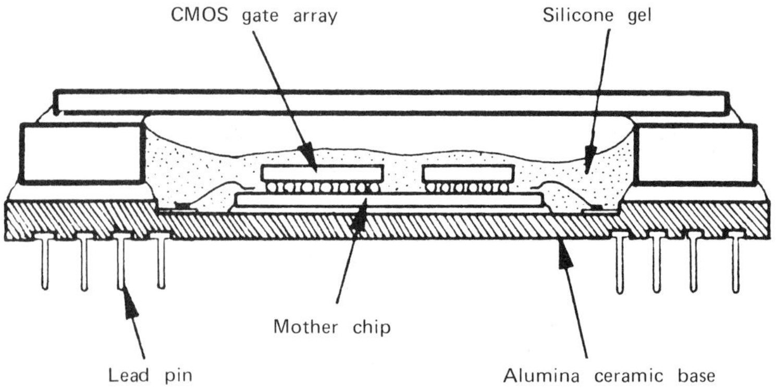

FIGURE 13.11 Microminiature multichip pin-grid array module. (Source: Hitachi, Ltd.)

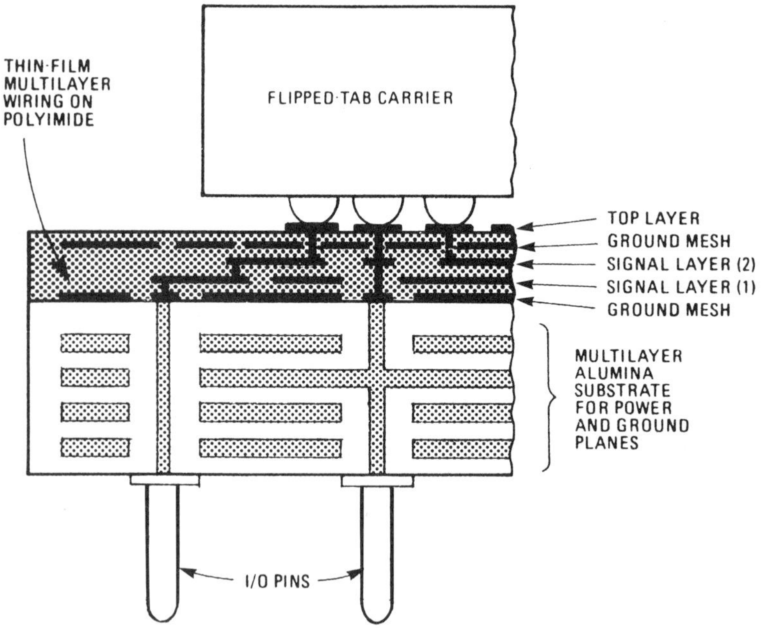

FIGURE 13.12 Microminiature multichip package, detail. (Source: NEC Corp.)

FIGURE 13.13 Multichip package with 2177 input/output pins. (Source: NEC Corp.)

combination with the alumina-fired ceramic with I/O pins. By utilizing this new technology, an MLS can be achieved with both high-speed/high-density wiring and high pin counts.

The low dielectric constant of polyimide (3.5) provides a signal delay time that is 60% of the conventional thick-film ceramic layers, where the dielectric constant is 9.0. Moreover, the distinctive features of polyimide, such as a smooth surface, fine via-hole formability, and low-temperature curability, make possible finer signal wiring on the ceramic substrate.

Wiring is 25-μm (1 mil) wide on a 75-μm (3-mil) grid with 50-μm (2-mil) square via holes. The signal lines utilize thin-film metallization. Thus, the maximum length of the signal lines on the 100-mm^2 MLS (Figure 13.14) is 110 m.

Also, the low curing temperature of the polyimide is very useful for the combination of fine-line circuitry with high pin counts and a high-power-distribution ceramic substrate. That is, the 400°C nitrogen curing of the polyimide does not cause oxidation of the tungsten-nickel-gold metallization on the ceramic substrate and the nickel-gold-plated brazed pins.

FIGURE 13.14 Multilayer ceramic substrate. (Source: NEC Corp.)

The complete multichip package (MCP) assembly consists of the multilayer substrate (MLS) with a maximum of 36 flipped-TAB carriers (FTCs) and a mounting flange (Figure 13.15). The 36 FTCs are attached simultaneously on the MLS by tin-lead reflow soldering.

FIGURE 13.15 Complete 36-FTC module. (Source: NEC Corp.)

13.2.5 Honeywell TAB Multichip Module Technology [13]

The Honeywell, Inc. (Bloomington, Minnesota) DPS-88 mainframe computer uses microminiature multichip modules that are packaged with tape automated bonding (TAB) integrated circuits. The TAB devices are bonded to 80-mm^2, 96% alumina multilayer substrates composed of five layers of thick-film copper circuitry (Figure 13.16). The substrates are hermetically sealed with a soldered cover and mounted flat against connectors on a printed wiring board that contacts the substrate's external peripheral pads.

The TAB assembly process employs Honeywell's frame-and-cannister technology. With this approach the various integrated circuit chips are inner-lead bonded to 35-mm tape. The tape is then severed and each device is mounted inside a frame, similar to mounting a photographic slide. The frames are stacked in a cannister and tested.

The cannisters of TAB frames are mounted on an automatic sequencing machine. Cannisters containing sequenced frames are then loaded onto an outer-lead bonder. The substrate is heated and the chips are die-bonded, active side down, onto a softened epoxy preform, aligned, and outer-lead bonded to the multilayer ceramic substrate using 100% tin reflow soldering.

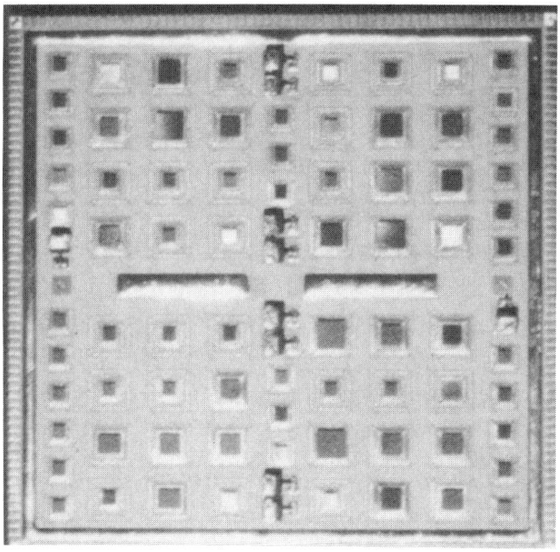

FIGURE 13.16 TAB multichip module. (Source: Honeywell, Inc.)

13.2.6 Cray Microminiature Packaging Technology [14]

The Cray Research (Chippewa Falls, Wisconsin) "number-crunching" computers have been known for their computing power; the present-day silicon-based Cray 2 is rated at 1200 megaflops. The use of gallium arsenide (GaAs) is one of the reasons that Cray's future-generation supercomputer, the Cray 3, is eight times faster than the Cray 2. But the apparent use of microminiature multichip packaging is probably the most significant factor in the Cray 3's projected 10,000-megaflop performance level.

The Cray 3, set for introduction in 1988, employs about 50,000 GaAs chips for about 55% of its total of 90,000 integrated circuits. (Using GaAs yields at least three times more speed than using ECL parts.) The remainder are mainly silicon-based CMOS RAM memories.

For the first time at Cray, the packaging consists of chips mounted directly to an eight-layer printed-wiring board substrate using a proprietary technology. These microminiature multichip modules are then mounted onto larger printed board assemblies (Figure 13.17).

There are about 200 of these assemblies, each of which holds up to 1024 (0.8 W average) chips. This packaging approach helps trim

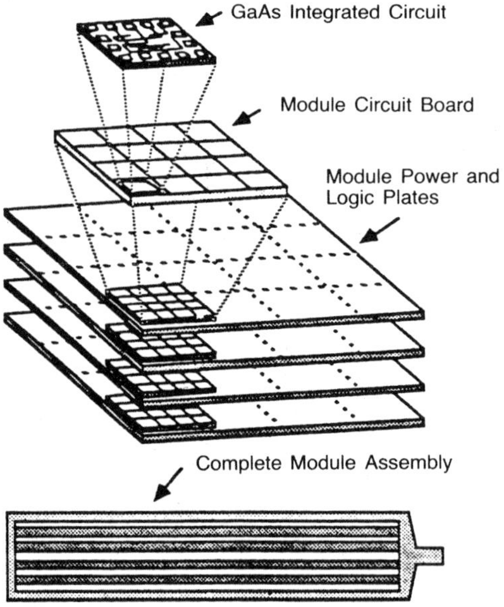

FIGURE 13.17 Multichip module packaging hierarchy. (Source: Cray Research)

the size of the Cray 3 to about one-half that of the Cray 2, that is, 32 in. octagon/34 in. high versus 53 in. diameter/45 in. high. (Both computers are liquid-cooled, with a system power of approximately 150,000 W.)

13.3 TELECOMMUNICATIONS APPLICATIONS [15]

A new packaging technology is being developed at AT&T Bell Laboratories (Murray Hill, New Jersey). This microminiature multichip module technology, which is referred to as advanced VLSI packaging (AVP), is designed to meet the needs of future VLSI-based telecommunications and data processing systems. It is expected that these systems will incorporate the use of mixed-device technologies, for instance, MOS and bipolar, silicon and gallium arsenide.

The AVP interconnection substrate is constructed to provide an excellent environment in which to route signal paths and distribute voltage reference levels. The substrate provides a microstrip configuration for two signal levels over a continuous power plane that acts as a high-frequency ground, as shown in Figure 13.18.

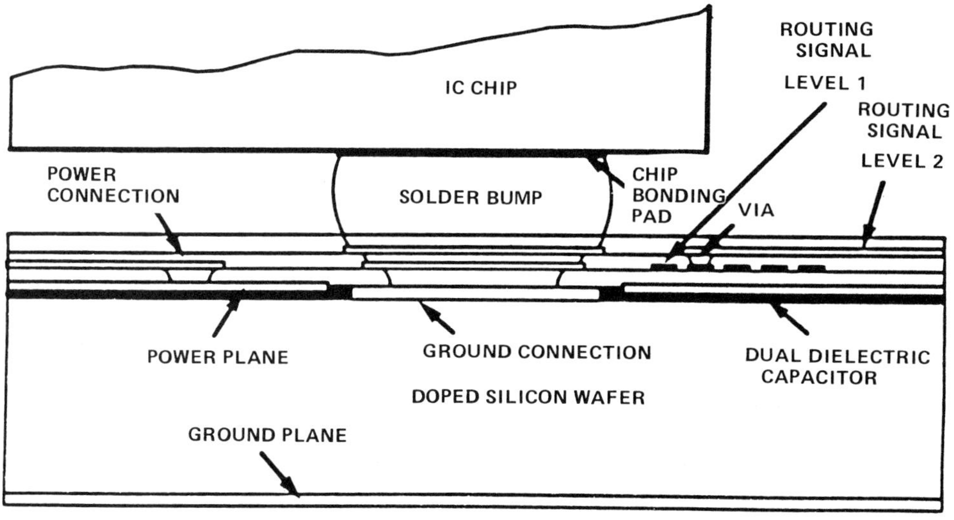

FIGURE 13.18 Cross section of AVP multichip module. (Source: AT&T Bell Labs)

The metallization of both signal lines and voltage planes is plated copper. The use of copper assures flexibility in altering line thickness and thus provides line resistivity values as designs require.

Good electrical contact between interconnection levels is achieved with solid nickel vias that are also plated. Signal-level "microvias" measure 10 μm (0.4 mil) in diameter, and power/ground vias measure 90 μm (3.6 mils) in diameter.

Polyimide forms the dielectric between metallized levels. Characteristic impedances of 50 to 70 Ω can be achieved with dielectric thicknesses of from 6 to 10 μm (0.24 to 0.4 mil) with line widths and spacings of 10 μm (0.4 mil).

The use of AVP technology enables the extension of the high level of integration of VLSI integrated circuits to the system level by allowing as many as 100 complex VLSI chips to be interconnected on a single silicon wafer. This ability to place large-scale systems or subsystems on AVP silicon wafers leads to minimal electrical parasitics, which, in turn, results in higher system operating frequencies and lower power dissipation.

Signal nets on multilayer boards may easily exceed 500 mm in total length, while nets on AVP substrates rarely exceed 100 mm. Thus, output-driver circuits typically have to drive board-level loads of more than 100 pf, whereas most interconnection paths for AVP circuits require drivers designed for only 25-pf loading. In addition, flip-chip solder-bump attachment removes the need to route internally generated signals to the perimeter of the chip, resulting in minimal integrated circuit parasitics.

What this leads to is that, by tailoring the size of the CMOS driver to the load, the total delay through the circuit is reduced about 50%. A large part of this delay reduction is accounted for within the smaller driver circuit.

Power dissipation levels are also reduced by shortening interconnection lengths and reducing capacitive loading. For high-I/O VLSI designs, output drivers can account for up to 40% of the total chip power dissipation. By reducing the capacitive loading by a factor of up to 4, total chip power can be reduced by up to 30%.

The first implementation of the AVP module was a three-chip laboratory test vehicle consisting of a central processing chip, a memory management chip, and a math accelerator chip. The chip I/Os range from 81 to 120. The 13-mm × 30-mm silicon interconnection substrate within the module is not much larger than the chips themselves.

The chips are spaced 0.5 mm apart, with most of the conductor routing taking place under the chips. Given abundant routing area, signal paths were widened to 20 μm (0.8 mil), and all signal nets were brought to the periphery of the wafer for full test access.

The complete multichip module (Figure 13.19) is in the form of a 160-I/O pin-grid array package with a four-layer printed-wiring

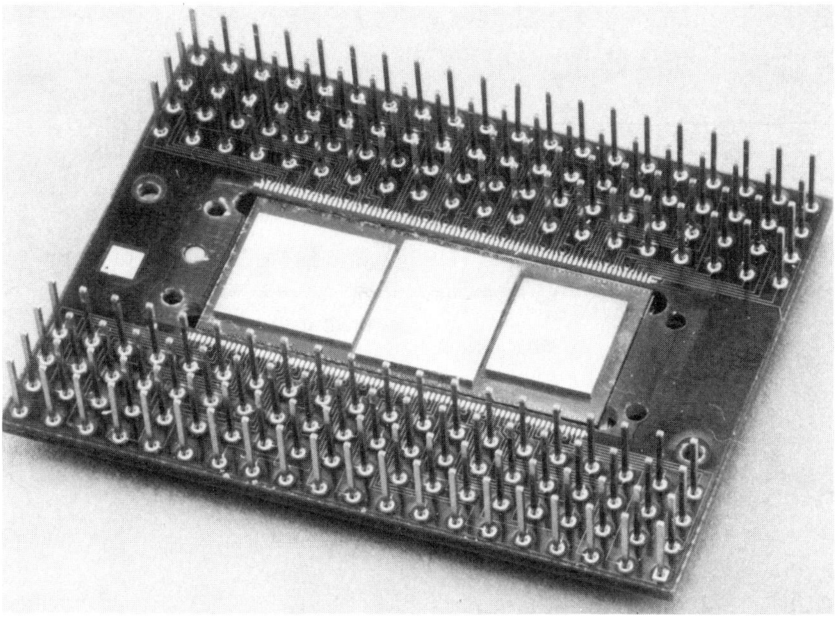

FIGURE 13.19 Three-chip AVP module without protective cover. (Source: AT&T Bell Labs.)

board overall substrate with minimum design features of 0.1-mm lines and 0.15-mm spaces. The board is fabricated using bismaleimide-triazine (BT) epoxy blends to accommodate thermosonic wire bonding at 150°C.

A rectangular cutout is routed in the center of the package to provide space for a metal heat sink/structural support member. The AVP interconnect substrate with the three chips flip-chip mounted is assembled to the aluminum heat sink using a thin layer of compliant silicone gel adhesive (Figure 13.20).

The heat sink structure is assembled to the printed wiring board and the assembly is then thermosonically ball/wedge-bonded using 0.001-in.-diameter gold wire. A protective plastic cover, designed with alignment tabs, provides the final mechanical protection for the wire-bonded assembly.

The final prototype modules were demonstrated at clock frequencies of 14 MHz. the operational limit of the integrated circuit chip set itself. Although this module is intended to be used only as a laboratory test vehicle, extensions of this packaging technology are being investigated.

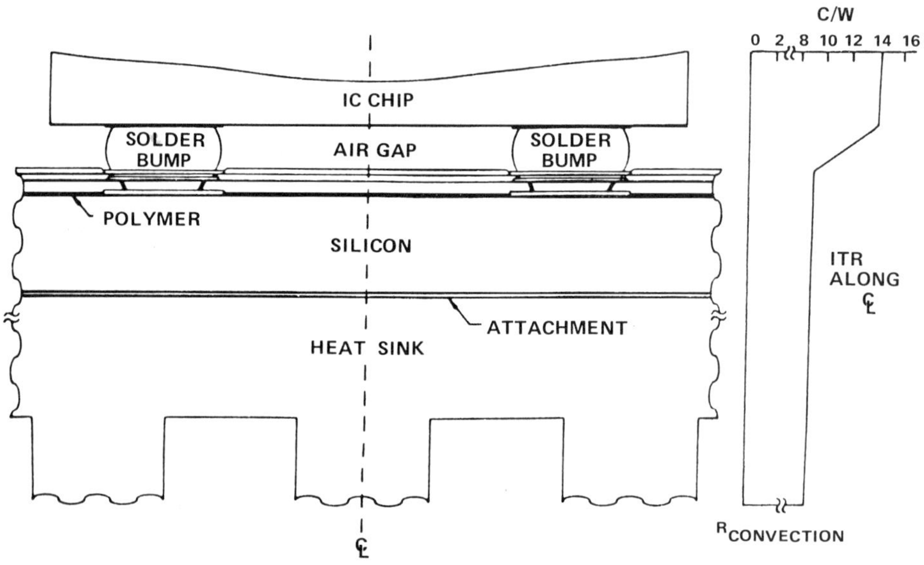

FIGURE 13.20 Internal thermal resistance (ITR) of AVP module. (Source: AT&T Bell Labs.)

13.4 COMMAND-CONTROL-COMMUNICATE APPLICATIONS

In the realm of command-control-communicate (C-3), primarily military applications, the success of the mission is of prime importance. Therefore, companies trying to satisfy C-3 requirements have a slightly different viewpoint as to what is cost-effective compared to the commercial computer and telecommunications applications just described. However, as the following technology description will show, there are not nearly as many differences in end results as one might expect.

13.4.1 IBM Federal Systems Division Multichip Modules [16]

A microminiature 500,000-gate multichip package (MCP) is being offered by IBM (Manassas, Virginia) for next-generation C-3 applications for use with 50-MHz CMOS integrated circuits that have feature sizes as small as 0.5 μm. The 64-mm^2 MCP module (Figure 13.21) will have a cofired multilayer ceramic substrate that will accommodate 16 chip sites.

FIGURE 13.21 MCP multichip module. (Source: IBM Federal Systems.)

Each module will have 236 signal I/Os. The overall module itself will have 625 I/O pins on 2.54-mm centers, including 498 I/Os for signals.

Package heremeticity will be established by using a ceramic cap, which is reflow solder-sealed to the substrate. Maximum power for the package is expected to be less than 30 W, with a 3-W maximum power for each chip.

13.4.2 Honeywell Bull Multichip Modules [17]

Honeywell Bull (Phoenix, Arizona) has developed a "silicon-on-silicon" multichip module concept for both C-3 and computer applications. The concept (Figure 13.22) combines flip-chips, a silicon substrate, a ceramic package or header, and conventional wire bonding or copper-polyimide beam-tape automated bonding between the silicon substrate and the package.

Low-cost packaging can be achieved, since the silicon substrate utilizes conventional integrated circuit fabrication processes and materials. Silicon substrate quality is not critical when active devices are not present in the interconnection substrate.

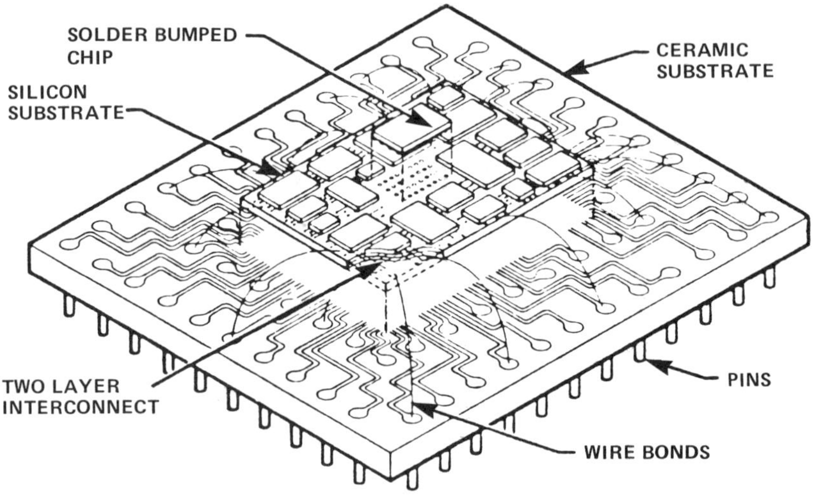

FIGURE 13.22 Silicon-on-silicon module. (Source: Honeywell Bull.)

Microminiature multichip packaging of this type also yields high-density interconnects. Using flip-chip technology, the chips can be placed relatively close together (Figure 13.23).

Another benefit of silicon-on-silicon packaging is repairability. The solder bumping and assembly processes can be repeated if chips have to be repaired or replaced. Also, since the integrated circuits are fabricated independently from the silicon substrate, it is possible, if not highly desirable, to mix chip technologies in the same module.

13.4.3 Texas Instruments TAB Multichip Modules [18]

Texas Instruments' (Dallas, Texas) multichip module consists basically of four TAB chips interconnected on a thick-film ceramic substrate in a 196-lead package. The four-chip module uses two TAB tapes assembled back to back. The TAB etched conductive pattern requires line widths and spaces of 50 μm (2 mil) in some areas.

The gold thick-film interconnections are printed on a 50-mm^2 (2-in.^{2})/0.9-mm (35-mil)-thick ceramic substrate. The final dimensions are achieved by laser drilling the chamfered corners and laser scribing to break the parallel edges. The 0.9-mm-thick ceramic substrate was used to prevent possible warpage problems, thus minimizing alignment and continuity problems in the upper layers.

The final interconnection between the thick-film conductors and the TAB leads on top of them is done by thermosonic bonding.

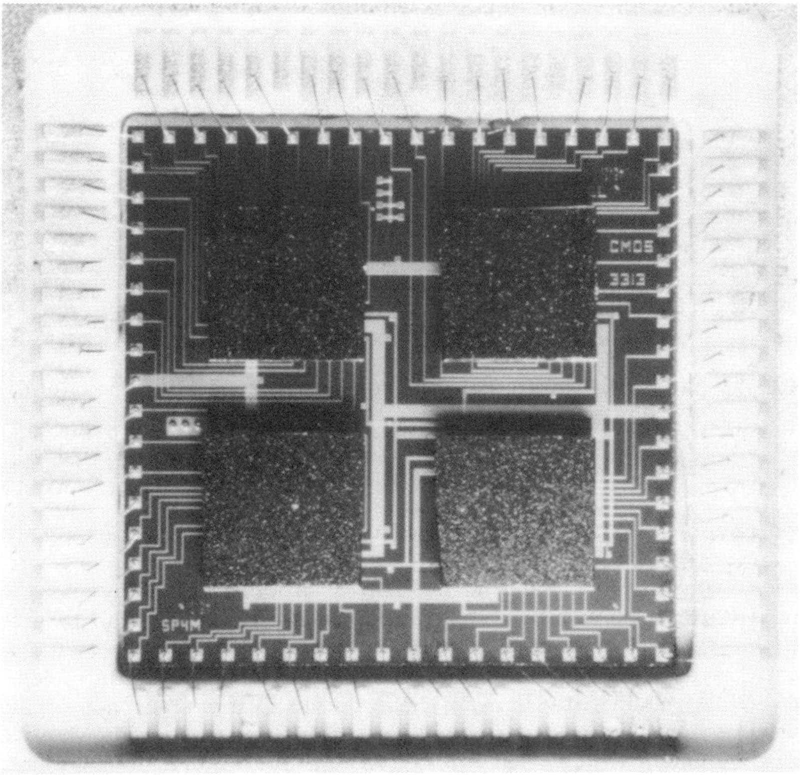

FIGURE 13.23 Closely spaced flip-chip integrated circuits. (Source: Honeywell Bull.)

13.4.4 Rockwell International Silicon-Substrate Multichip Module [19]

Studies of various interconnect concepts led the Collins Avionics Division of Rockwell International (Cedar Rapids, Iowa) to the conclusion that a concept consisting of multichip modules with silicon interconnect substrates (SISs) was a viable approach. Their packaging concept (Figure 13.24) provides direct connection of an integrated circuit die to a silicon substrate, thereby eliminating one level of packaging and interconnection.

The SIS fabrication process begins with an aluminum conductor material being deposited on silicon wafers using conventional thin-film

FIGURE 13.24 Silicon substrate microminiature multichip module. (Source: Rockwell Collins.)

thermal evaporation techniques. Conductor lines of 25-μm (1-mil) width and spacing and 2-μm thickness are formed by a subtractive approach, whereby a pattern is defined by photolithography using standard chrome/glass masks. Unwanted material is removed by wet etching.

Metal layers are separated by 2- to 4-μm-thick polyimide dielectric material. When fully cured, these polyimide films are heat-resistant, mechanically tough, chemically resistant, possess good electrical properties, and provide good step coverage by planarizing surface topography. A wet chemical process was developed for forming 75-μm (3-mil)-diameter via holes.

Both digital and RF circuits have been adapted to SIS. Functional testing to prove the validity of the concept was completed on a 32-bit reprogrammable memory module (Figure 13.25) with excellent results.

The principal mechanical design issues are related to providing large 50-mm^2 (2-in.^{2}) silicon substrates with adequate structural support to survive vibration, shock, and thermal cycling environments while maintaining the significant SIS advantages of small size, low weight, and high reliability. Fortunately, although unsupported silicon wafers are quite fragile, after bonding to a support plane they become quite rugged.

Matched thermal expansion and high thermal conductivity of the support material are important. Thus, molybdenum, aluminum nitride, silicon carbide, and graphite composites are candidate materials.

Figure 13.26 shows a miniaturized version of a digital data processor interface assembly. The relatively large copper-Invar-copper

FIGURE 13.25 Thirty-two-bit reprogrammable memory module. (Source: Rockwell Collins.)

constraining-core leadless chip carrier assembly on the right is replaced by the much smaller 61-chip SIS module on the left.

The integrated circuit chips, which are either wire bonded, tape automated bonded, or flip-chip soldered to the silicon substrate, are subsequently encapsulated with a silicone gel coating.

The wide range of packaging flexibility of the SIS approach is also demonstrated in Figure 13.27, which shows a large, hermetically sealed, eight-SIS module assembly. Four of the SIS multichip modules are mounted to each side of the standard electronics module (SEM-E) assembly. A rail-mounted thermal core provides high heat removal capacity, and a new SEM-E connector provides 250-I/O capability.

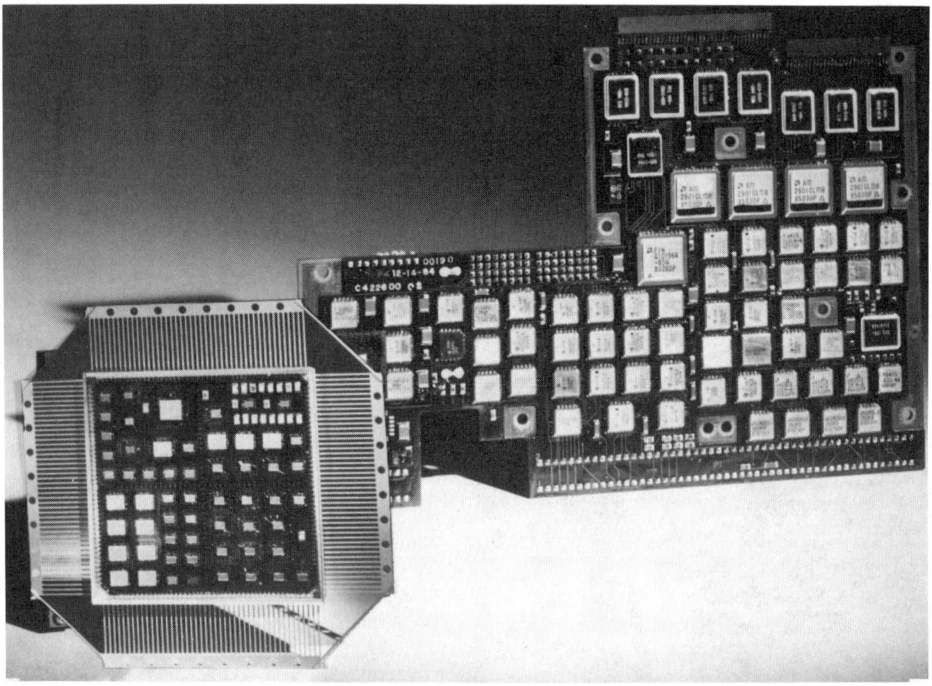

FIGURE 13.26 Digital data processor interface assemblies. (Source: Rockwell Collins.)

13.5 RELATED TECHNOLOGIES

The work being done by original equipment manufacturers to package multichip modules use, for the most part, existing fabrication and assembly technologies. However, work is also being done by independent suppliers to develop more sophisticated interconnecting structures for these applications. Some of the most promising of these interconnection technologies will be described in this section.

13.5.1 Mosaic Systems Silicon Circuit Boards [20]

The foundation of Mosaic Systems' (Troy, Michigan) new multichip module interconnection substrate is a high-density silicon circuit board (SCB). The basic element of the SCB is a "sea of bonding pads" vertically connected to a matrix of conducting lines separated

FIGURE 13.27 Eight-module SEM-E assembly. (Source: Rockwell Collins.)

at over 100,000 intersections by an electrically programmable amorphous silicon "antifuse" (Figure 13.28).

To make an SCB, a metal layer is deposited on the silicon substrate, followed by a silicon dioxide dielectric layer, which in turn is followed by an additional metal layer. The two metal layers are vertically connected to 840 bonding pads distributed over the surface of the segment.

The "antifuses" between the intersections of the conductor lines are made of a material that normally has a high resistance. However, when a voltage above a certain threshold is applied, the material

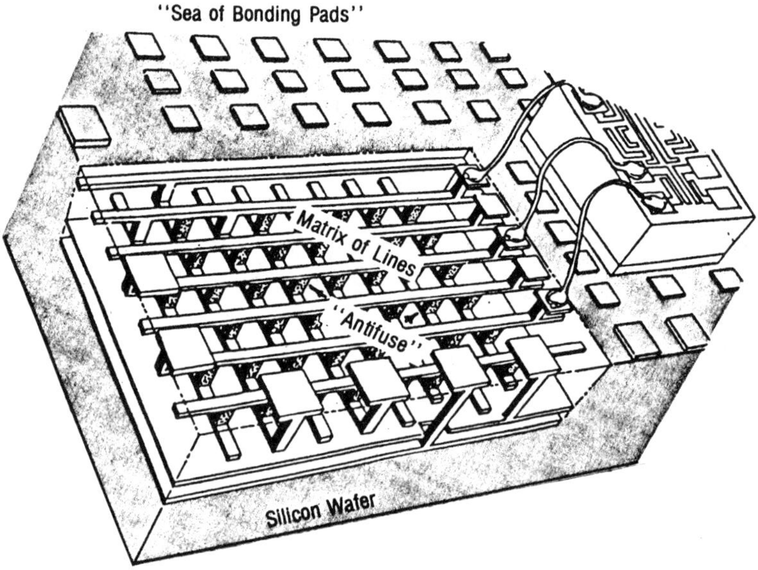

FIGURE 13.28 Basic elements of silicon circuit board. (Source: Mosaic Systems.)

switches to a low resistance. This action is opposite to that of a fuse. Thus, it forms an overall interconnection pattern by connecting the conductor lines at preprogrammed intersections.

The small, 0.3-mm (12-mil) bond pitch permits the SCB to handle some relatively high padout die in a very small footprint. Subsequent wire bonding interconnects the chips to the silicon circuit board.

Figure 13.29 shows two 25-mm^2 (1-in.^{2}) SCBs assembled together to hold a 1.1-Mbit SRAM. It is composed of seventeen 64K × 1 SRAM integrated circuit chips plus six buffers to drive the memory.

13.5.2 Raychem High-Density Interconnect Technology [21]

The Raychem Corporation (San Jose, California) has combined the use of integrated circuit manufacturing processes with specialty materials to make custom designs for multichip modules using thin-film aluminum multilayer techniques. Called HDI, for high-density interconnects, the multichip module interconnection substrate has 25-μm (1-mil) lines and spaces. Figure 13.30 shows a partial close-up view of a four-chip module.

A cross section through the structure (Figure 13.31) shows HDI's special features. The conductors are 5 μm (0.2 mil) thick;

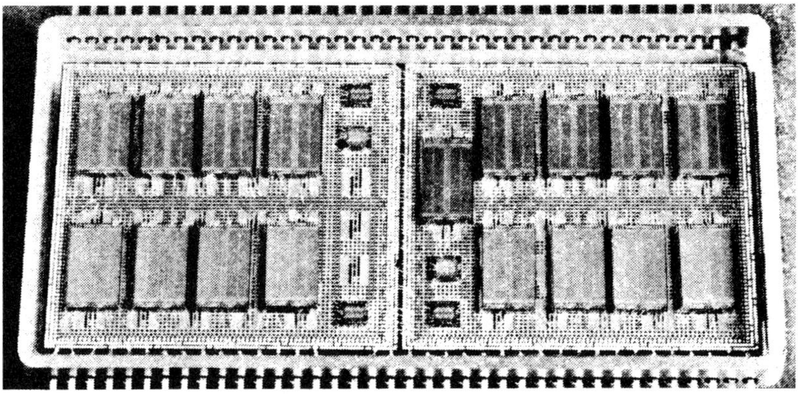

FIGURE 13.29 A 1.1-Mbit SRAM silicon circuit board assembly. (Source: Mosaic Systems.)

the dielectric is 10 μm (0.4 mil) thick; and vias are 30 μm (1.2 mil) in diameter.

Significantly, the vias are not stacked one above the other. Rather, they are staggered, so that the one above is offset from the one below, either in a linear chain or in a spiral staircase. These staggered vias are considered by Raychem to be more robust to thermal cycling than stacked vias.

Designs with up to five metal layers have been fabricated by alternating aluminum metal conductor patterns with polyimide dielectric on a ceramic base substrate. Power and ground planes can be fabricated where needed for controlled impedance or for low coupling noise on signal traces.

13.6 ADVANCED WAFER-SCALE PACKAGING [23]

The quest continues for a practical way to put entire systems on a single wafer, even though earlier attempts at wafer-scale integration (WSI) were plagued by intractable problems with yield and discretionary wiring. To date the approaches taken by Rockwell International and Mosaic Systems represent the closest attempts to achieve full WSI. However, packaging engineers are not giving up, because system performance now depends as much on the connections between the integrated circuit chips as it does on what goes on in the chips themselves.

FIGURE 13.30 Close-up view of high-density interconnect technology. (Source: Raychem Corp.)

13.6.1 Auburn University Planar Hybrid Interconnection Technology [24]

Research scientists at the Microelectronics Science and Technology Center of Auburn University are developing a silicon-based, hybrid wafer-scale packaging technology that they say promises to solve WSI problems. Although their interconnection concept (Figure 13.32) uses a silicon wafer as the master interconnection medium, the semiconductor chips are not mounted on its surface. Instead, they are placed into the wafer.

The process steps begin with chemically etching rectangular cutouts (holes) into the silicon wafer and having interchip metallization paths deposited and patterned on its surface. Pretested chips are

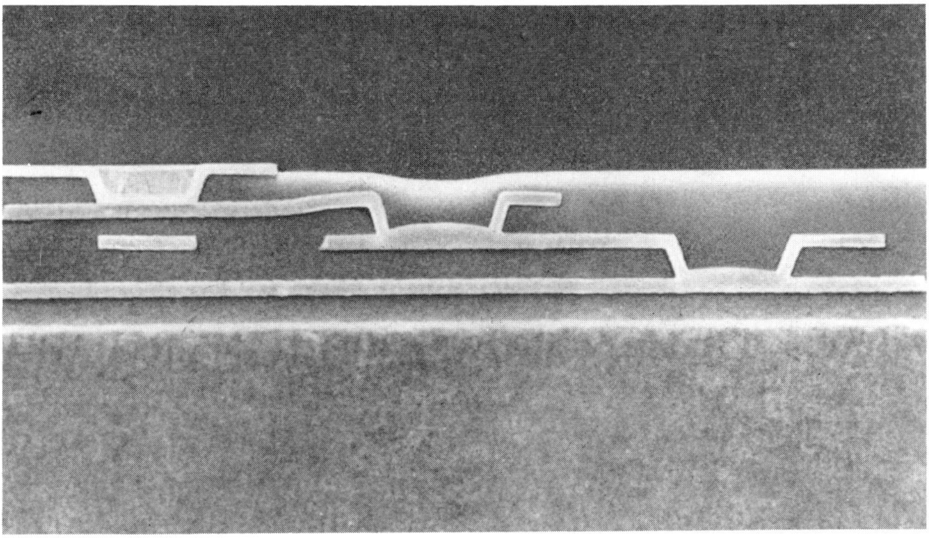

FIGURE 13.31 Cross section of HDI technology. (Source: Raychem Corp.)

then mounted in the holes with their top surface coplanar with the wafer's surface. Electrical connection from chip to wafer is made by either wire bonding or with thin-film metal links.

13.6.2 Rensselaer Polytechnic Institute Wafer Transmission Module [25]

Recently, research efforts at Rensselaer Polytechnic Institute have been directed toward developing cost-effective, multilayer metal-polyimide structures for use with gallium arsenide (GaAs) devices at a wiring pitch of only a few micrometers. An advantage of this approach, which is variously termed the wafer transmission module (WTM) or the wafer-scale hybrid (WSH), is that all the wiring capacitance can be made small as a result of the low dielectric constant of polyimide.

Figure 13.33 shows a prototype of the wafer transmission module. With a fully developed WTM bonding technology, the die spacing could be made extremely tight, resulting in an effective gate density approximating that of an ultralarge-scale integrated (ULSI) circuit.

The resulting multilayer structure resembles a miniaturized printed circuit board, with integrated circuit chips and other components bonded directly to the surface. Such a hybrid package can also be

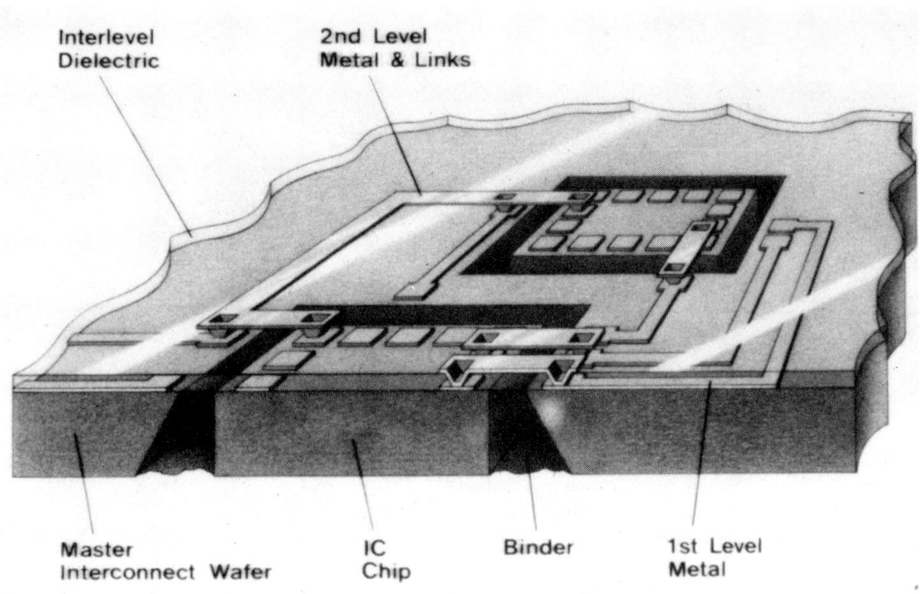

FIGURE 13.32 Planar hybrid interconnection technology. (Source: Auburn University.)

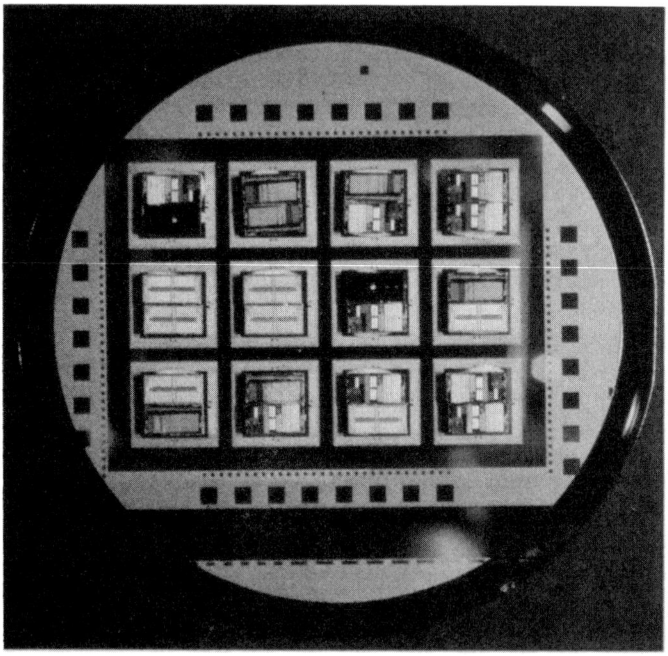

FIGURE 13.33 Wafer transmission module, prototype. (Source: Rensselaer Polytechnic Institute.)

regarded as supplying the additional layers of interconnections needed to improve the active device density to better exploit the speed of gallium arsenide devices.

Unfortunately, the yield of GaAs is currently so low that utilizing the maximum achievable device density is extraordinarily expensive. Hence, until the yield/cost problems associated with this approach are solved, the more traditional microminiature multichip module packaging schemes will be more attractive alternatives.

REFERENCES

1. Dieter Bergman, Institute for Interconnecting and Packaging Electronic Circuits, personal correspondence, August 12, 1987.
2. J. F. McDonald, A. J. Stecki, C. A. Neugebauer, R. O. Carlson, and A. S. Bergendahl, "Multilevel Interconnections for Wafer Scale Integration," *Journal of Vacuum Science and Technology*, November/December 1986, pp. 3127-3138.
3. C. J. Bartlett, "Advanced Packaging for VLSI," *Solid State Technology*, June 1986, pp. 119-123.
4. G. Messner, "Cost-Density Analysis of Interconnections," Printed Circuit World Conference IV, Tokyo, Japan, June 1987.
5. A. Bar-Cohen, "Thermal Management of Air and Liquid Cooled Multi-Chip Modules," Control Data Corporation, IPC Technical Paper, IPC-TP-632.
6. H. Nakahara, "Complex Multilayer Boards Vie for Space in Japanese Computers," *Electronic Packaging and Production*, February 1987.
7. L. S. Goldman and P. A. Totta, "Area Array Solder Interconnections for VLSI," *Solid State Technology*, June 1983, pp. 91-97.
8. J. Gow and A. R. Leach, "Memory Packaging at IBM," *Circuits Manufacturing*, May 1983, pp. 72-97.
9. D. Lammers, "Fujitsu Shoehorns Mainframe CPU onto Single Board," *Electronic Engineering Times*, January 26, 1987, p. 8.
10. T. Yamada, K. Otsuka, K. Okutani, and K. Sahara, "Low Stress Design of Flip Chip Technology for Si on Si Multichip Modules," Proceedings, International Electronic Packaging Conference, Orlando, Florida, October 1985.
11. F. Kobayashi, K. Ogiue, G. Toda, and M. Wajima, "Packaging Technology for the Supercomputer Hitachi S-810 Array-Pro-

cessor," Proceedings, Electronic Components Conference, May 1984, pp. 379-382.

12. T. Watari and H. Murano, "Packaging Technology for the NEC SX Supercomputer," *IEEE Transactions on Components, Hybrids and Manufacturing Technology*, December 1985, pp. 462-467.

13. Thomas J. Moravec, Micro Interconnects, Honeywell, Inc., personal correspondence, August 8, 1987.

14. T. Costlow, "Designers Give Packaging the Credit for Cray 3 Speed," *Electronic Engineering Times*, October 12, 1987, p. 10.

15. C. J. Bartlett, J. M. Segelken, and N. A. Teneketges, "Multi-Chip Packaging Design for VLSI-Based Systems," Proceedings, Electronic Components Conference, May 1987, pp. 518-525.

16. R. H. Estrada, "VHSIC Technology Provides a Technical Base for Spaceborne Processor Requirements," *Technical Directions*, November 1, 1987, pp. 6-13.

17. R. K. Spielberger, C. D. Huang, W. H. Nunne, A. H. Mones, D. L. Fett, and F. L. Hampton, "Silicon-on-Silicon Packaging," *IEEE Transactions on Components, Hybrids and Manufacturing Technology*, June 1984, pp. 193-196.

18. S. Brown, M. Kressley, R. Natali, R. Rath, and P. Rima, "Multilayer Interconnects and TAB for VHSIC Chips,"

19. A. A. Evans and J. K. Hagge, "Advanced Packaging Concepts—Microelectronics Multiple Chip Modules Utilizing Silicon Substrates,"

20. A. A. Bogdan, "Electrically Programmable Silicon Circuit Board Expands Package Options," *Hybrid Circuit Technology*, April 1987.

21. E. Bogatin, "Beyond Printed Wiring Board Densities: A New Commercial Multichip Packaging Technology," Proceedings, National Electronic Packaging and Production Conference, Anaheim, California, February 1987, pp. 208-224.

22. S. Lebow, "High Density/High Speed Multi-Chip Packaging," Proceedings, International Electronic Packaging Conference, San Diego, California, November 1986, pp. 417-423.

23. J. Lyman, "The Latest Wafer-Scale Design is a Hybrid," *Electronics*, March 17, 1986, p. 28.

24. R. W. Johnson, M. Cornelius, J. L. Davidson, and R. C. Jaeger, "Planar Hybrid Interconnection Technology," *International Journal for Hybrid Microelectronics*, Vol. 10, no. 1, 1987.

25. J. F. McDonald, H. J. Greub, R. H. Steinvorth, B. J. Donlan, and A. S. Bergendahl, "Wafer Scale Interconnection for GaAs Packaging—Applications to RISC Architecture," *Computer*, April 1987, pp. 21-34.

14

Summary and Conclusions

14.1 SUMMARY [1, 2]

The scope of this book does not allow for an in-depth treatment of all of the specific technologies that are actively under development throughout the electronics industry. While the focus has been primarily on the hardware aspects of surface mounting implementation, many other disciplines are also involved.

The potentials of using chip carriers and other surface-mounted components are great (Table 14.1). Yet there exist some critical problems to be solved in successful applications. Key among these are the following:

Thermal management, involving both the thermal expansion mismatch at the component/substrate level and the overall dissipation of heat at the system level.

Electrical performance, with respect to achieving the high circuit performance of the VLSI components without undue degradation caused by the packaging and interconnecting structure.

Cost-effective fabrication, in conjunction with the development of new fabrication, assembly, and testing concepts and equipment.

TABLE 14.1 A Comparison of Integrated Circuit Packaging Technologies

Characteristics	Through-hole	Leaded surface mount	Leadless surface mount	Direct deposit	Bare chip
Packaging density	Low	Moderate	Good	Good	High
Standardization	Very good	Good	Good	Good	Limited
Thermal performance	Moderate	Good	Very good	Good	Fair
Substrate choices	Very good	Very good	Good	Fair	Limited
Fab investment	Low	Low	Moderate	Moderate	High
Assembly investment	Moderate	Low	Moderate	Moderate	High
Support investment	Low	Moderate	Moderate	High	High
External assembly services	Very high	High	Moderate	Moderate	Limited
Maintenance skills	Low	Moderate	Moderate	High	High

Change risk	Very low	Low	Moderate	Moderate	High
Assembly test	Very easy	Easy	Moderate	Moderate	Complex
Documentation	Easy	Easy	Moderate	Complex	Complex
Logistics support	Field change	Field change	Field change	Field change	Factory only
Inspect (circuit)	100% test	100% test	100% test	100% test	Lot accept
Component burn-in	Easy	Easy	Easy	Easy	Impractical
Pretest	Very easy	Very easy	Easy	Easy	Impractical
Change and repair	Easy	Easy	Easy	Easy	Difficult
Chip availability	Excellent	Very good	Good	Good	Limited
Multiple sourcing	Excellent	Very good	Good	Good	Limited
Footprint commonality	Excellent	Very good	Good	Good	Poor
Profile	High	High	Moderate	Low	Low

14.2 CONCLUSIONS [3]

The following points can be made about electronic packaging in general, and surface mounting in particular.

1. Although new, high-density, discrete packaged component technologies, such as multichip modules, are gaining in importance, they do not represent a long-term solution for VLSI and ULSI applications.
2. New multichip (and surface mount) packaging technologies can be successfully introduced into products only if equivalent progress is made on the associated system technologies, including design, interconnection, assembly, and testing.
3. As the breadth of applications of VLSI-based electronics increases, so will the diversity of packaging designs and technologies.
4. Packaging development activity is rapidly extending beyond the needs of VLSI. New device technologies, for instance, gallium arsenide, optical interconnects, and cryogenic operating temperatures, are imposing new requirements on package design and technology.

The extent that surface mounting will be used in future VLSI-based products will depend on both the performance gains that can be achieved and the costs. However, a cost-effective technology will not guarantee a successful introduction of surface mounting into system designs. When companies develop the needed infrastructure to meet these challenges successfully, the electronics industry will take a major step toward the fabrication of more sophisticated electronics equipment.

REFERENCES

1. "Guidelines for Component Packaging and Interconnection with Emphasis on Surface Mounting," IPC-SM-780, The Institute for Interconnecting and Packaging Electronic Circuits, 1988.

2. C. A. Harper and W. W. Staley, "Some Critical Materials Factors in the Application of Leadless Chip Carrier Packages," *Electronic Packaging and Production*, August 1981, pp. 137-140.

3. C. J. Bartlett, J. M. Segelken, and N. A. Teneketges, "Multi-Chip Packaging Design for VLSI-Based Systems," Proceedings, Electronic Components Conference, May 1987, pp. 518-525.

Appendix A

Terms and Definitions

The following definitions from IPC-T-50D apply to the terms used in this book.

Aspect Ratio A ratio of the length of a hole to its preplated diameter.

Blind Via A via extending to only one surface of a printed board.

Buried Via A via not extending to the surface of the printed board.

Castellation A metallized feature that is recessed on the edge of a chip carrier, used to interconnect conducting surfaces or planes within or on the chip carrier.

Chip Carrier A low-profile surface mount component package, usually square, whose semiconductor chip cavity or mounting area is a large fraction of the package size and whose external connections are usually on all four sides of the package (may be either leaded or leadless).

Coefficient of Thermal Expansion (CTE) The linear dimensional change of a material per unit change in temperature.

Constraining Core A supporting plane that is internal to a packaging and interconnecting structure.

Double-Sided Assembly A packaging and interconnecting structure with components mounted on both the primary and secondary sides.

Dual-Inline Package (DIP) A component package having two rows of leads extending from its sides and formed at right angles from a plane parallel to its base.

Flat Pack A component package having two rows of leads extending from its sides and parallel to its base.

Land A portion of a conductive pattern usually, but not exclusively, used for electrical connection, component attachment, or both.

Land Pattern A combination of lands intended for the mounting, interconnection, and testing of a particular component.

Leaded Chip Carrier A chip carrier whose external connections consist of leads around and down the sides of the package.

Leaded Surface Mount Component (Device) See **Surface Mount Component (Leaded)**.

Leadless Chip Carrier A chip carrier whose external connections consist of metallized terminations.

Mixed-Component Mounting Technology A component mounting technology that uses both through-hole and surface-mounting technologies on the same packaging and interconnecting structure.

Packaging and Interconnecting Assembly The general term for an assembly that has electronic components mounted on either or both sides of a packaging and interconnecting structure.

Packaging and Interconnecting Structure The general term for a completely processed combination of base materials, supporting planes or constraining cores, and interconnection wiring used for the purpose of mounting components.

Primary Side That side of a packaging and interconnecting structure so defined on the master drawing. (It usually contains the most complex or greatest number of components.)

Printed Board The general term for completely processed printed circuit or printed wiring configurations. It includes single-sided, double-sided, and multilayer boards with rigid, flexible, and rigid-flex base materials.

Repair(ing) The act of restoring the functional capability of a defective article in a manner that precludes compliance with applicable drawings or specifications.

Rework(ing) The act of reprocessing noncomplying articles, through the use of original or alternative equivalent processing, in order to bring the article into compliance with applicable drawings and specifications.

Secondary Side That side of a packaging and interconnecting structure that is opposite the primary side. (The same as the "solder side" in through-hole component mounting technology.)

Sequentially Laminated Multilayer Printed Board A multilayer printed board formed by laminating together through-hole-plated double-sided and/or multilayer printed boards. (Thus, some of its circuitry layers are interconnected with blind and/or buried vias.)

Single-Sided Assembly A packaging and interconnecting assembly with components mounted only on the primary side.

Solder Ball A small sphere of solder adhering to a laminate, resist, or conductor surface. (This generally occurs after wave or reflow soldering.)

Solder Fillet A normally concave surface of solder at the intersection of the metal surfaces of the solder connection.

Solder Spatter Extraneous irregular-shaped solder fragments.

Solder Webbing A continuous film or curtain of solder parallel to, but not necessarily adhering to, a surface or between separate conductive patterns that should be free of solder.

Step Soldering The technique for making solder connections by sequentially using solder alloys with successively lower melting temperatures.

Straight-Through Lead A component lead that extends through a hole in a printed board without subsequent forming of the lead.

Supporting Plane A planar structure that is part of a packaging and interconnecting structure in order to provide mechanical support, thermomechanical constraint, thermal conduction, and/or electrical characteristics. (It may be either internal or external to the packaging and interconnecting structure.)

Surface Mount Component (Device) A component (device) capable of being attached to a printed board by surface mounting (may be either leaded or leadless).

Leaded Surface Mount Component A surface mount component whose external connections consists of formed leads.

Leadless Surface Mount Component A surface mount component whose external connections consist of metallized terminations that are an integral part of the component body.

Surface mounting The electrical connection of components to the surface of a conductive pattern that does not utilize component holes.

Thermal Coefficient of Expansion (TCE) See **Coefficient of Thermal Expansion (CTE)**.

Thermal Expansion Mismatch The absolute difference in the thermal expansion of two components.

Wave Soldering A process wherein printed boards are brought in contact with the surface of continuously flowing and circulating solder.

Appendix B

Surface Mounting Standards, Specifications, and Guidelines

The following documents center on surface mount technology. These documents were/are being developed by standards organizations in the United States and internationally. The letters of each document indicate the organization that has responsibility for the document:

EIA indicates documents prepared by the Electronic Industries Association
JEDEC indicates documents prepared by the Joint Electronic Device Engineering Council of EIA
IPC indicates documents prepared by the Institute for Interconnecting and Packaging Electronic Circuits
MIL indicates documents prepared by the U.S. military
DOD indicates standardized documents prepared by the U.S. Department of Defense
IEC indicates documents prepared by the International Electrotechnical Commission

B.1 COMPONENTS, GENERAL

EIA-481-A	Taping of Surface Mount Components for Automatic Placement
EIA-PDP-100	Mechanical Outlines for Registered and Standard Electronic Parts
EIA-JEP-95	JEDEC Registered and Standard Outlines for Semiconductor Devices
EIA-JESD-11	Chip Carrier Pinouts Standardized for CMOS 4000, HC and HCT Series of Logic Circuits

B.2 COMPONENTS, PASSIVE

B.2.1 Capacitors

EIA-CB-11	Guidelines for the Surface Mounting of Multilayer Ceramic Chip Capacitors
EIA-IS-28	Fixed Tantalum Chip Capacitor Style 1—Protected—Standard Capacitance Range
EIA-IS-29	Fixed Tantalum Chip Capacitor Style 1—Protected—Extended Capacitance Range
EIA-IS-36	Chip Capacitors, Multilayer (Ceramic Dielectric)
EIA-IS-37	Multiple Layer High Voltage Capacitors (Radial Lead Chip Capacitor)

IEC-384-3	Sectional Specification, Tantalum Chip Capacitors
IEC-384-10	Sectional Specification, Fixed Multilayer Ceramic Chip Capacitors
IECQ Draft	Blank Detail Specification, Fixed Multilayer Ceramic Chip Capacitors
IECQ-PQC-31	Sectional Specification, Fixed Tantalum Chip Capacitors with Solid Electrolyte
IECQ-PQC-32	Blank Detail Specification, Fixed Tantalum Chip Capacitor

B.2.2 Resistors

EIA-IS-30-A	Resistors, Rectangular Surface Mount Thick Film
EIA-IS-34	Leaded Surface Mount Resistor Networks Fixed Film

B.3 COMPONENTS, ACTIVE

B.3.1 Microelectronic Ceramics

EIA-JEDEC-MS-002-009	Ceramic Chip Carriers
EIA-JEDEC-MO-044	(50 mil) Ceramic j-lead
EIA-JEDEC-MS-014	(40 mil) Ceramic Chip Carrier
EIA-JEDEC MO-056	(45 mil) Ceramic Chip Carrier
EIA-JEDEC-MO-057	(20 mil) Ceramic Chip Carrier
EIA-JEDEC-MO-062	(25 mil) Ceramic Chip Carrier
EIA-JEDEC-MO-070	(375 mil) Ceramic Flat Pack

B.3.2 Microelectronic Plastics

EIA-JEDEC-MO-046	(200 mil) Gullwing SOIC
EIA-JEDEC-MS-012	(150 mil) Gullwing SOIC
EIA-JEDEC-MS-013	(300 mil) Gullwing SOIC
EIA-JEDEC-MS-059	(330 mil) Gullwing SOIC
EIA-JEDEC-MO-053	(25 mil) Leaded Square PLCC
EIA-JEDEC-MO-060	(40 mil) Ceramic Quad Flat Pack
EIA-JEDEC-MO-069	(25 mil) Plastic Quad Flat Pack Gullwing

EIA-JEDEC-MO-052	(50 mil) j-lead Rectangular PLCC
EIA-JEDEC-MO-047	(50 mil) j-lead Square PLCC
EIA-JEDEC-MO-041	Type E Rectangular Leadless Chip Carrier
EIA-JEDEC-MO-042	Type F Rectangular Leadless Chip Carrier
EIA-JEDEC-MO-061	Plastic Small Outline j-lead (SOJ) .400 Body Family
EIA-JEDEC-MO-063	Plastic Small Outline j-lead (SOJ) .350 Body Family
EIA-JEDEC-MO-065	Plastic Small Outline j-lead (SOJ) .300 Body Family
EIA-JEDEC-MO-071	Plastic Chip Carrier Gullwing (20 mil)
EIA-JEDEC-MO-075	Leadless Quad 50 mil Centers
EIA-JEDEC-MO-076	Leadless Chip Carrier 50 Series 50 mil Centers

B.3.3 Discrete Devices

EIA-JEDEC-TO-236	(similar to SOT-23) Transistor
EIA-JEDEC-TO-243	(similar to SOT-89) Transistor
EIA-JEDEC-TO-253	(similar to SOT-143) Transistor
EIA-JEDEC-DO-207	Leadless Diodes
EIA-JEDEC-DO-212	Round Lead Axial Diodes
EIA-JEDEC-DO-213	Leadless Diodes
EIA-JEDEC-DO-214	C-Bend Plastic Surface Mount Diode
EIA-JEDEC-DO-215	Gullwing Plastic Surface Mount
MIL-M-38510	Military Specification on Microelectronic Devices
EIA-JESD 21-A	Configurations for Solid State Memories (1985)
EIA-JESD 21-A-1	Addendum to JEDEC Standard 21A (1986)

B.4 COMPONENTS, ELECTROMECHANICAL

B.4.1 Connectors

EIA-506	Dimensional and Functional Characteristics Defining Sockets for Leadless Type A Chip Carriers (0.050 Spacing)
EIA-507	Dimensional Characteristics Defining Edge Clips for Use with Hybrids and Chip Carriers
EIA-IS-47	Contact Termination Finish Standard for Surface Mount Devices

B.4.2 Switches

IECQ-PQC-41-US0003	Detail Specification, Dual-in-Line Switch, Surface Mountable, Slide Actuated

B.5 BASE MATERIALS

IPC-L-108	Specification for Thin Laminate Metal Clad Primarily for High Temperature Multilayer Printed Boards
IPC-L-109	Specification for Glass Cloth, Resin Preimpregnated, (B Stage) for High-Temperature Multilayer Printed Boards
IPC-L-115	Specification for Plastic Sheet Laminated Metal Clad for High Temperature Performance Printed Boards
IPC-CF-148	Resin Coated Metal for Multilayer Printed Boards
IPC-FC-150	Copper Foil for Printed Wiring Applications
IPC-CF-152	Metallic Foil Specification for Copper/Invar/Copper (CIC) for Printed Wiring and Other Related Applications
IPC-SM-817	General Requirements for SMT Adhesives
IPC-SF-818	General Requirements for Electronic Soldering Fluxes
IPC-SP-819	General Requirements for Electronic Grade Solder Paste
IPC-CC-830	Qualification and Performance of Electrical Insulation Compounds for Printed Board Assemblies
IPC-SM-840	Qualification and Performance of Permanent Polymer Coating (Solder Mask) for Printed Boards
MIL-F-14256E	Flux, Soldering, Liquid (Rosin Base)

B.6 PRINTED BOARDS

IPC-FC-250	Performance Specification for Single- and Double-Sided Flexible Printed Boards
IPC-SD-320	Performance Specification for Rigid Single- and Double-Sided Printed Boards
IPC-ML-950	Performance Specification for Multilayer Printed Boards
MIL-P-50884	Military Specification Printed Wiring, Flexible and Rigid Flex

MIL-P-55110	Military Specification Printed Wiring Boards, General Specification Form

B.7 INTERCONNECTING SUBSTRATES

IPC-RF-245	Performance Specification for Rigid-Flex Multilayer Printed Boards
IPC-MC-324	Performance Specification for Metal Core Boards
IPC-HM-860	Performance Specification for Hybrid Multilayer

B.8 DESIGN ACTIVITIES

IPC-T-50	Terms and Definitions for Electronic Interconnections
IPC-CM-78	Surface Mount and Interconnecting Chip Carrier Guidelines
IPC-H-855	Hybrid Microcircuit Design Guide
IPC-D-249	Design Standard for Flexible Single- and Double-Sided Printed Boards
IPC-D-317	Design Standard for Electronic Packaging Utilizing High-Speed Techniques
IPC-D-319	Design Standards for Rigid Single- and Double-Sided Printed Boards
IPC-SM-782	Surface Mount Land Patterns (Configurations and Design Rules)
IPC-D-859	Design Standard for Multilayer Hybrid Circuits
IPC-D-949	Design Standard for Rigid Multilayer Printed Boards
MIL-STD-275	Military Standard Printed Wiring for Electronic Equipment
MIL-STD-2118	Design Standard for Flexible Printed Wiring

B.9 COMPONENT MOUNTING

EIA Bulletin CB-11	Guidelines for the Surface Mounting of Multilayer Ceramic Chip Capacitors
IPC-CM-770	Guidelines for Printed Board Component Mounting
IPC-SM-784	Guidelines for Direct Chip Attachment

DOD-STD-2002-2	Part and Component Mounting for High Quality/High Reliability Soldered Electrical and Electronic Assemblies

B.10 ATTACHMENT TECHNIQUES

IPC-SM-780	Electronic Component Packaging and Interconnection with Emphasis on Surface Mounting

B.11 SOLDERING AND SOLDERABILITY

EIA-IS-46	Test Procedure for Resistance to Soldering: (Vapor Phase Technique) for Surface Devices
EIA-IS-49	Solderability Test Procedures
EIA-448-19	Test Standard for Electromechanical Components, Environmental Effects of Machine Soldering Using A Vapor Phase System—Method 19
IPC-S-804	Solderability Test Method for Printed Wiring Boards
IPC-S-805	Solderability Test for Component Leads and Terminations
IPC-S-815	General Requirements for Soldering Electronic Interconnections
DOD-STD-2000-1	Soldering Technology, High Quality/High Reliability
DOD-STD-2000-4	General Soldering Requirements for Electrical and Electronic Equipment

B.12 QUALITY ASSESSMENT

EIA-469-A	Standard Test Method for Destructive Physical Analysis of High Reliability Ceramic Monolithic Capacitors (embraces chips)
EIA-510	Standard Test Method for Destructive Physical Analysis of Industrial Grade Ceramic Monolithic Capacitors (embraces chips)
IPC-A-600	Acceptability of Printed Boards
IPC-A-610	Acceptability of Printed Board Assemblies
MIL-STD-883	Methods and Procedures for Microelectronics

B.13 RELIABILITY

IPC-A-38	Fine Line Round Robin Test Pattern
IPC-A-48	Surface Mount Artwork
IPC-SC-60	Post Solder Solvent Cleaning Handbook
IPC-AC-62	Post Solder Aqueous Cleaning Handbook
IPC-AI-640	User Requirements for Automatic Inspection of Unpopulated Thick Film Hybrid Substrates
IPC-AI-641	User Guidelines for Automated Solder Joint Inspection Systems
IPC-AI-642	User Guidelines for Automated Inspection of Artwork and Innerlayers
IPC-AI-643	User Guidelines for Automatic Optical Inspection of Populated Packaging and Interconnection Assemblies
DOD-STD-2000-3	Criteria for High Quality/High Reliability Soldering Technology

B.14 NUMERICAL CONTROL STANDARDS

IPC-NC-349	Computer Numerical Formatting for Drilling and Routing Equipment
IPC-D-350	Printed Board Description in Digital Form
IPC-D-351	Printed Wiring Documentation in Digital Form
IPC-D-352	Electronic Design Data Base Description for Printed Boards
IPC-D-354	Library Format Description for Printed Boards in Digital Form
EIA-224-B	Character Code for Numerical Machine Control Perforated Tape
EIA-227-A	One-Inch Perforated Tape
EIA-267-B	Axis and Motion Nomenclature for Numerically Controlled Machines
EIA-274-D	Interchangeable Variable Block Data Format for Positioning, Contouring and Contouring/Positioning Numerically Controlled Machines
EIA-281-B	Electrical and Construction Standards for Numerical Machine Control

EIA-358-B	Subset of American National Standard Code for Information Interchange for Numerical Machine Control Perforated Tape
EIA-408	Interface Between Numerical Control Equipment on Data Terminal Equipment Employing Parallel Binary Data Interchange
EIA-431	Electrical Interface Between Numerical Control and Machine Tools
EIA-441	Operator Interface Function of Numerical Controls
EIA-474	Flexible Disk Format for Numerical Control Equipment Information Interchange
EIA-484	Electrical and Mechanical Interface Characteristics and Line Control Protocol Using Communication Control Characters for Serial Data Link Between a Direct Numerical Control System and Numerical Control Equipment Employing Asynchronous Full Duplex Transmission
EIA-494	32 BIT Binary CL Exchange (BCL) Input Format for Numerically Controlled Machines

B.15 TEST METHODS

EIA-JEDEC Method B 105-A	Lead Integrity—Plastic Leaded Chip Carrier (PLCC) Packages
EIA-JEDEC Method B 102	Surface Mount Solderability Test (JESD22-B)
EIA-JEDEC Method B 108	Coplanarity (intended for inclusion into JESD 22-C)
IPC-TM-650	Test Methods Manual

B.16 REPAIR

IPC-R-700	Guidelines for Repair and Modification of Printed Board Assemblies

Documents May Be Obtained from:

Electronic Industries Association (EIA)
2001 Eye Street NW
Washington, DC 20006

Institute for Interconnecting and Packaging
Electronic Circuits (IPC)
7380 North Lincoln Avenue
Lincolnwood, IL 60646

Naval Publications and Forms Center
5801 Tabor Road
Philadelphia, PA 19120

Central Office of the IEC:

International Electrotechnical Commission
3 Rue de Varembe
1211 Geneva 20, Switzerland

IEC documents are available from ANSI:

American National Standards Institute
1430 Broadway
New York, NY 10018

Appendix C

Surface Mount Land Patterns from IPC-SM-782

Appendix C Surface Mount Land Patterns from IPC-SM-782 provides information on land pattern geometries used for the surface attachment of electronic components. The intent of the following extracts from IPC-SM-782 is to provide the appropriate size, shape, and tolerance of the surface mount land patterns to ensure sufficient area for the appropriate solder fillet and also to allow for inspection and testing of those solder joints.

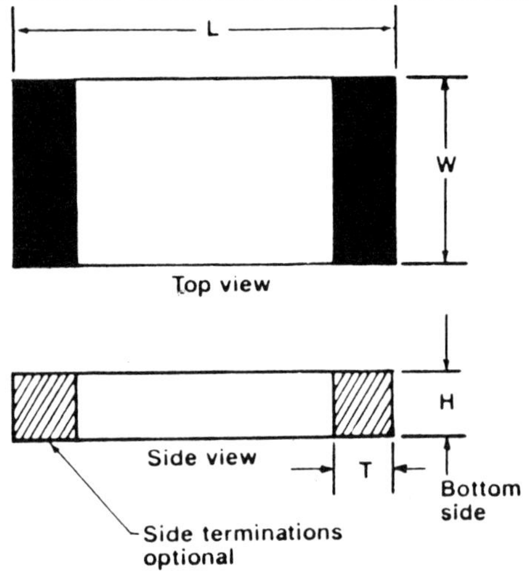

Resistor Number	L	W	H	T[a]
RC0805	1.8-2.2 [.070-.087]	1.0-1.4 [.040-.055]	0.3-0.7 [.012-.028]	0.3-0.6 [.012-.024]
RC1206	3.0-3.4 [.118-.134]	1.4-1.8 [.055-.070]	0.4-0.7 [.016-.028]	0.4-0.7 [.016-.028]
RC1210	3.0-3.4 [.118-.134]	2.3-2.7 [.090-.106]	0.4-0.7 [.016-.028]	0.4-0.7 [.016-.028]

[a] Dimensions have been tailored to facilitate standard land pattern design

FIGURE C.1 Rectangular chip resistor feature dimensions. (*Note*: All dimensions apply after solder coating.)

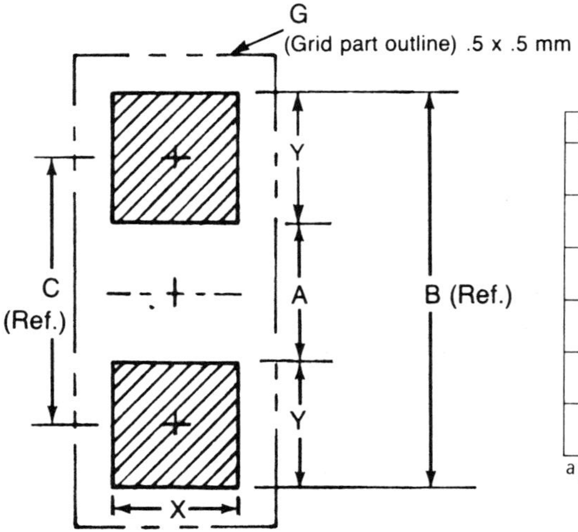

Code Letter	RC 0805	RC 1206[a]	RC 1210
A	0.8 [.032]	1.8 [.070]	1.8 [.070]
B	3.8 [.150]	5.0 [.200]	5.0 [.200]
C	2.3 [.090]	3.4 [.134]	3.4 [.134]
X	1.4 [.055]	1.6 [.063]	2.6 [.102]
Y	1.5 [.060]	1.6 [.063]	1.6 [.063]
Grid part outline	4 x 8	4 x 12	6 x 12

[a] Example shown in graphics

FIGURE C.2 Land patterns for rectangular chip resistors (see Figure C.1).

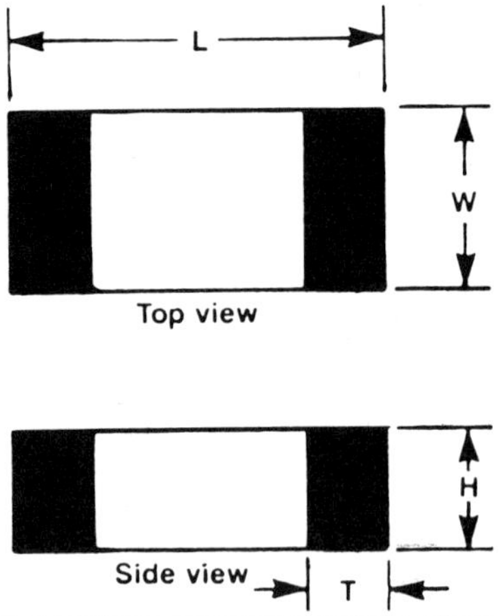

Capacitor Number	L	W	H max	T[a]
CC0805	1.8-2.2 [.070-.087]	1.0-1.4 [.040-.055]	1.3 [.05]	0.3-0.6 [.012-.024]
CC1206	3.0-3.4 [.118-.134]	1.4-1.8 [.055-.070]	1.5 [.06]	0.4-0.7 [.016-.028]
CC1210	3.0-3.4 [.118-.134]	2.3-2.7 [.090-.106]	1.7 [.067]	0.4-0.7 [.016-.028]
CC1812	4.2-4.8 [.166-.190]	3.0-3.4 [.118-.134]	1.7 [.067]	0.4-0.7 [.016-.028]
CC1825	4.2-4.8 [.166-.190]	6.0-6.8 [.236-.268]	1.7 [.067]	0.4-0.7 [.016-.028]

[a] Dimensions have been tailored to facilitate standard land pattern design

FIGURE C.3 Rectangular ceramic chip capacitor feature dimensions.

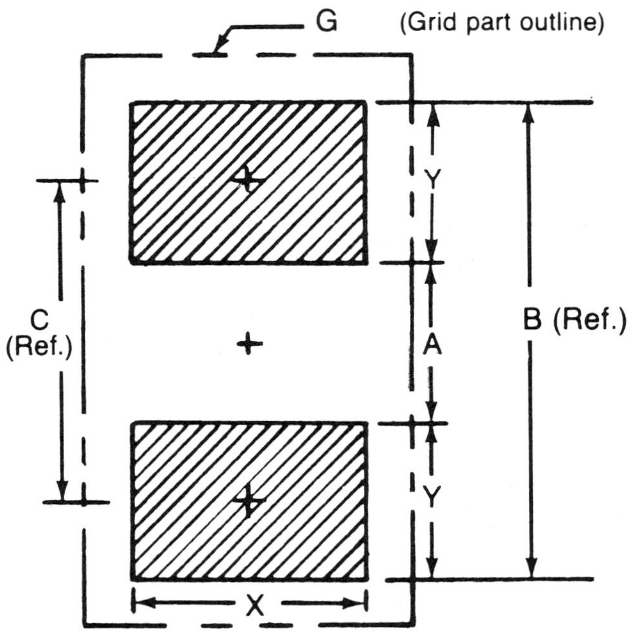

Code Letter	CC 0805	CC 1206	CC 1210[a]	CC 1812	CC 1825
A	0.8 [.032]	1.8 [.070]	1.8 [.070]	3.2 [.126]	3.2 [.126]
B	3.8 [.150]	5.0 [.200]	5.4 [.213]	6.8 [.268]	6.8 [.268]
C	2.3 [.090]	3.4 [.134]	3.6 [.142]	5.0 [.200]	5.0 [.200]
X	1.4 [.055]	1.6 [.063]	2.6 [.102]	3.2 [.126]	6.6 [.260]
Y	1.5 [.060]	1.6 [.063]	1.8 [.070]	1.8 [.070]	1.8 [.070]
Grid part outline	4 x 8	4 x 12	6 x 12	8 x 16	14 x 16

[a] Example shown in graphics

FIGURE C.4 Land patterns for rectangular ceramic chip capacitors (see Figure C.3).

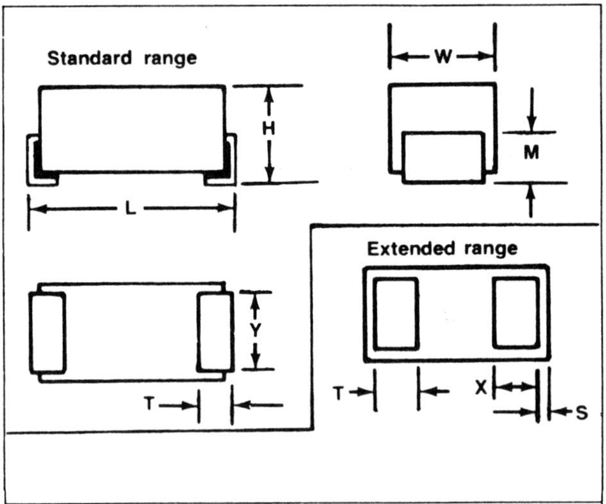

Dim.	Size Code/Standard Capacitance Range			
	[A] 3216	[B] 3528	[C] 6032	[D] 7243
L	3.0-3.4 [.118-.134]	3.3-3.7 [.130-.146]	5.7-6.3 [.224-.248]	6.8-7.6 [.268-.299]
H[a]	1.4-1.8 [.055-.070]	1.7-2.1 [.067-.083]	2.2-2.8 [.087-.110]	2.5-3.1 [.098-.122]
W	1.4-1.8 [.05-.070]	2.6-3.0 [.102-.118]	2.9-3.5 [.114-.138]	4.0-4.6 [.157-.181]
M Min	0.7 [.028]	0.7 [.028]	1.0 [.040]	1.0 [.040]
Y	1.1-1.3 [.043-.051]	2.1-2.3 [.083-.090]	2.1-2.3 [.083-.09]	2.3-2.5 [.090-.098]
T[a]	0.5-1.1 [.020-.043]	0.5-1.1 [.020-.043]	0.5-1.1 [.020-.043]	0.5-1.1 [.020-.043]
Dim.	Size Code/Extended Capacitance Range			
	3518	3527	7227	7257
L	3.3-3.7 [.130-.146]	3.3-3.7 [.130-.146]	6.9-7.5 [.272-.295]	6.9-7.5 [.272-.295]
H[a]	1.7-2.1 [.067-.083]	1.7-2.1 [.067-.083]	2.5-3.1 [.098-.122]	3.0-3.7 [.118-.146]
W	1.6-2.0 [.063-.079]	2.4-3.0 [.095-.118]	2.4-3.0 [.095-.118]	5.2-6.2 [.205-.244]
M Min	0.7 [.028]	0.7 [.028]	1.0 [.040]	1.2 [.047]
Y[a]	1.6-1.8 [.063-.071]	2.4-2.6 [.095-.102]	2.4-2.6 [.095-.102]	5.4-5.8 [.213-.228]
X[a]	0.6-1.0 [.024-.040]	0.6-1.0 [.024-.040]	0.8-1.2 [.031-.047]	0.8-1.2 [.031-.047]
T[a]	0.7-1.1 [.028-.043]	0.7-1.1 [.028-.043]	1.3-1.7 [.051-.067]	1.3-1.7 [.051-.067]
S[a]	0.4-0.6 [.016-.024]	0.4-0.6 [.016-.024]	0.6-0.8 [.024-.032]	0.6-0.8 [.024-.032]

[a] Dimensions are tailored to facilitate standard land pattern design

FIGURE C.5 Tantalum capacitor feature dimensions. (*Note*: All dimensions apply after solder coating.)

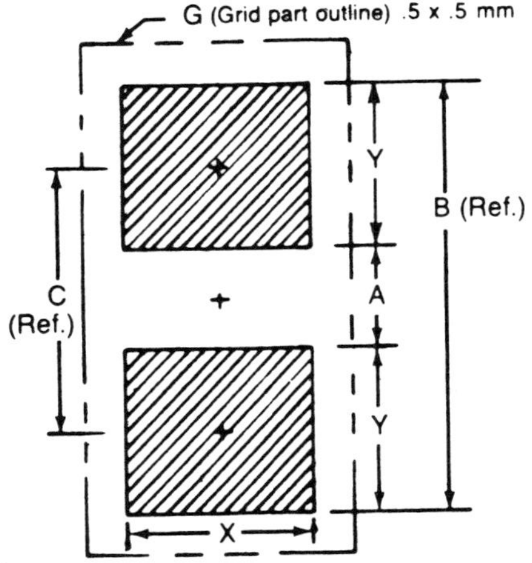

Code Letter	Standard Capacitance Range/Land Patterns			
	[A] 3216	[B] 3528[a]	[C] 6032	[D] 7243
A	0.8 [.032]	1.2 [.047]	3.2 [.126]	4.2 [.166]
B	4.8 [.190]	5.2 [.205]	8.0 [.315]	9.4 [.370]
C	2.8 [.110]	3.2 [.126]	5.6 [.220]	7.0 [.276]
X	1.4 [.055]	2.4 [.095]	2.4 [.095]	2.6 [.102]
Y	2.0 [.080]	2.0 [.080]	2.4 [.095]	2.4 [.095]
G	4 x 10	8 x 12	8 x 18	10 x 20
Code Letter	Extended Capacitance Range/Land Patterns			
	3518	3527	7227	7257
A	0.8 [.032]	0.8 [.032]	3.2 [.126]	3.2 [.126]
B	4.8 [.190]	4.8 [.190]	9.2 [.362]	9.2 [.362]
C	2.8 [.110]	2.8 [.110]	6.2 [.244]	6.2 [.244]
X	2.0 [.080]	2.8 [.110]	2.8 [.110]	6.0 [.236]
Y	2.0 [.080]	2.0 [.080]	3.0 [.118]	3.0 [.118]
G	4 x 10	6 x 10	6 x 20	14 x 20

[a] Example shown in graphics

FIGURE C.6 Land patterns for tantalum capacitors (see Figure C.5).

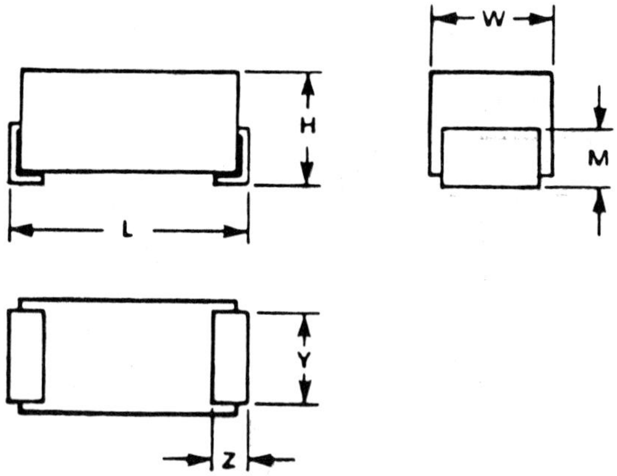

Dim.	Size Code A	Size Code B
L	3.0–3.4 [0.118–0.134]	42–4.8 [0.166–0.190]
H	1.9–2.5 [0.075–0.098]	2.9–3.5 [0.114–0.138]
W	2.0–2.8 [0.079–0.110]	3.0–3.4 [0.118–0.134]
M Min	0.2 [0.008]	0.2 [0.008]
Y	1.6–2.0 [0.063–0.079]	2.3–2.7 [0.091–0.106]
Z	0.2–0.8 [0.008–0.031]	0.2–0.8 [0.008–0.031]

FIGURE C.7 Wirewound inductor feature dimensions. (*Note*: All dimensions apply after solder coating.)

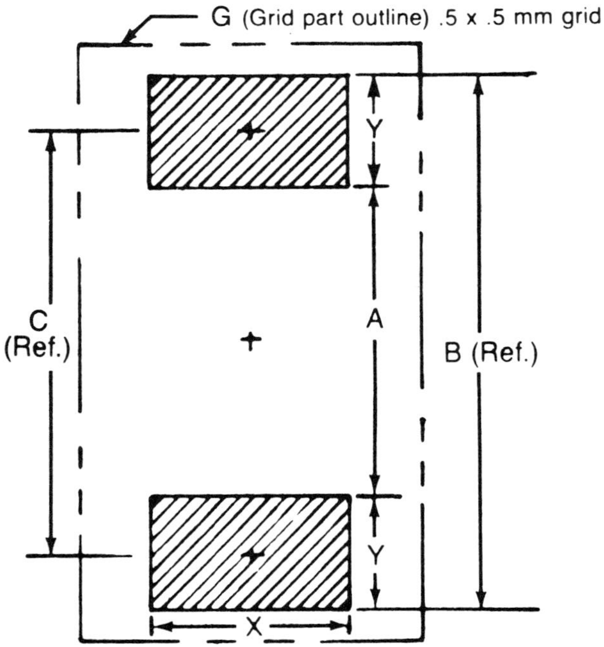

Code Letter	Size (A)	Size (B)[a]
A	1.4 [.055]	2.4 [.095]
B	4.2 [.165]	5.6 [.220]
C	2.8 [.110]	4.0 [.158]
X	2.4 [.095]	3.0 [.118]
Y	1.4 [.055]	1.6 [.063]
G	6 x 10	8 x 12

[a] Example shown in graphics

FIGURE C.8 Land pattern for wirewound inductors (see Figure C.7).

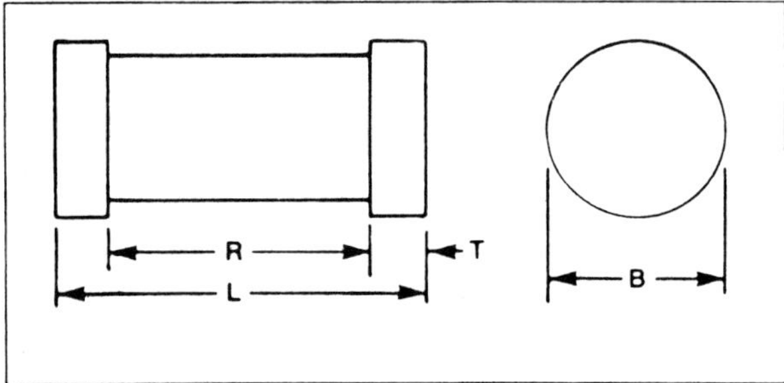

Dim	Case Style	
	MLL34 (SOD-80)	MLL41
L	3.30-3.70 [0.130-0.146]	4.80-5.20 [0.189-0.205]
B	1.50-1.70 [0.059-0.067]	2.44-2.54 [0.096-0.100]
R	2.49-2.59 [0.098-0.102]	3.71-4.59 [0.146-0.181]
T	0.29-0.55 [0.011-0.022]	0.35-0.51 [0.014-0.020]

FIGURE C.9 Tubular MELF component feature dimensions. (*Note*: All dimensions apply after solder coating.)

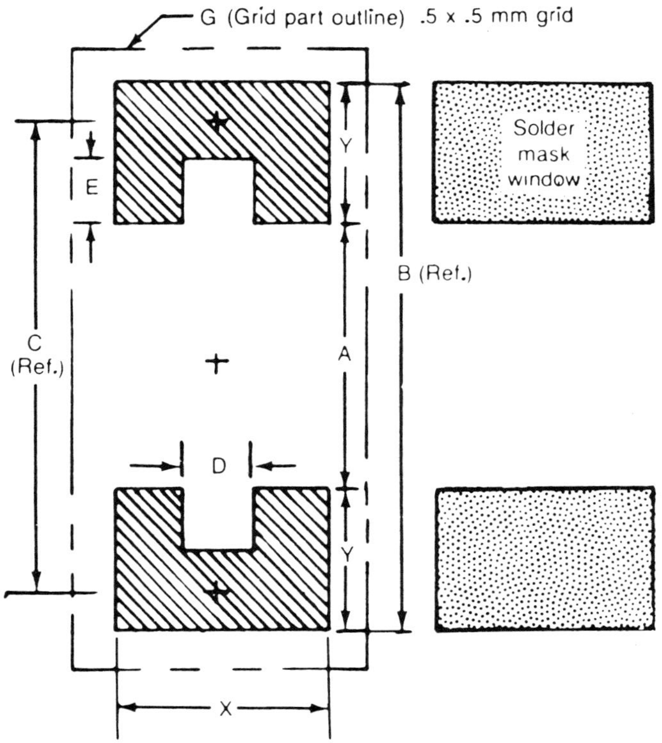

Code Letter	MELF 1/4 W	MLL34	SOD80	MLL41[a]
A	4.4 [.174]	2.2 [.087]	2.2 [.087]	3.4 [.134]
B	8.4 [.330]	5.0 [.200]	5.4 [.213]	7.0 [.276]
C	6.4 [.252]	3.6 [.142]	3.8 [.150]	5.2 [.205]
D[b]	0.3 [.012]	0.3 [.012]	0.3 [.012]	0.3 [.012]
E[b]	0.8 [.032]	0.7 [.028]	0.8 [.032]	1.0 [.040]
X	2.5 [.100]	2.0 [.080]	2.0 [.080]	2.8 [.110]
Y	2.0 [.080]	1.4 [.055]	1.6 [.063]	1.8 [.070]
G	6 x 18	6 x 12	6 x 12	6 x 16

a Example shown in graphics

b The notch shown by dimensions "D" and "E" are intended to reduce skewing and are optional. The depth of this notch may be determined by the following:

$$E = Y - \left(\frac{B - L_{MAX}}{2}\right)$$

Where "L" equals the maximum body length.

FIGURE C.10 Land patterns for tubular MELF components (see Figure C.9).

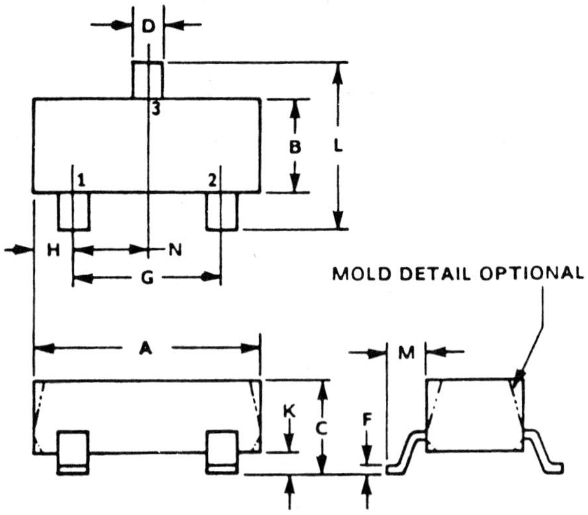

Dimension	Min.-Max Dimensions
A	2.80-3.05 mm [0.110-0.120 inches]
B	1.20-1.40 mm [0.048-0.055 inches]
C	0.85-1.20 mm [0.033-0.048 inches]
D	0.38-0.46 mm [0.015-0.018 inches]
F	0.085-0.015 mm [0.003-0.006 inches]
G	1.76-2.04 mm [0.070-0.081 inches]
H	0.51-0.60 mm [0.020-0.024 inches]
K	(see Figure 3-14)
L	2.10-2.50 mm [0.083-0.098 inches]
M	0.45-0.60 mm [0.018-0.024 inches]
N	0.89-1.02 mm [0.035-0.040 inches]

FIGURE C.11 SOT-23 transistor feature dimensions. (*Note*: All dimensions apply after solder coating.)

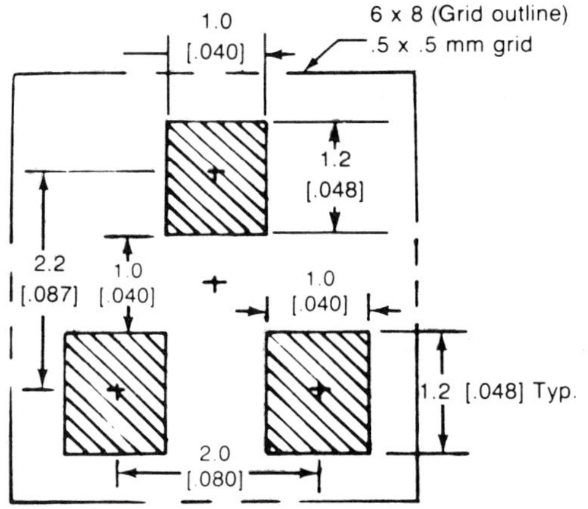

FIGURE C.12 Land pattern for SOT-23 transistors (see Figure C.11).

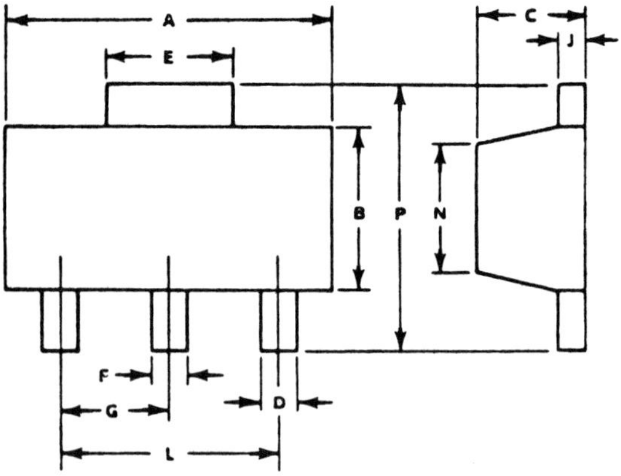

Dimension	Min.-Max Dimensions
A	4.40-4.60 mm [0.174-0.182 inches]
B	2.29-2.60 mm [0.090-0.102 inches]
C	1.40-1.60 mm [0.056-0.062 inches]
D	0.38-0.46 mm [0.015-0.018 inches]
E	1.62-1.80 mm [0.064-0.070 inches]
F	0.46-0.51 mm [0.018-0.020 inches]
G	1.50 BSC [0.059 BSC]
J	0.35-0.44 mm [0.014-0.017 inches]
L	3.00 BSC [0.118 BSC]
N	2.04-2.28 mm [0.081-0.089 inches]
P	3.94-4.25 mm [0.156-0.167 inches]

FIGURE C.13 SOT-89 transistor feature dimensions. (*Note*: All dimensions apply after solder coating.)

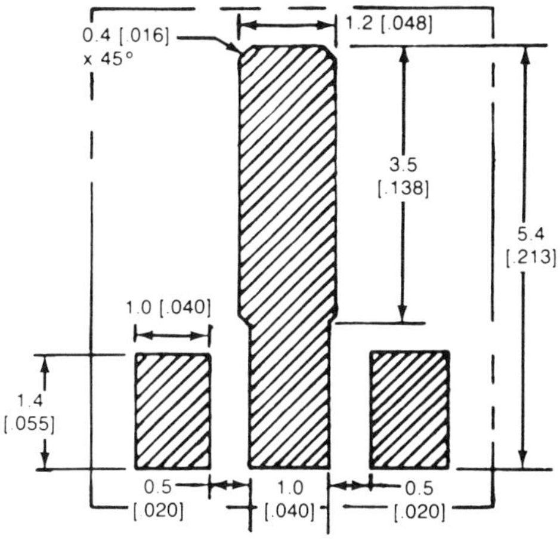

FIGURE C.14 Land pattern for SOT-89 transistors (see Figure C.13).

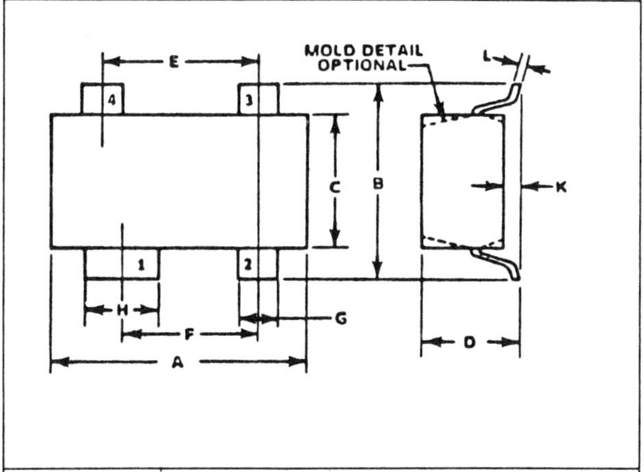

Dimension	*Min.-Max Dimensions*
A	2.80-3.10 mm [0.110-0:122 inches]
B	2.10-2.60 mm [0.083-0.102 inches]
C	1.20-1.70 mm [0.048-0.067 inches]
D	0.85-1.20 mm [0.033-0.048 inches]
E	1.80-2.00 mm [0.071-0.079 inches]
F	1.60-1.80 mm [0.063-0.071 inches]
G	0.37-0.48 mm [0.015-0.019 inches]
H	0.78-0.88 mm [0.031-0.035 inches]
K	0.05-0.25 mm [0.002-0.010 inches]
L	0.09-0.15 mm [0.004-0.006 inches]

FIGURE C.15 SOT-143 transistor feature dimensions. (*Note*: All dimensions apply after solder coating.)

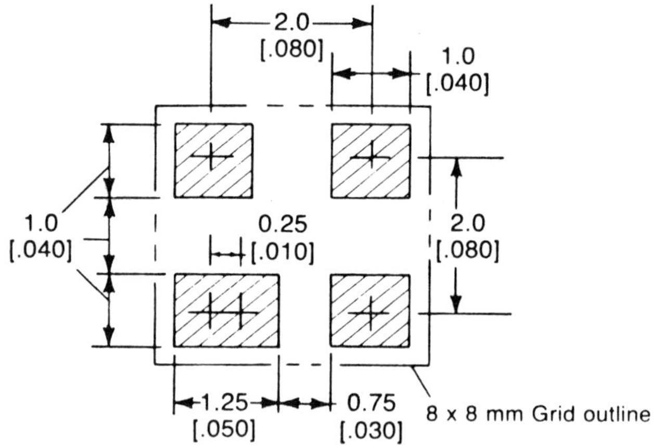

FIGURE C.16 Land pattern for SOT-143 transistors (see Figure C.15).

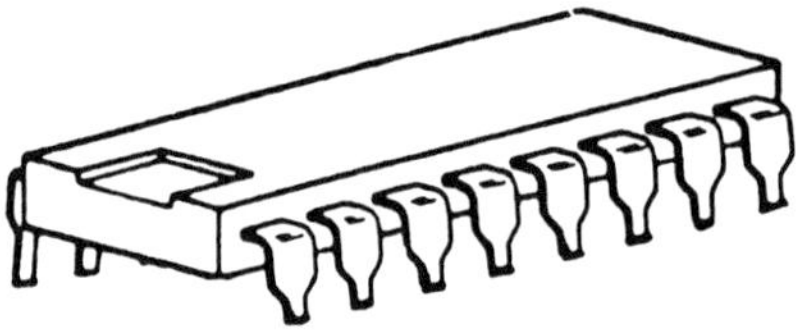

FIGURE C.17 Modified I-lead dual-inline package.

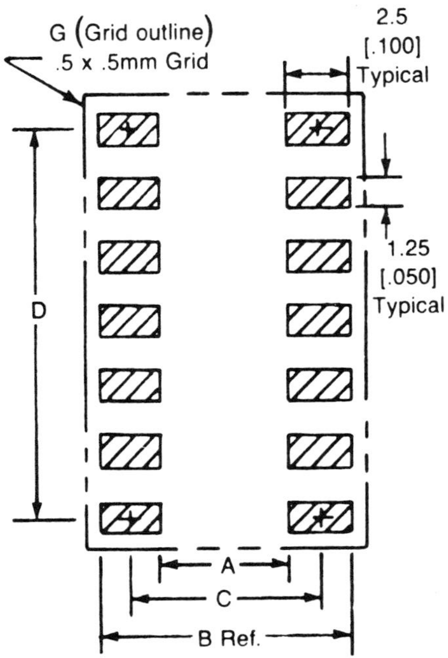

Package Lead Count	A	B	C	D	G
8	5.2 [.200]	10.2 [.400]	7.7 [.300]	7.62 [.300]	22 x 12
14	5.2 [.200]	10.2 [.400]	7.7 [.300]	15.24 [.600]	22 x 36
16	5.2 [.200]	10.2 [.400]	7.7 [.300]	17.78 [.700]	22 x 42
18	5.2 [.200]	10.2 [.400]	7.7 [.300]	20.32 [.800]	22 x 46
20	5.2 [.200]	10.2 [.400]	7.7 [.300]	22.86 [.900]	22 x 52
22	7.7 [.300]	12.7 [.500]	10.2 [.400]	25.40 [1.000]	26 x 56
24	12.7 [.500]	17.7 [.700]	15.2 [.600]	27.94 [1.100]	36 x 62
28	12.7 [.500]	17.7 [.700]	15.2 [.600]	33.02 [1.300]	36 x 72

FIGURE C.18 Land patterns for modified I-lead dual-inline packages (see Figure C.17).

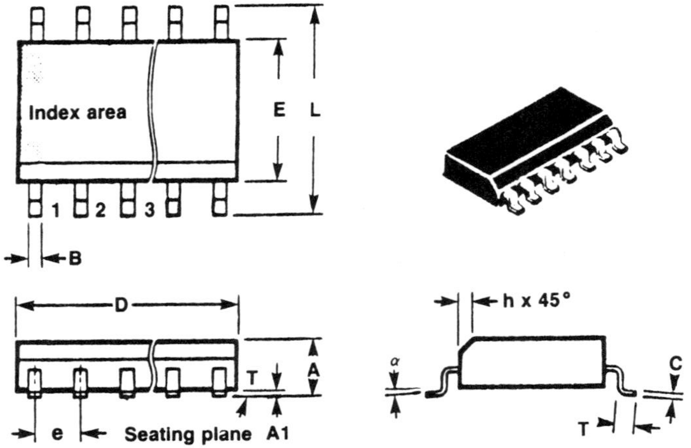

Symbol	8 Pin	14 Pin	16 Pin	16L Pin	20L Pin	24L Pin	28L Pin
A	1.3-1.75 [.058-.069]	1.3-1.75 [.058-.069	1.3-1.75 [.058-.069]	2.4-2.6 [.094-1.02]	2.4-2.6 [.094-1.02]	2.4-2.6 [.094-1.02]	2.4-2.6 [.094-1.02]
A_1	0.01-0.02 [.004-.008]	0.01-0.02 [.004-.008]	0.01-0.02 [.004-.008]	0.01-0.02 [.004-.008]	0.01-0.02 [.004-.008]	0.01-0.02 [.004-.008]	0.01-0.02 [.004-.008]
B	0.36-0.45 [.014-.018]	0.36-0.45 [.014-.018]	0.36-0.45 [.014-.018]	0.36-0.48 [.014-.019]	0.36-0.48 [.014-.019]	0.36-0.48 [.014-.019]	0.36-0.48 [.014-.019]
C	0.18-0.23 [.007-.009]	0.18-0.23 [.007-.009]	0.18-0.23 [.007-.009]	0.23-0.32 [.009-.013]	0.23-0.32 [.009-.013	0.23-0.32 [.009-.013]	0.23-0.32 [.009-.013
D	4.75-5.0 [.188-.197	8.53-8.74 [.336-.344]	9.78-10.0 [.385-.394]	10.11-10.41 [.398-.413]	12.6-13.00 [.496-.512]	15.21-15.6 .599-.614	17.7-18.11 .697-.713
E	3.8-4.0 [.150-.158]	3.8-4.0 [.150-.158	3.8-4.0 [.150-.158]	7.4-7.6 [2.91-2.99]	7.4-7.6 [2.91-2.99]	7.4-7.6 [2.91-2.99]	7.4-7.6 [2.91-2.99]
e	1.27 BSC [.050]	1.27 BSC [.050]	1.27 BSC [.050]	1.27 BSC [.050]	1.27 BSC [.050]	1.27 BSC [.050]	1.27 BSC [.050]
L	5.8-6.2 [.228-.244]	5.8-6.2 [.228-.244]	5.8-6.2 [.228-.244]	10.0-10.6 [.394-.419]	10.0-10.6 [.394-.419]	10.0-10.6 [.394-.419]	10.0-10.6 [.394-.419]
h	0.25-0.5 [.01-.02]	0.25-0.5 [.01-.02]	0.25-0.5 [.01-.02]	0.25-0.5 [.01-.02]	0.25-0.5 [.01-.02]	0.25-0.5 [.01-.02]	0.25-0.5 [.01-.02]
T	0.5-1.15 [.02-.045]	0.5-1.15 [.02-.045]	0.5-1.15 [.02-.045]	0.58-1.25 [.023-.050]	0.58-1.25 [.023-.050]	0.58-1.25 [.023-.050]	0.58-1.25 [.023-.050]
a	0-8°	0-8°	0-8°	0-8°	0-8°	0-8°	0-8°

FIGURE C.19 Small-outline gullwing integrated circuit feature dimensions.

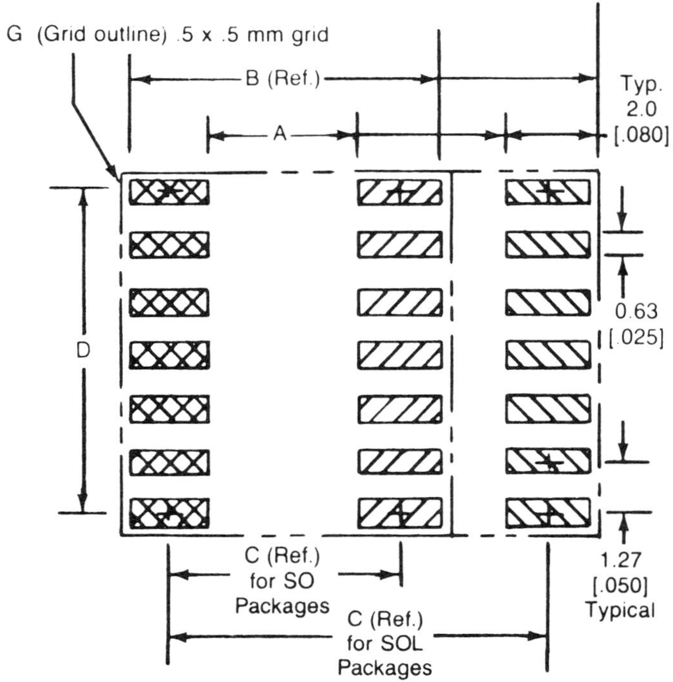

SOIC Lead Count	A	B	C	D	G
SO-8	3.6 [.140]	7.6 [.300]	5.6 [.220]	3.8 [.150]	16 x 10
SO-14[a]	3.6 [.140]	7.6 [.300]	5.6 [.220]	7.6 [.300]	16 x 18
SO-16	3.6 [.140]	7.6 [.300]	5.6 [.220]	8.9 [.350]	16 x 20
SOL-14[a]	7.6 [.300]	11.6 [.460]	9.6 [.380]	7.6 [.300]	24 x 18
SOL-16	7.6 [.300]	11.6 [.460]	9.6 [.380]	8.9 [.350]	24 x 20
SOL-20	7.6 [.300]	11.6 [.460]	9.6 [.380]	11.4 [.450]	24 x 26
SOL-24	7.6 [.300]	11.6 [.460]	9.6 [.380]	14.0 [.550]	24 x 30
SOL-28	7.6 [.300]	11.6 [.460]	9.6 [.380]	16.5 [.650]	24 x 36

[a] Example shown in graphics

FIGURE C.20 Land patterns for small-outline gullwing integrated circuits (see Figure C.19).

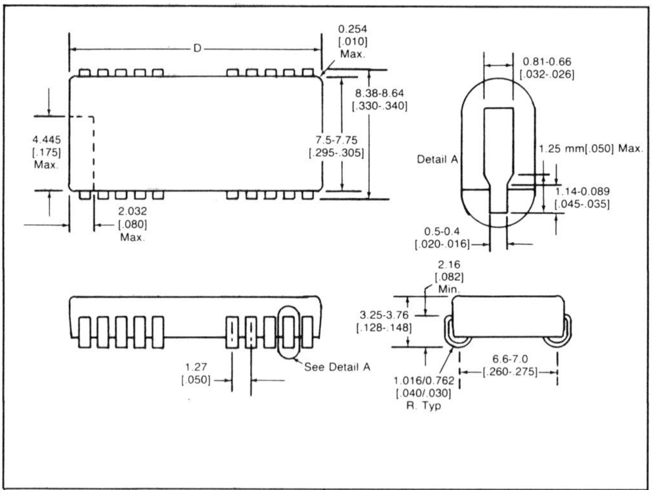

No. of Terminals	D
14	9.4-9.65 [.370-.380]
16	10.67-10.9 [.420-.430]
18	11.94-12.19 [.470-.480]
20	13.20-13.46 [.520-.530]
22	14.48-14.73 [.570-.580]
24	15.75-16.00 [.620-.630]
26	17.02-17.27 [.670-.680]
28	18.29-18.54 [.720-.730]

FIGURE C.21 Small-outline J-lead integrated circuit feature dimensions.

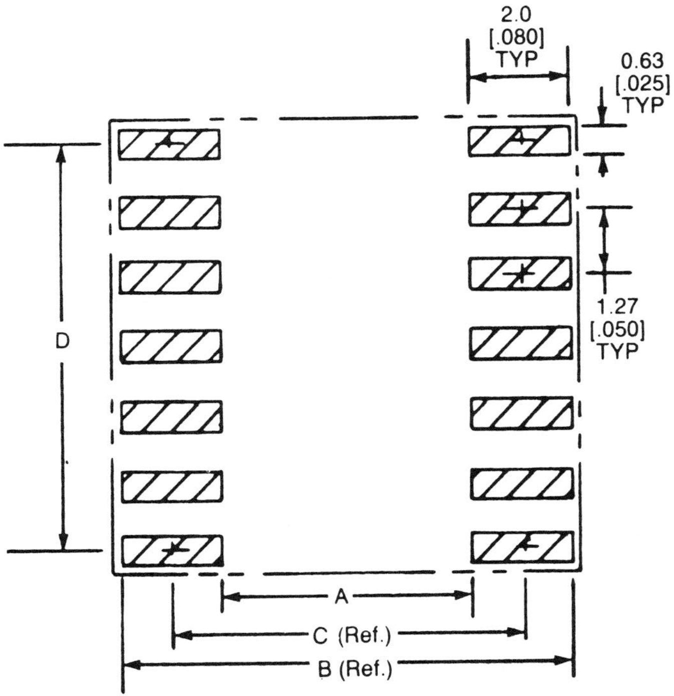

Package Lead Count	A	B	C	D	G
14 [a]	4.9 [.190]	8.9 [.350]	6.9 [.270]	7.6 [.300]	18 x 18
16	4.9 [.190]	8.9 [.350]	6.9 [.270]	8.9 [.350]	18 x 20
18	4.9 [.190]	8.9 [.350]	6.9 [.270]	10.2 [.400]	18 x 24
20	4.9 [.190]	8.9 [.350]	6.9 [.270]	11.4 [.450]	18 x 26
22	4.9 [.190]	8.9 [.350]	6.9 [.270]	12.7 [.500]	18 x 28
24	4.9 [.190]	8.9 [.350]	6.9 [.270]	14.0 [.550]	18 x 30
26	4.9 [.190]	8.9 [.350]	6.9 [.270]	15.2 [.600]	18 x 34
28	4.9 [.190]	8.9 [.350]	6.9 [.270]	16.5 [.650]	18 x 36

[a] Example shown in graphics

FIGURE C.22 Land patterns for small-outline J-lead integrated circuits (see Figure C.21).

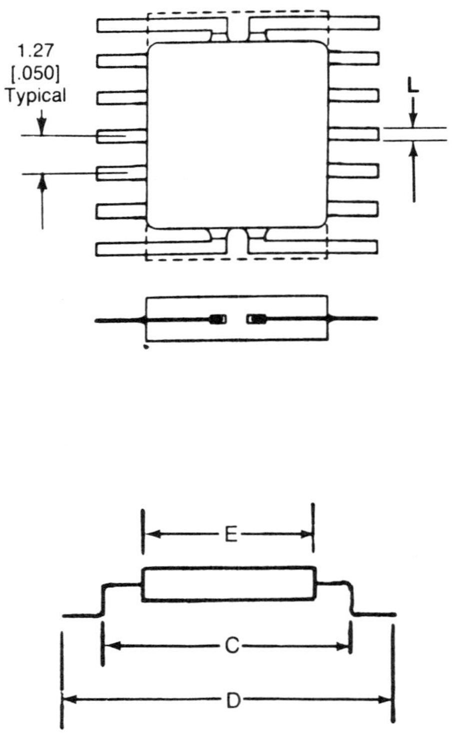

Designation	No. Leads	L	E	C	D
MO-003	10	.25-.35 [.010-.014]	2.5-5.0 [.100-.200]	6.2-6.8 [.244-.268]	9.0-9.6 [.354-.378]
	14	.25-.35 [.010-.014]	2.5-5.0 [.100-.200]	6.2-6.8 [.244-.268]	9.0-9.6 [.354-.378]
MO-004	10	.38-.48 [.015-.018]	5.0-7.6 [.200-.300]	8.2-8.8 [.323-.347]	11.0-11.6 [.433-.457]
	14	.38-.48 [.015-.019]	5.0-7.6 [.200-.300]	8.2-8.8 [.323-.347]	11.0-11.6 [.433-.457]
	16	.38-.48 [.015-.019]	5.0-7.6 [.200-.300]	8.2-8.8 [.323-.347]	11.0-11.6 [.433-.457]
MO-018	40	.25-.36 [.010-.014]	5.0-7.6 [.300-.400]	8.2-8.8 [.323-.347]	11.0-11.6 [.433-.457]
MO-019	24	.38-.48 [.015-.019]	7.6-10.1 [.300-.400]	12.2-12.8 [.480-.504]	15.0-15.6 [.590-.614]
	28	.46-.56 [.018-.022]	7.6-10.1 [.300-.400]	12.2-12.8 [.480-.504]	15.0-15.6 [.590-.614]
MO-020	36	.38-.48 [.015-.019]	10.1-12.7 [.400-.500]	14.2-14.8 [.559-.583]	17.0-17.6 [.669-.693]
	40	.33-.43 [.013-.017]	10.1-12.7 [.400-.500]	14.2-14.8 [.559-.583]	17.0-17.6 [.669-.693]
MO-021	16	.28-.48 [.015-.019]	12.7-15.2 [.500-.600]	17.2-17.8 [.677-.701]	20.0-20.6 [.787-.811]
	24	.38-.48 [.015-.019]	12.7-15.2 [.500-.600]	17.2-17.8 [.677-.701]	20.0-20.6 [.787-.811]
	36	.38-.48 [.015-.019]	12.7-15.2 [.500-.600]	17.2-17.8 [.677-.701]	20.0-20.6 [.787-.811]
MO-022	20	.38-.48 [.015-.019]	15.2-17.8 [.600-.700]	19.2-19.8 [.756-.780]	22.0-22.6 [.866-.890]
	42	.46-.56 [.018-.022]	15.2-17.8 [.600-.700]	19.2-19.8 [.756-.780]	22.0-22.6 [.866-.890]
MO-023	36	.38-.48 [.015-.019]	20.3-22.9 [.800-.900]	24.2-24.8 [.953-.976]	27.0-27.6 [1.063-1.087]
	50	.38-.48 [.015-.019]	20.3-22.9 [.800-.900]	24.2-24.8 [.953-.976]	27.0-27.6 [1.063-1.087]

FIGURE C.23 MO-type flatpack feature dimensions.

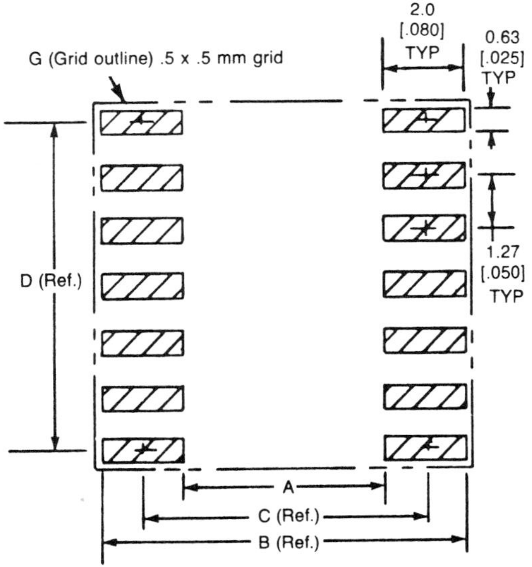

	A	B	C	D	G
MO-003 10	6.0 [.235]	10.0 [.395]	8.0 [.315]	7.6 [.300]	22 x 12
14	6.0 [.235]	10.0 [.395]	8.0 [.315]	7.6 [.300]	22 x 18
MO-004 10	8.0 [.315]	12.0 [.475]	10.0 [.395]	5.1 [.200]	26 x 12
14	8.0 [.315]	12.0 [.475]	10.0 [.395]	5.1 [.200]	26 x 12
16	8.0 [.315]	12.0 [.475]	10.0 [.395]	8.9 [.350]	24 x 20
MO-018 40	8.0 [.315]	12.0 [.475]	10.0 [.395]	10.2 [.400]	26 x 22
MO-019 24	12.0 [.470]	16.0 [.630]	14.0 [.550]	14.0 [.550]	34 x 30
28	12.0 [.470]	16.0 [.630]	14.0 [.550]	16.5 [.650]	34 x 36
MO-020 36	14.0 [.550]	18.0 [.710]	16.0 [.630]	21.6 [.850]	38 x 46
40	14.0 [.550]	18.0 [.710]	16.0 [.630]	24.1 [.950]	38 x 50
MO-021 16	17.0 [.670]	21.0 [.830]	19.0 [.750]	8.9 [.350]	44 x 20
24	17.0 [.670]	21.0 [.830]	19.0 [.750]	14.0 [.550]	44 x 30
36	17.0 [.670]	21.0 [.830]	19.0 [.750]	21.6 [.850]	44 x 46
MO-022 20	19.0 [.745]	23.0 [.905]	21.0 [.825]	11.4 [.450]	40 x 26
42	19.0 [.745]	23.0 [.905]	21.0 [.825]	25.4 [10.0]	48 x 50
MO-023 36	24.0 [.945]	28.0 [1.105]	26.0 [1.025]	21.6 [.850]	58 x 46
50	24.0 [.945]	28.0 [1.105]	26.0 [1.025]	30.1 [1.185]	58 x 124

FIGURE C.24 Land patterns for MO-type flatpacks (see Figure C.23).

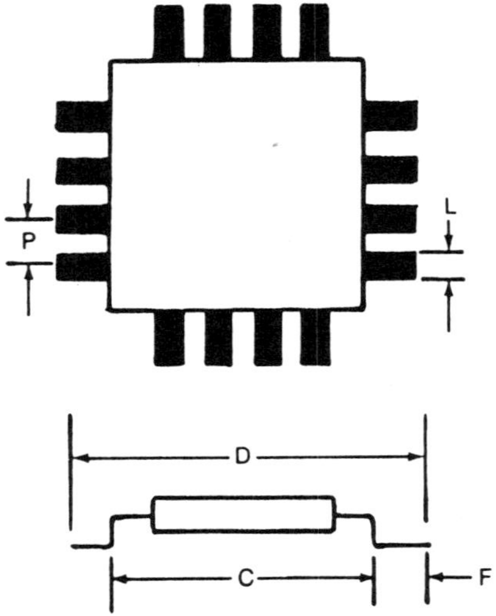

Terminal Count	L	P	C	D	F
44 (11 per side)	0.3-0.4 [.012-.016]	0.8 [.031]	11.2-11.8 [.441-.465]	14.0-14.8 [.551-.583]	1.4-1.7 [.043-.067]
48 (12 per side)	0.24-0.36 [.009-.014]	0.8 [.031]	13.2-13.8 [.520-.544]	17.0-17.6 [.670-.693]	1.5-2.1 [.059-.083]
52 (13 per side)	0.3-0.4 [.012-.016]	1.0 [.04]	17.2-17.8 [.677-.701]	21.0-21.6 [.827-.851]	1.5-2.1 [.059-.083]
54 (13/14 per side)	0.3-0.4 [.012-.016]	0.65 [.026]	11.2-11.8 [.441-.465]	14.0-14.8 [.551-.583]	1.1-1.7 [.043-.047]
64 (13/19 per side)	.03-.05 [.012-.020]	1.0 [.04]	15.4-16.0 [.606-.630] 21.4-22.0 [.842-.866]	18.3-19.1 [.720-.751] 24.3-25.1 [.957-.988]	1.0-1.4 [.04-.055]
70 (11/24 per side)	0.2-0.4 [.008-.016]	0.8 [.031]	12.0-12.6 [.472-.496] 24.3-24.9 [.950-.980]	16.0-16.8 [.630-.661] 28.2-29.0 [1.11-1.142]	1.7-2.3 [.067-.091]
80 (16/24 per side)	0.25-0.45 [.010-.018]	0.8 [.031]	15.7-16.0 [.598-.630] 21.2-22.0 [.835-.866]	17.5-18.3 [.689-.720] 23.5-24.3 [.925-.957]	.6-1.0 [.024-.040]
100 (20-30 per side)	0.2-0.4 [.008-.016]	0.65 [.026]	15.2-16.0 [.598-.630] 21.2-22.0 [.835-.866]	17.5-18.3 [.689-.720] 23.5-24.3 [.925-.957]	.6-1.0 [.024-.040]

FIGURE C.25 Gullwing quad flatpack feature dimensions.

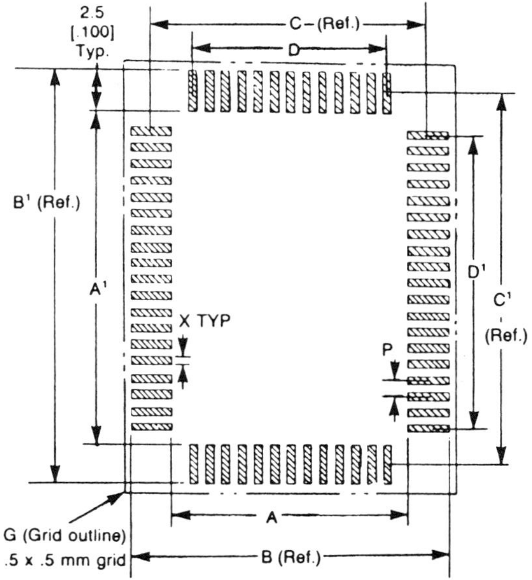

Quad Packs	A/A_1	B/B_1	C/C_1	D/D_1	P	X	G
44 Pin	10.0 [.390]	15.0 [.590]	12.5 [.490]	8.0 [.315]	0.8 [.031]	0.4 [.016]	32 x 32
48 Pin	13.0 [.510]	18.0 [.710]	15.5 [.610]	8.8 [.346]	0.8 [.031]	0.4 [.016]	38 x 38
52 Pin	17.0 [.668]	22.0 [.868]	19.5 [.768]	12.0 [.472]	1.0 [.394]	0.5 [.020]	46 x 46
54 Pin (13 pin)	10.0 [.390]	15.0 [.590]	12.5 [.490]	8.5 [.335]	0.65 [.026]	0.35 [.014]	32 x 32
(14 pin)	10.0 [.390]	15.0 [.590]	12.5 [.490]	8.5 [.335]	0.65 [.026]	0.35 [.014]	32 x 32
64 pin (13 pin)	14.4 [.565]	19.4 [.765]	16.9 [.665]	12.0 [.472]	1.0 [.394]	0.5 [.020]	41 x 53
(19 pin)	20.4 [.800]	25.4 [1.000]	22.9 [.900]	18.0 [.709]	1.0 [.394]	0.5 [.020]	41 x 53
70 pin (11 pin)	12.0 [.470]	17.0 [.670]	14.5 [.570]	8.0 [.315]	0.8 [.031]	0.4 [.016]	36 x 60
(24 pin)	24.2 [.950]	29.2 [1.150]	26.7 [1.050]	10.4 [.409]	0.8 [.031]	0.4 [.016]	36 x 60
80 pin (16 pin)	13.5 [.530]	18.5 [.730]	16.0 [.630]	12.0 [.472]	0.8 [.031]	0.4 [.016]	38 x 50
(24 pin)	19.5 [.766]	24.5 [.966]	22.0 [.866]	10.4 [.409]	0.8 [.031]	0.4 [.016]	38 x 50
100 pin (20 pin)	13.5 [.530]	18.5 [.730]	16.0 [.630]	12.35 [.486]	0.65 [.026]	0.35 [.016]	38 x 50
(30 pin)	19.5 [.766]	24.5 [.966]	22.0 [.866]	18.85 [.742]	0.65 [.026]	0.35 [.016]	38 x 50

FIGURE C.26 Land patterns for gullwing quad flatpacks (see Figure C.25).

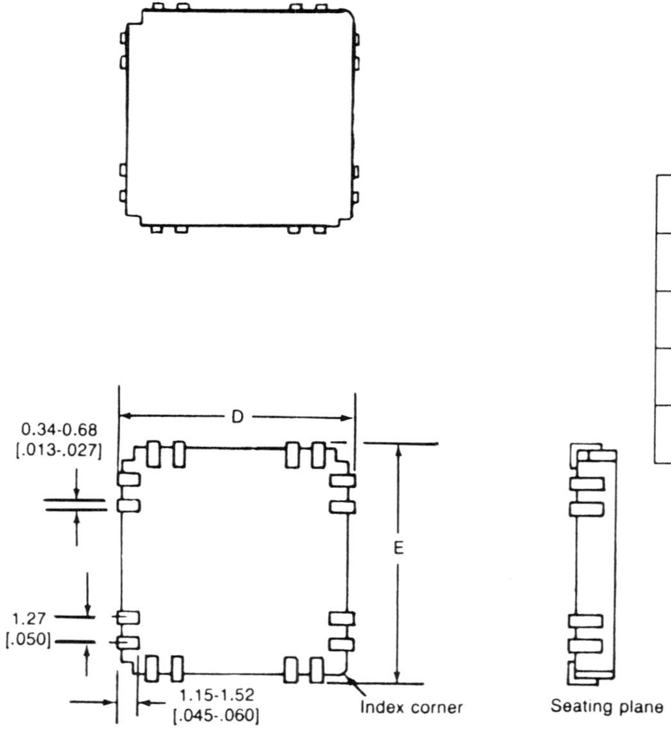

Terminal Count	D	E
28	11.94-12.95 [.470-.510]	11.94-12.95 [.470-.510]
44	17.02-18.03 [.670-.710]	17.02-18.03 [.670-.710]
52	19.56-20.57 [.770-.810]	19.56-20.57 [.770-.810]
68	24.64-25.65 [.970-1.010]	24.64-25.65 [.970-1.010]

FIGURE C.27 JEDEC type A premolded plastic-leaded chip carrier feature dimensions.

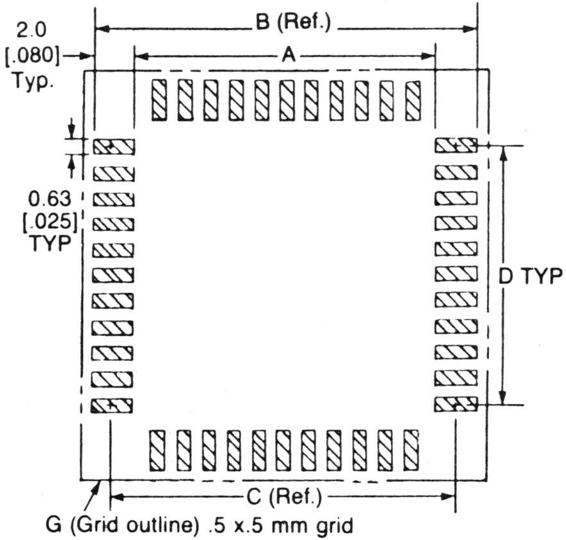

Type A Pin Count	A	B	C	D	G
28	10.0 [.390]	14.0 [.550]	12.0 [.470]	7.6 [.300]	30 x 30
44[a]	15.0 [.590]	19.0 [.750]	17.0 [.670]	12.7 [.500]	40 x 40
52	17.6 [.690]	21.6 [.850]	19.6 [.770]	15.2 [.600]	46 x 46
68	22.6 [.890]	26.6 [1.050]	24.6 [.970]	20.3 [.800]	58 x 56

[a] Example shown in graphics

FIGURE 2.8 Land patterns for JEDEC type A premolded plastic-leaded chip carriers (see Figure C.27).

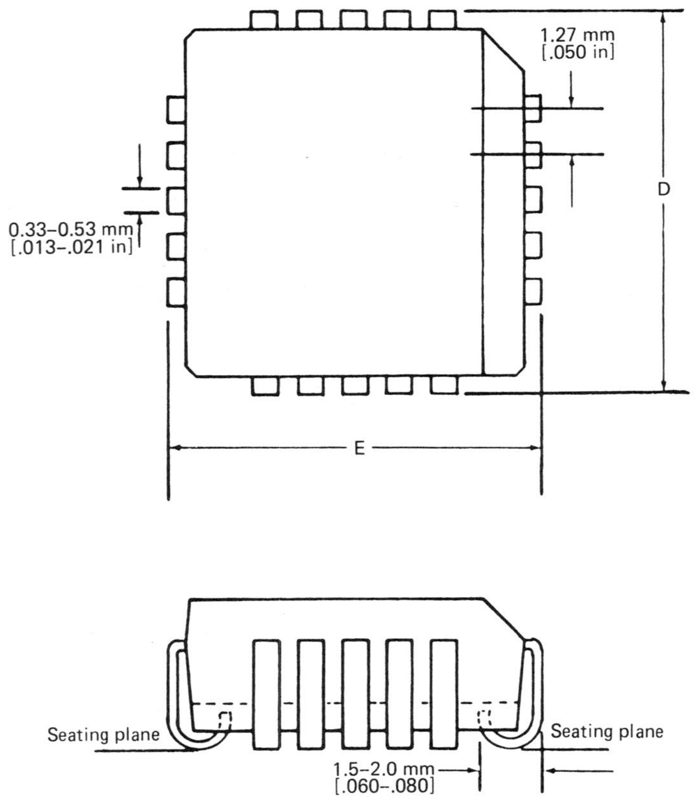

Terminal Count	D	E
20	9.78-10.03 [.385-.395]	9.78-10.03 [.385-.395]
28	12.32-12.57 [.485-.495]	12.32-12.57 [.485-.495]
44	17.40-17.65 [.685-.695]	17.40-17.65 [.685-.695]
52	19.94-20.19 [.785-.795]	19.94-20.19 [.785-.795]
68	25.02-25.27 [.985-.995]	25.02-25.27 [.985-.995]
84	30.10-30.35 [1.185-1.195]	30.10-30.35 [1.185-1.195]
100	35.18-35.43 [1.385-1.395]	35.18-35.43 [1.385-1.395]
124	42.80-43.05 [1.685-1.695]	42.80-43.05 [1.685-1.695]

FIGURE C.29 JEDEC square-outline postmolded plastic J-lead chip carrier feature dimensions.

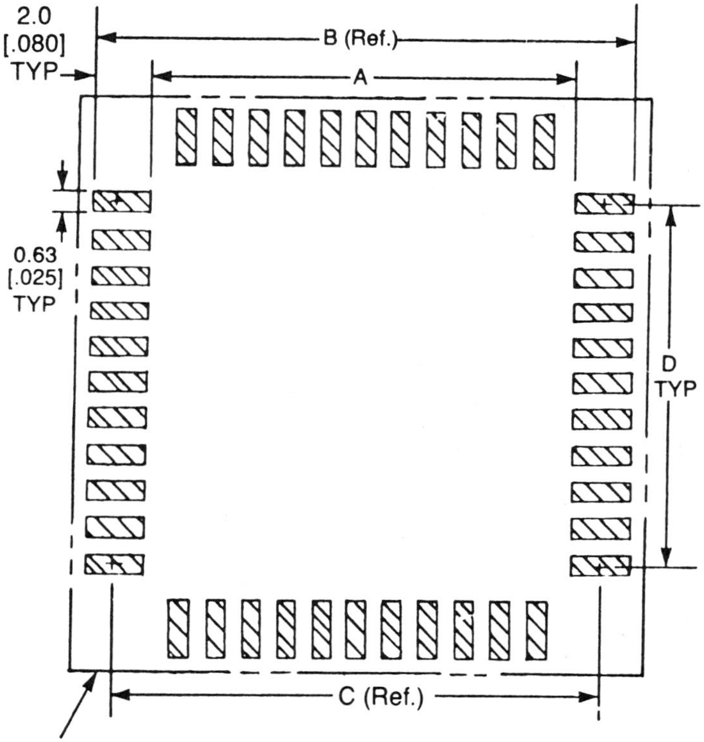

J Lead Pin Count	A	B	C	D	G
20	6.8 [.265]	10.8 [.425]	8.8 [.345]	5.08 [.200]	22 x 22
28	9.3 [.365]	13.3 [.525]	11.3 [.445]	7.6 [.300]	28 x 28
44[a]	14.4 [.565]	18.4 [.725]	16.4 [.645]	12.7 [.500]	38 x 38
52	17.0 [.665]	21.0 [.825]	19.0 [.745]	15.2 [.600]	44 x 44
68	22.0 [.865]	26.0 [1.025]	24.0 [.945]	20.3 [.800]	54 x 54
84	27.1 [1.065]	31.1 [1.225]	29.1 [1.145]	25.4 [1.000]	64 x 64
100	32.2 [1.265]	36.2 [1.425]	34.2 [1.345]	30.5 [1.200]	74 x 74
124	39.8 [1.565]	43.8 [1.725]	41.8 [1.645]	38.1 [1.500]	90 x 90

[a] Example shown in graphics

FIGURE C.30 Land patterns for JEDEC square-outline postmolded plastic J-lead chip carriers (see Figure C.29).

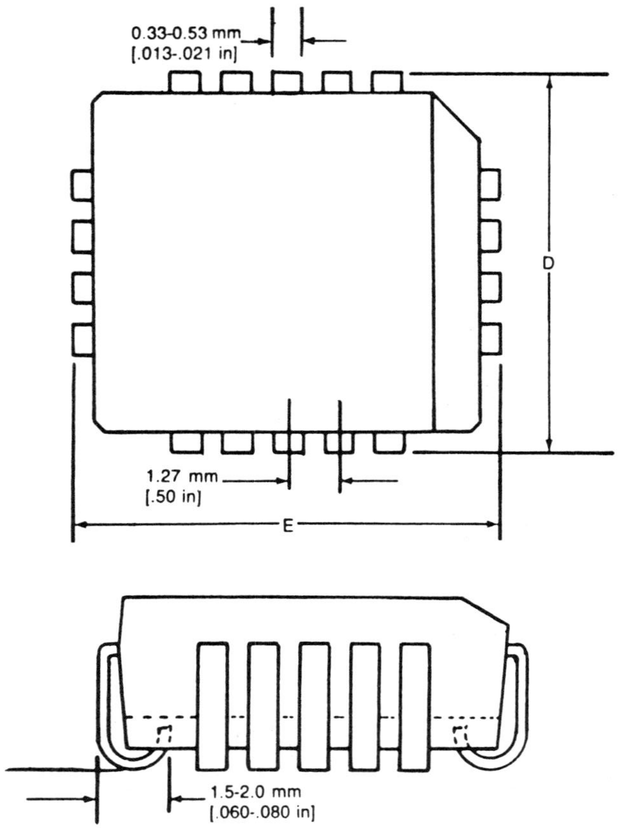

Terminal Count	D	E
18	8.05-8.3 [.317-.327]	11.61-11.86 [.457-.467]
18L	8.13-8.51 [.320-.335]	13.21-13.59 [.520-.535]
22	8.13-8.51 [.320-.335]	13.21-13.59 [.520-.535]
28	9.28-10.03 [.385-.395]	14.86-15.11 [.585-.595]
32	12.32-12.57 [.485-.495]	14.86-15.11 [.585-.595]

FIGURE C.31 JEDEC rectangular-outline postmolded plastic J-lead chip carrier feature dimensions.

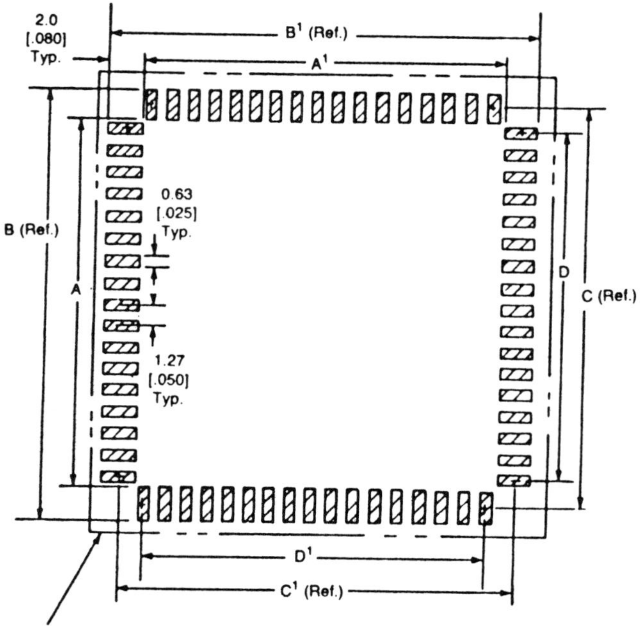

J Lead Pin Count		A/A_1	B/B_1	C/C_1	D/D_1	G
18	4	5.0 [.195]	9.0 [.355]	7.0 [.275]	3.8 [.150]	20 x 26
	5	8.6 [.335]	12.6 [.495]	10.6 [.415]	5.08 [.200]	
18L	4	5.4 [.210]	9.4 [.370]	7.4 [.290]	3.8 [.150]	20 x 30
	5	10.6 [.415]	14.6 [.575]	12.6 [.495]	5.08 [.200]	
22	4	5.4 [.210]	9.4 [.370]	7.4 [.290]	3.8 [.150]	20 x 30
	7	10.6 [.415]	14.6 [.575]	12.6 [.495]	7.6 [.300]	
28[a]	5	6.8 [.265]	10.8 [.425]	8.8 [.345]	5.08 [.200]	22 x 34
	9	12.0 [.470]	16.0 [.630]	14.0 [.550]	10.2 [.400]	
32	7	9.4 [.370]	13.4 [.530]	11.4 [.450]	7.6 [.300]	28 x 34
	9	12.0 [.470]	16.0 [.630]	14.0 [.550]	10.2 [.400]	

[a] Example shown in graphics

FIGURE C.32 Land patterns for JEDEC rectangular-outline postmolded plastic J-lead chip carriers (see Figure C.31).

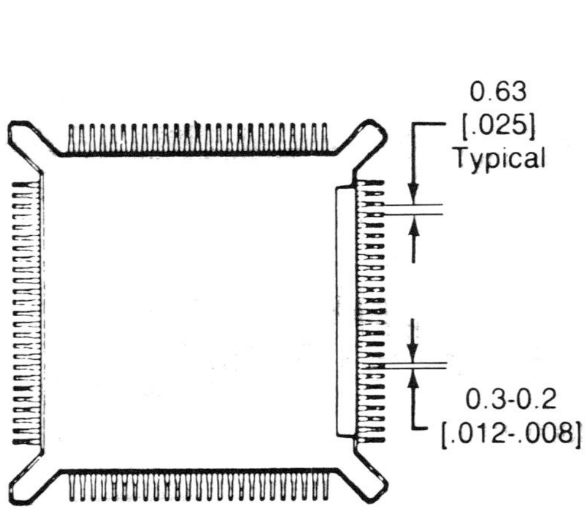

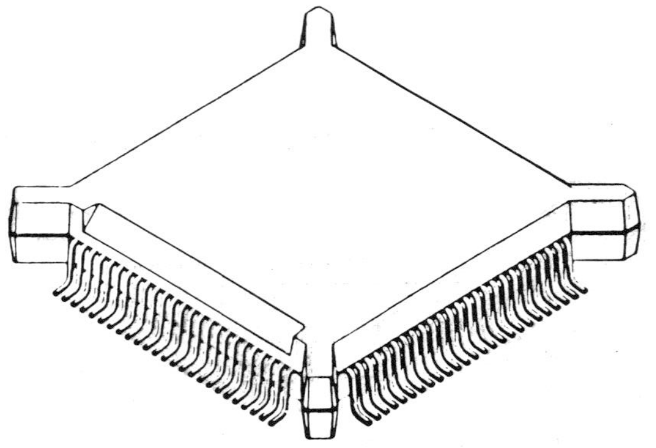

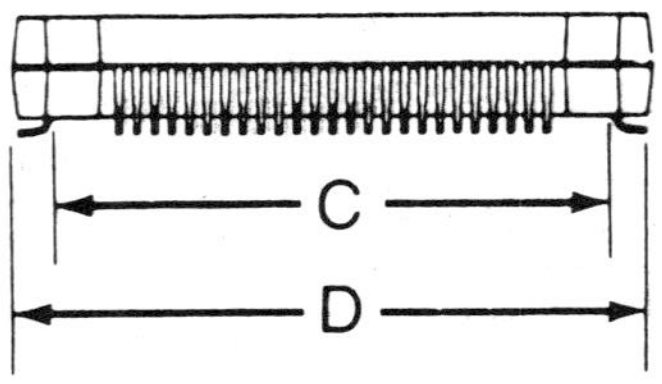

Terminal Count	C	D
84	18.05-19.05 [.710-.750]	19.56-20.05 [.770-.790]
100	20.60-21.60 [.810-.850]	22.10-22.60 [.870-.890]
132	25.65-26.65 [1.010-1.050]	27.20-27.70 [1.070-1.090]
164	30.75-31.75 [1.210-1.250]	32.25-32.75 [1.270-1.290]
196	35.80-36.80 [1.410-1.450]	37.35-37.85 [1.470-1.490]
244	40.15-41.15 [1.580-1.620]	41.65-42.15 [1.640-1.660]

FIGURE C.33 JEDEC gullwing plastic quad flatpack feature dimensions.

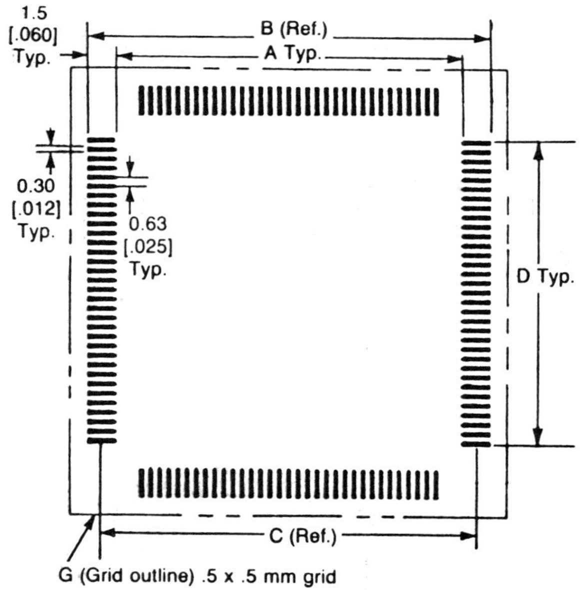

Terminal Count	A	B	C	D	G
84	17.5 [.690]	20.5 [.810]	19.0 [.750]	12.7 [.500]	42 x 42
100	20.0 [.785]	23.0 [.905]	21.5 [.845]	15.24 [.600]	48 x 48
132[a]	25.0 [.985]	28.0 [1.105]	26.5 [1.045]	20.32 [.800]	58 x 58
164	30.2 [1.190]	33.2 [1.310]	31.7 [1.250]	25.4 [1.0]	68 x 68
196	35.2 [1.385]	38.2 [1.505]	36.7 [1.445]	30.48 [1.20]	78 x 78
244	39.5 [1.555]	42.5 [1.675]	41.0 [1.615]	38.10 [1.50]	86 x 86

[a] Example shown in graphics

FIGURE C.34 Land patterns for JEDEC gullwing plastic quad flatpack (see Figure C.33),

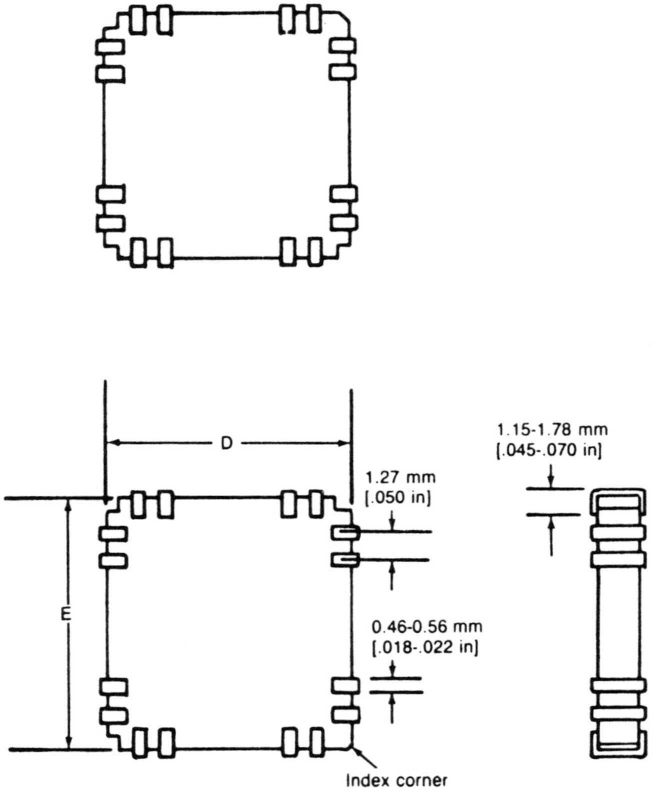

Terminal Count	D	E
28	11.30-11.56 [.445-.455]	11.30-11.56 [.445-.455]
44	16.34-16.68 [.643-.657]	16.34-16.68 [.643-.657]
52	18.85-19.25 [.742-.758]	18.85-19.25 [.742-.758]
68	23.88-24.38 [.940-.960]	23.88-24.38 [.940-.960]
84	28.91-29.51 [1.139-1.161]	28.91-29.51 [1.139-1.161]
100	34.02-34.56 [1.339-1.361]	34.02-34.56 [1.339-1.361]
124	41.64-42.18 [1.639-1.661]	41.64-42.18 [1.639-1.661]
156	51.80-52.34 [2.039-2.061]	51.80-52.34 [2.039-2.061]

FIGURE C.35 JEDEC square-outline leaded type B ceramic chip carrier feature dimensions.

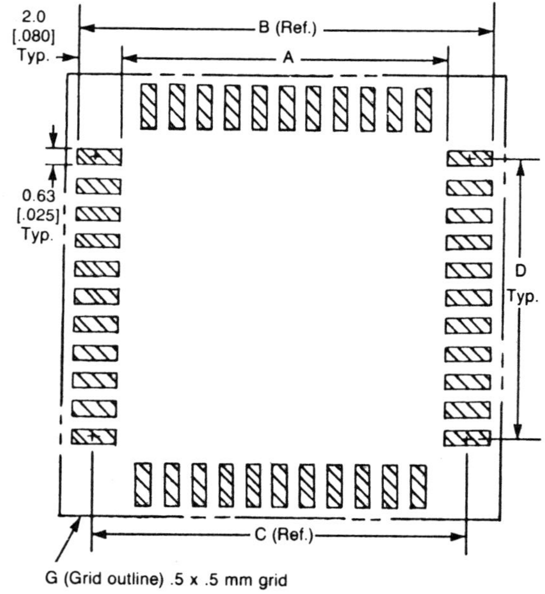

Pin Count	A	B	C	D	G
28	8.6 [.335]	12.6 [.495]	10.6 [.415]	7.6 [.300]	26 x 26
44[a]	13.6 [.535]	17.6 [.695]	15.6 [.615]	12.7 [.500]	36 x 36
52	16.0 [.630]	20.0 [.790]	18.0 [.710]	15.2 [.600]	42 x 42
68	21.2 [.835]	25.2 [.995]	23.2 [.915]	20.3 [.800]	52 x 52
84	26.4 [1.040]	30.4 [1.200]	28.4 [1.120]	25.4 [1.000]	62 x 62
100	31.4 [1.235]	35.4 [1.395]	33.4 [1.315]	30.5 [1.200]	76 x 76
124	39.0 [1.535]	43.0 [1.695]	41.0 [1.615]	38.1 [1.500]	88 x 88
156	49.2 [1.935]	53.2 [2.095]	51.2 [2.015]	48.3 [1.900]	108 x 108

[a]Example shown in graphics

FIGURE C.36 Land patterns for JEDEC square-outline leaded type B ceramic chip carriers (see Figure C.35).

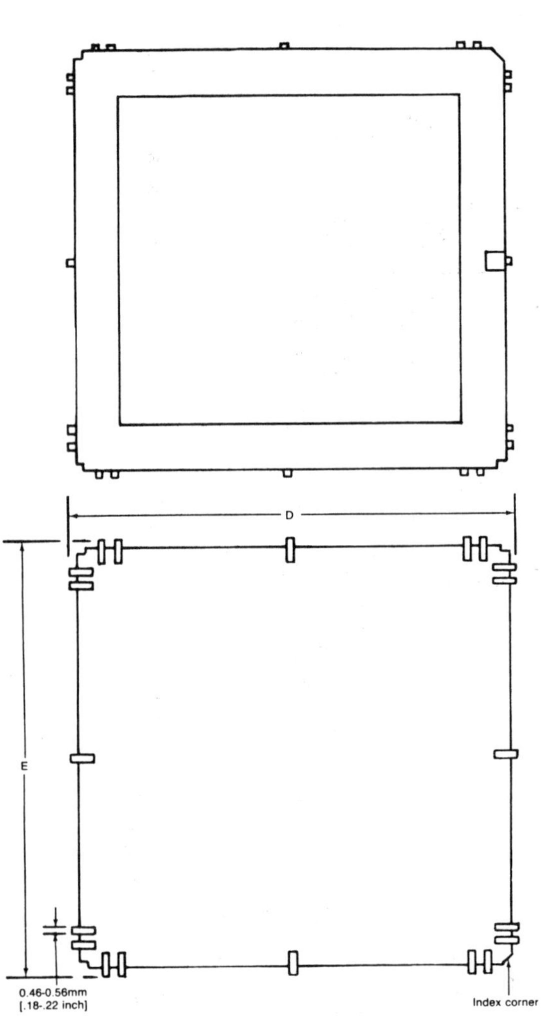

Terminal Count	D	E	H Typ
68	25.02-25.27 [.985-.995]	25.02-25.27 [.985-.995]	23.12-24.12 [.910-.950]
84	30.10-30.35 [1.185-1.195]	30.10-30.35 [1.185-1.195]	28.2-29.2 [1.11-1.15]

FIGURE C.37 JEDEC MS-044 ceramic-leaded chip carrier with CERDIP construction feature dimensions.

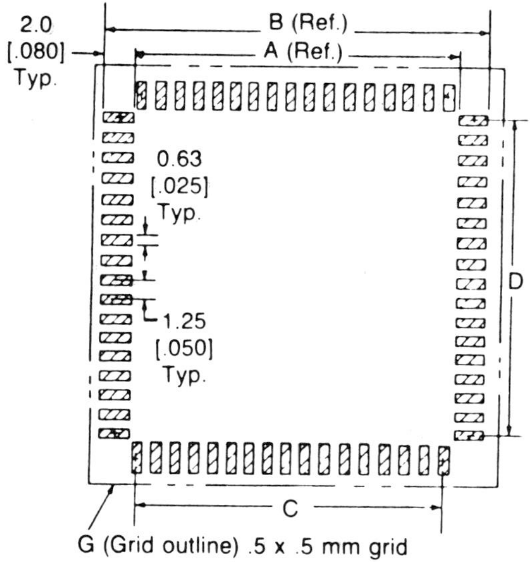

Pin Count	A	B	C	D	G
68[a]	22.4 [.880]	26.4 [1.040]	24.4 [.960]	20.3 [.800]	54 x 54
84	27.0 [1.060]	31.0 [1.220]	29.0 [1.140]	25.4 [1.000]	64 x 64

[a] Example shown in graphics

FIGURE C.38 Land patterns for JEDEC MS-044 ceramic-leaded chip carrier with CERDIP construction (see Figure C.37).

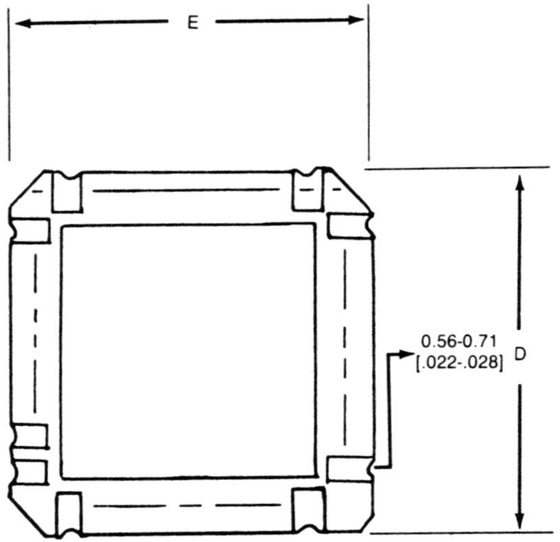

Terminal Count	D	E
16	7.4-7.8 [.291-.307]	7.4-7.8 [.291-.307]
20	8.7-9.1 [.342-.358]	8.7-9.1 [.342-.358]
24	10.0-10.4 [.390-.410]	10.0-10.4 [.390-.410]
28	11.2-11.7 [.441-.460]	11.2-11.7 [.441-.460]
44	16.3-16.8 [.640-.660]	16.3-16.8 [.640-.660]
52	18.8-19.3 [.740-.760]	18.8-19.3 [.740-.760]
68	23.8-24.4 [.940-.960]	23.8-24.4 [.940-.960]
84	28.8-29.6 [1.135-1.165]	28.8-29.6 [1.135-1.165]
100	34.0-34.7 [1.335-1.365]	34.0-34.7 [1.335-1.365]
124	41.3-42.1 [1.625-1.655]	41.3-42.1 [1.625-1.655]
156	51.7-52.5 [2.035-2.065]	51.7-52.5 [2.035-2.065]

FIGURE C.39 Features common to the JEDEC 50-mil-center chip carriers.

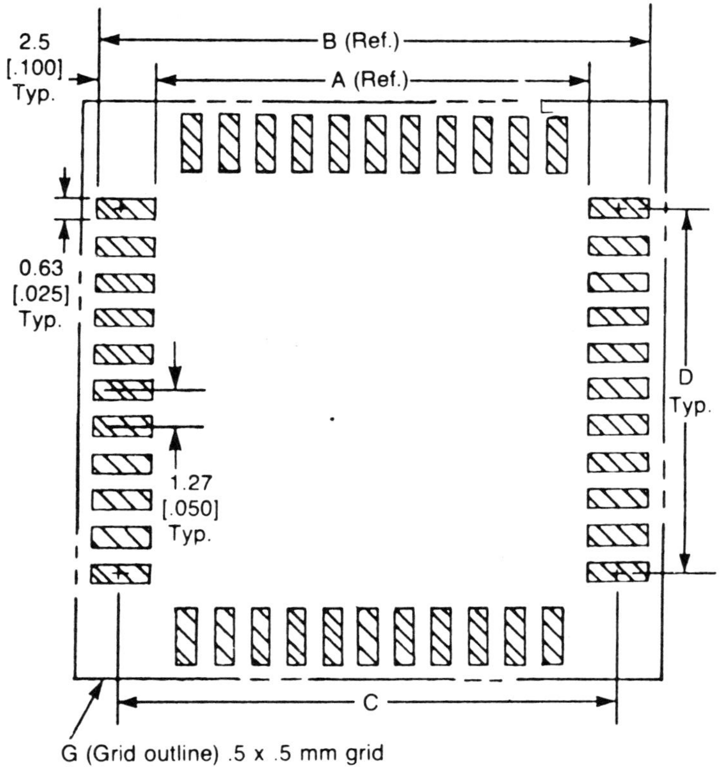

Pin Count	A	B	C	D	G
16	4.8 [.185]	9.8 [.385]	7.3 [.285]	3.8 [.150]	20 x 20
20	6.1 [.245]	11.1 [.435]	8.6 [.335]	5.08 [.200]	24 x 24
24	7.4 [.290]	12.4 [.490]	9.9 [.390]	6.35 [.250]	26 x 26
28	8.6 [.335]	13.6 [.535]	11.1 [.435]	7.62 [.300]	28 x 28
44[a]	13.8 [.540]	18.8 [.740]	16.3 [.640]	12.7 [.500]	38 x 38
52	16.3 [.640]	21.3 [.840]	18.8 [.740]	15.24 [.600]	44 x 44
68	21.4 [.840]	26.4 [1.040]	23.9 [.940]	20.3 [.800]	54 x 54
84	26.5 [1.040]	31.5 [1.240]	29.0 [1.140]	25.4 [1.000]	64 x 64
100	31.5 [1.240]	36.5 [1.440]	34.0 [1.340]	30.5 [1.200]	74 x 74
124	39.2 [1.540]	44.2 [1.740]	41.7 [1.640]	38.1 [1.500]	90 x 90
156	49.3 [1.940]	54.3 [2.140]	51.8 [2.040]	48.3 [1.900]	110 x 110

[a] Example shown in graphics

FIGURE C.40 Land patterns for the JEDEC 50-mil-center chip carriers (see Figure C.39).

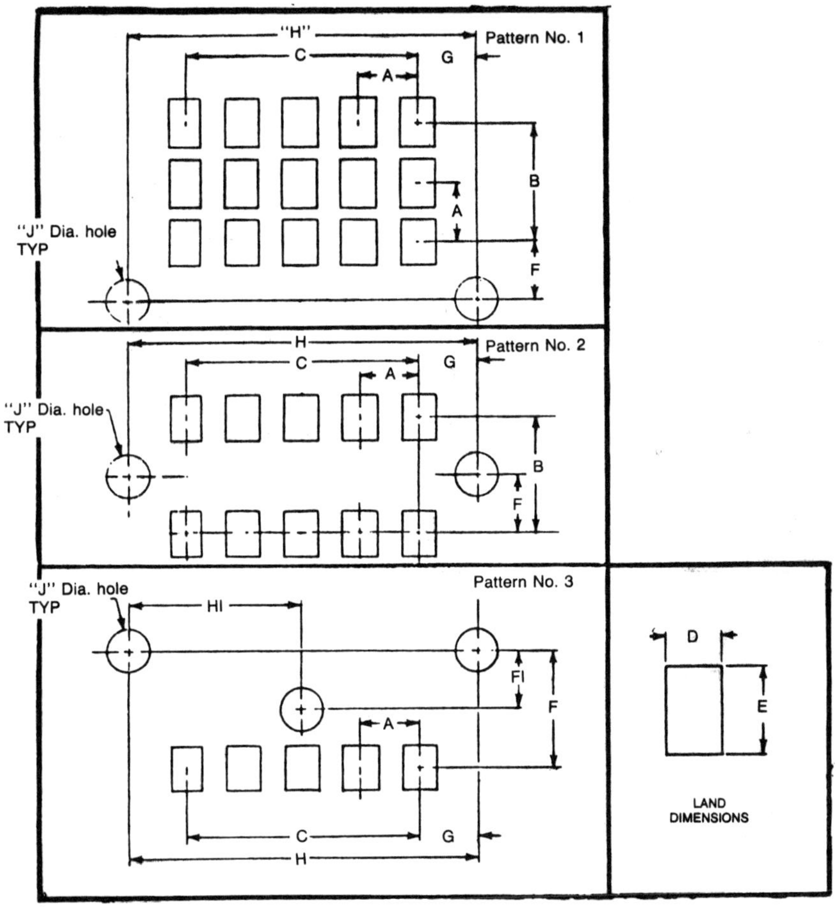

J	HI	H	G	FI	F	E	D	C	B	A	Product Family	Pat. No.
3.13 [0.125]	N/A	88.9 [3.50]	5.0 [0.200]	N/A	2.5 [0.100]	1.38 [0.055]	DIA	76.2 [3.10]	5.0 [0.20]	2.5 [0.100]	Eurocard, Receptacle Ass'y	1
2.5 [0.100]	N/A	88.9 [3.50]	5.0 [0.200]	N/A	2.5 [0.100]	1.38 [0.055]	DIA	76.2 [3.10]	5.0 [0.20]	2.5 [0.100]	Eurocard, Pin Ass'y	1
2.5 [0.100]	N/A	15.0-88.9 [0.60-3.50]	2.5 [0.100]	N/A	3.13 [0.125]	1.38 [0.055]	DIA	5.0-76.2 [0.20-3.10]	6.3 [0.25]	2.5 [0.10]	Mod IV	2
3.13 [0.125]	N/A	13.75-76.2 [0.55-3.0]	4.38-8.8 [0.175-0.350]	N/A	4.38 [0.175]	2.0 [0.08]	1.5 [0.06]	10-76.2 [0.40-3.10]	8.8 [0.35]	2.5 [0.10]	Vertical Header	2
3.13 [0.125]	N/A	13.75-90.4 [0.55-3.56]	4.38-5.75 [0.175-0.230]	N/A	1.25 [0.050]	2.0 [0.08]	1.5 [0.06]	5-76.2 [0.20-3.10]	2.5 [0.10]	2.5 [0.10]	Right Angle Header	2
3.13 [0.125]	11.4-36.8 [0.45-1.45]	13.75-82.5 [0.55-3.25]	4.38 [0.175]	6.3 [0.25]	3.13 [0.125]	2.0 [0.08]	1.5 [0.06]	5-73.5 [0.20-2.90]	N/A	2.5 [0.10]	Vertical Header, Single Row	2
3.13 [0.125]	N/A	11.4 [0.45]	3.75 [0.15]	N/A	8.25 [0.33]	1.75 [0.070]	0.75 [0.030]	3.75 [0.15]	N/A	1.25 [0.050]	SDL Receptacle	3
2.5 [0.100]	10.0 [0.40]	20.0 [0.80]	2.5 [0.10]	8.8 [0.35]	12.7 [0.50]	3.75 [0.15]	2.5 [0.10]	15.0 [0.60]	N/A	0.5 [0.20]	Mate-N-Lok Header	3

FIGURE C.41 Typical connector land patterns.

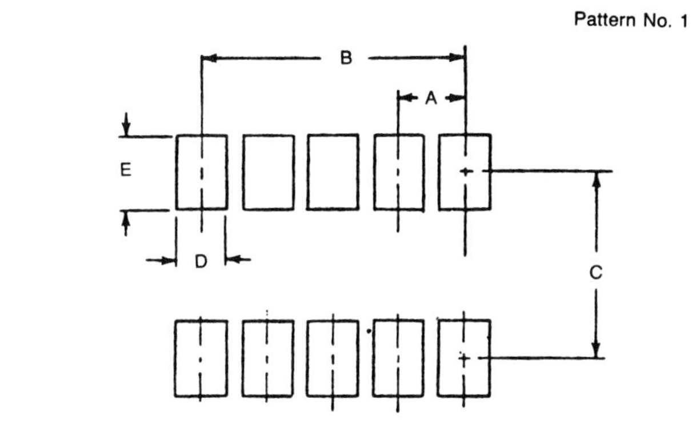

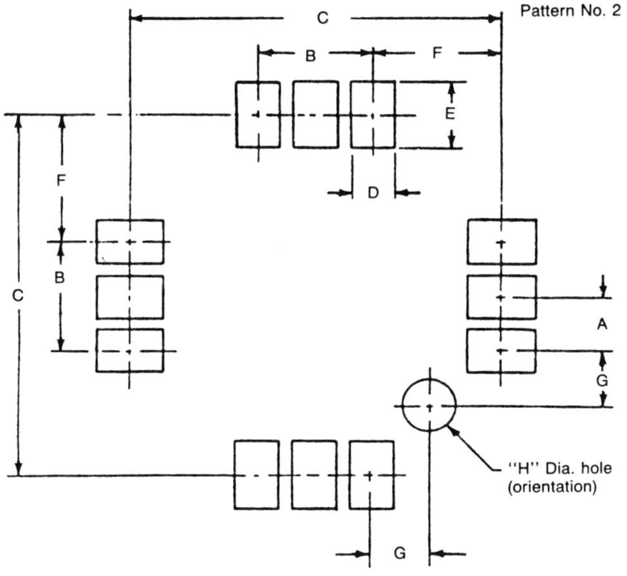

H	G	F	E	D	C	B	A	Product Family	Pat. No.
N/A	N/A	N/A	2.5 [0.100]	1.4 [.056]	8.8-16.5 [0.350-0.650]	15.2-45.7 [0.60-1.80]	2.5 [0.100]	DIP Sockets	1
N/A	N/A	N/A	1.25 [0.050]	0.75 [0.030]	5.0 [0.20]	5.0-11.4 [0.20-0.45]	1.25 [0.050]	Shunts—1.25 [0.050] Center	1
N/A	N/A	N/A	1.5 [0.060]	0.95 [0.038]	9.5 [0.375]	5.0-27.9 [0.20-1.100]	2.5 [0.100]	Switches/Shunts—2.5 [0.100] Center	1
1.5 [0.060]	2.5 [0.100]	6.3 [0.250]	1.5 [0.060]	0.9 [0.036]	25.4-43.2 [1.00-1.70]	12.7-30.1 [0.50-1.20]	1.25 [0.050]	Chip Carrier Sockets	2

FIGURE C.42 Typical chip carrier socket and switch land patterns.

Appendix D

General Requirements for Soldering Electronic Interconnections from IPC-S-815B

IPC-S-815B prescribes general requirements for high-quality soldering of electronic interconnections. It also defines the approved materials, methods, and inspection criteria for producing the quality of electrical soldering workmanship necessary for soldering of electrical connections and printed board assemblies. The intent of the following abstracts from IPC-S-815B is to illustrate solder joint features that generally result in obtaining an acceptable solder connection.

Certain solder joint features are unspecified in size, and the only requirement is that a properly wetted fillet be visible; this parameter is designated as "F." Solder joint thickness values, parameter "X," may be defined by cleaning and reliability requirements.

Per IPC-S-815B, one percent of the joints of Class 3 (High Reliability) product may meet Class 2 (General Industrial) requirements. One percent of Class 2 product may meet the requirements of Class 1

(Consumer), except as described in paragraph D.1. Exceptions to the alignment requirements are permitted in IPC-S-815 if the vendor demonstrates to the user's satisfaction that the solder joints are sound both mechanically and electrically.

D.1 CONNECTOR AND SOCKET LEADS

Solder joints formed to connector, socket, and other leads or terminations subject to stress from insertion and withdrawal of components or printed wiring boards shall not be rated as Class 1. Otherwise, the solder joints are characterized according to paragraphs D.2 through D.9.

D.2 FLAT, RIBBON, "L," AND GULLWING LEADS

Solder joints formed from flat, ribbon, "L-," and gullwing-shaped leads of both stiff and flexible materials shall meet the alignment and solder fillet requirements of Table D.1 and Figures D.1 and D.2 for each product classification. For Class 1 and 2 products leads not having wettable sides by design, such as leads stamped or sheared from preplated stock, are not required to have side fillets per Figure D.1. However, the design should permit easy inspection of wetting to the wettable surfaces. Class 3 product requires side fillets.

D.3 ROUND, FLATTENED, OR COINED LEADS

Solder joints formed to round, flattened, or coined leads shall meet the dimensional and fillet requirements of Table D.2 and Figures D.3 and D.4 for each product classification. The dimension D is the lead diameter before flattening or coining.

D.4 "J" AND "V" LEADS

Solder joints formed to leads having a "J" or "V" shape at the joint site shall meet the dimensional and fillet requirements of Table D.3 and Figures D.5 and D.6 for each product classification. Solder shall not fill the gap between the lead and the package body. For Class 1 and 2 product, leads not having wettable sides by design, such as leads stamped or sheared from preplated stock, are not required to have side fillets. However, the design should permit easy inspection of the wetting to the wettable surfaces. Class 3 product requires side fillets.

TABLE D.1 Dimensional Criteria—Flat, Ribbon, "L," and Gullwing Leads. Dimensions in mm and [in.] (Minimum Values Apply)

Feature	Dim.	Class 1 W < 0.5 [0.02]	Class 2 W > 0.5 [0.02]	Class 2	Class 3
Max side overhang	A	0.5W	0.5W or 0.5 [0.02]	0.5W or 0.5 [0.02]	0.25W or 0.5 [0.02]
Max toe overhang	Z	U[b]	U[b]	U[b]	0.25W or 0.5 [0.02]
Min end joint width	B	F[c]	0.5W	(W-A)	(W-A)
Min side joint length	L	F[c]	W or 0.5 [0.02]	W or (D-Z)	(D-Z)
Heel fillet max[a]	H	F[c]	F[c]	To top bend[d]	To top bend[d]
Heel fillet min[a]	M	F[c]	F[c]	F[c]	Mid. pt.[e]
Max thickness	X	F[c]	F[c]	F[c]	2T or 0.5 [0.02]

[a]For Class 1 and 2 product, the heel fillet requirements are not applicable when the heel of the lead is not positioned over the pad area.
[b]Unspecified parameter.
[c]Properly wetted fillet evident.
[d]Fillet should not enter the top bend (if it exists) or touch the package body.
[e]Midpoint of heel.

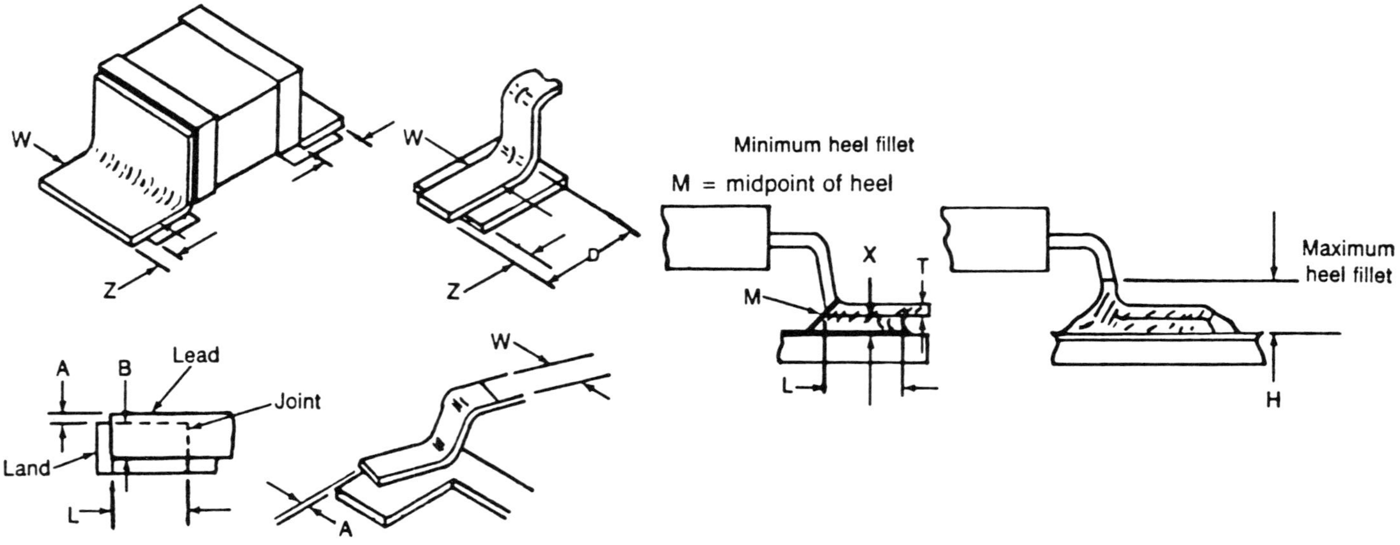

FIGURE D.1 Flat, ribbon, "L," and gullwing—solder joint.

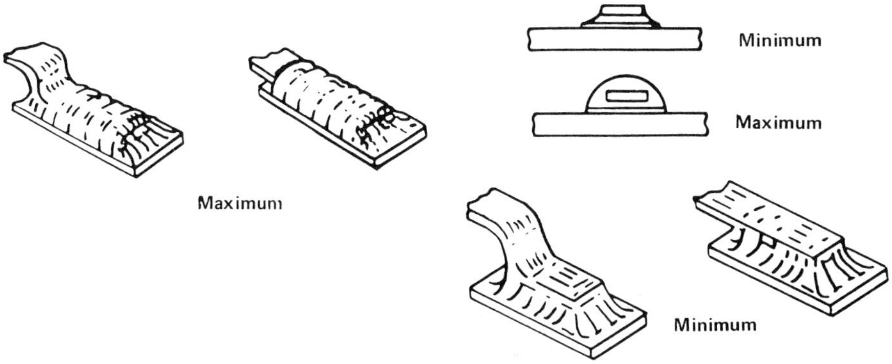

FIGURE D.2 Flat, ribbon, "L," and gullwing—solder joint. Minimum: Solder fillet has flowed up to the top of the lead and has covered the lead. Maximum: Solder covers lead and entire pad, but does not overhang. Toe of lead still discernible. No evidence of dewetting.

TABLE D.2 Dimensional Criteria—Round, Flattened, or Coined Leads. Dimensions in mm and [in.]

Feature	Dim.	Class 1	Class 2	Class 3
Max side overhang	A	0.5D	0.5D	0
Max end overhang	Z	U[a]	U[a]	0.25D
Min side joint length	L	D	D	2D
Min joint hgt.	Q	F[c]	0.25D	0.25D
Heel fillet max[b]	H	U[a]	To top bend[d]	To top bend[d]
Heel fillet min[b]	M	F[c]	F[c]	Mid. pt.[e]
Max thickness	X	F[c]	F[c]	0.75 [0.03]

[a] Unspecified parameter.
[b] For Class 1 and 2 product, the heel fillet requirements are not applicable when the heel of the lead is not positioned over the pad area.
[c] Properly wetted fillet evident.
[d] Fillet should not enter the top bend (if it exists) or touch the package body.
[e] Midpoint of heel.

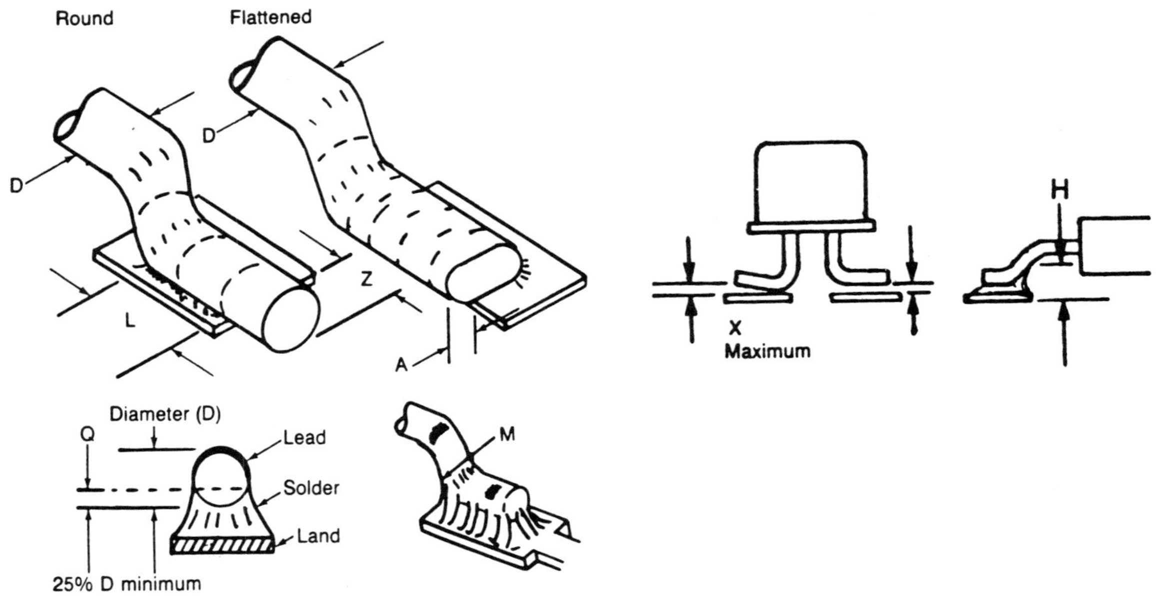

FIGURE D.3 Round, flattened, or coined leads—solder joint geometrical description. D: lead diameter prior to flattening.

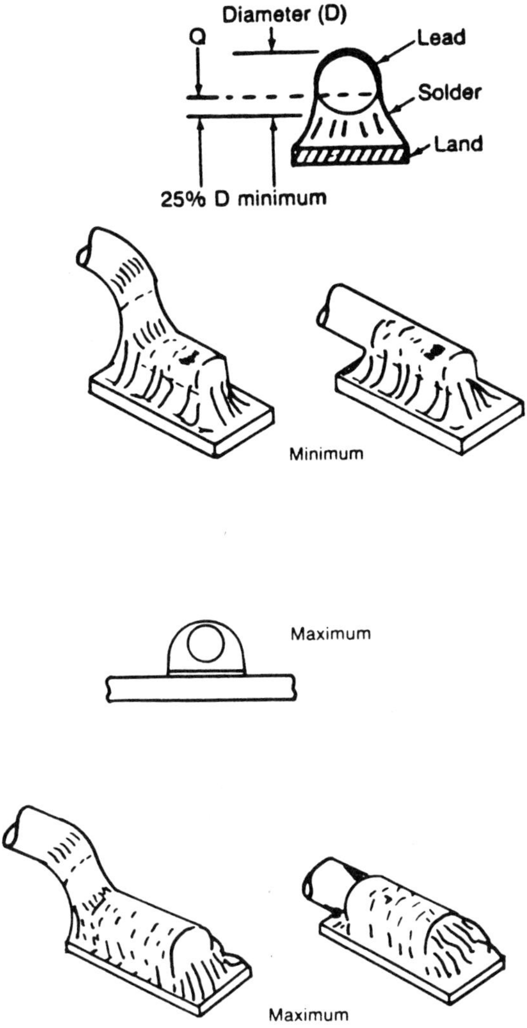

FIGURE D.4 Round, flattened, or coined leads—solder joint fillet appearance. Minimum: Solder fillet 25% of lead diameter. Solder may cover lead. Maximum: Solder covers lead and entire pad, but does not overhang. Toe of lead still discernible. No evidence of dewetting.

TABLE D.3 Dimensional Criteria—"J" and "V" Leads. Dimensions in mm and [in.]

Feature	Dim.	Class 1	Class 2	Class 3
Max side overhang	A	0.5W	0.5W	0.5W
Min side joint length	L	F[a]	1.5W	2W
Min joint width	B	F[a]	0.5W	(W-A)
Min heel fillet	M	F[a]	F[a]	Mid. pt.[b]
Max thickness	X	F[a]	F[a]	0.75 [0.03]

[a]Properly wetted fillet evident.
[b]Midpoint of heel as shown.

D.5 RECTANGULAR OR SQUARE END COMPONENTS WITH END, "3," AND "5" FACE TERMINATIONS

Solder joints to components having end, "3," and "5" face terminations of a square or rectangular configuration shall meet the dimensional and solder fillet requirements of Table D.4 and Figures D.7 and D.8 for each product classification. The width of the component and pad are W and P, respectively. The termination overhang describes the condition whereby the smaller extends beyond the larger termination, i.e., W or P.

D.6 CYLINDRICAL END CAP TERMINATIONS

Solder joints to components having cylindrical and cap terminations shall meet the dimensional solder fillet requirements of Table D.5 and Figures D.9 and D.10 for each product classification.

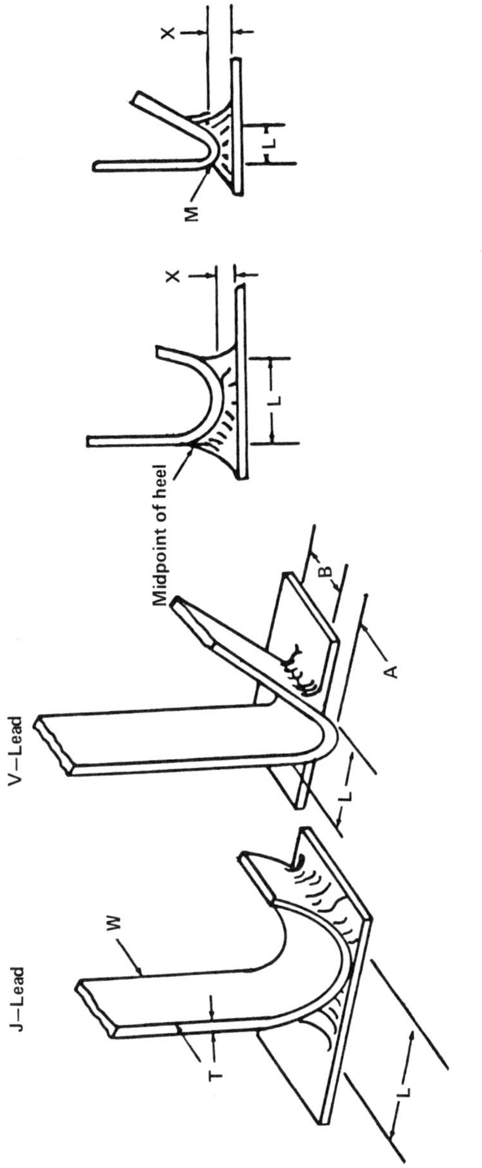

FIGURE D.5 "J" and "V" leads—solder joint geometrical description.

FIGURE D.6 "J" and "V" leads—solder joint fillet appearance. Class 3: Wettable sides only—side fillets required. Class 1 and 2: Nonwettable sides by design, such as leads stamped or sheared from preplated material, may have side fillets.

D.7 BOTTOM-ONLY TERMINATIONS

Discrete chip components, leadless chip carriers, and other devices having metallized terminations on the bottom side only shall meet the dimensional and solder fillet reqirements of Table D.6 and Figures D.11, D.12, and D.13. The width of the component and pad are W and P, respectively. The termination overhang describes the condition whereby the smaller extends beyond the larger termination, i.e., W or P.

D.8 LEADLESS CHIP CARRIERS WITH CASTELLATED TERMINATIONS

Solder joints formed to castellated terminations of leadless chip carriers shall meet the dimensional and solder fillet requirements of Table D.7 and Figures D.14 and D.15.

TABLE D.4 Dimensional Criteria—Rectangular or Square End Components with End, "3," and "5" Face Terminations. Dimensions in mm and [in.] (Minimum Values Apply)

Feature	Dim.	Class 1	Class 2	Class 3
Max side overhang	A	0.5 (W or P)	0.5 (W or P) or 1.5 [0.06]	0.25 (W or P) or 0.5 [0.02]
Min end overlap	L	NR[a]	R[b]	R[b]
Min end joint width	B	0.5 (W or P)	0.5 (W or P)	(W or P)−A
Min joint height	M	F[c]	F[c]	0.3H or 0.5 [0.02]
Side fillet	S	NR[a]	NR[a]	NR[a]
Max thickness	X	F[c]	F[c]	0.5 [0.02]

[a]Not required.
[b]Required across width of fillet B.
[c]Properly wetted fillet evident.

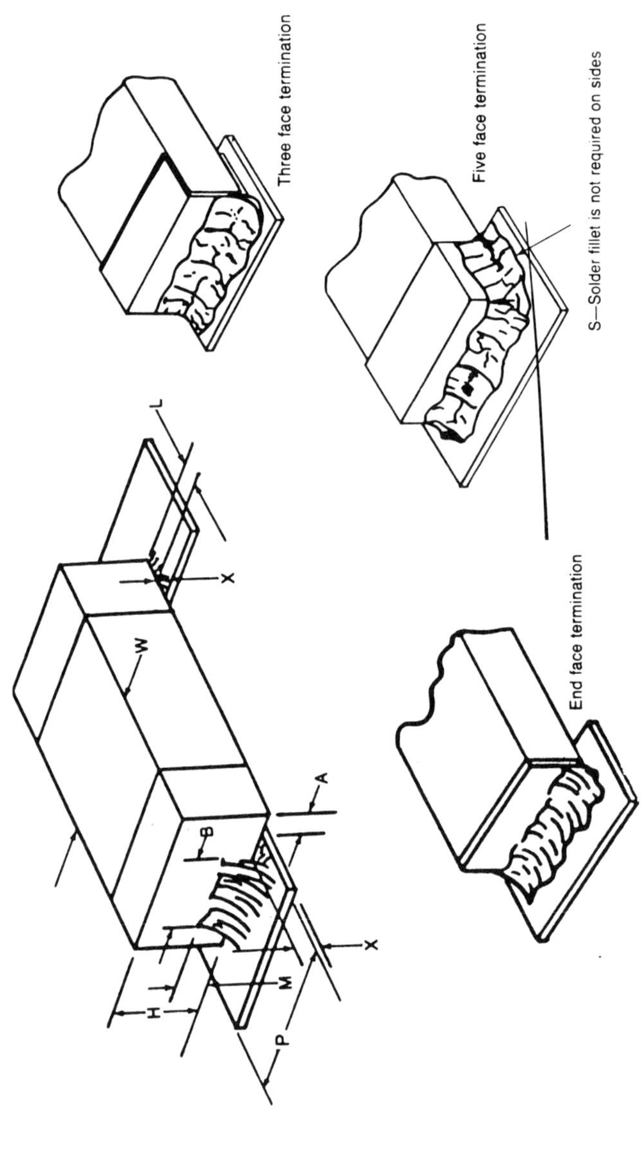

FIGURE D.7 Rectangular or square end components with end, "3," and "5" face terminations—solder joint geometrical description.

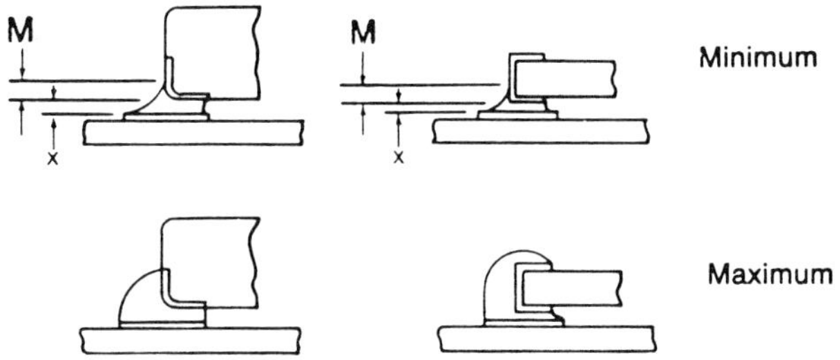

FIGURE D.8 Rectangular or square end components with end, "3," and "5" face terminations—solder joint fillet appearance. Minimum: Solder fillet flows up end more than 25% or 0.51 mm (0.020 in.), whichever is less (for Class 2 and 3 product). Maximum: Solder covers entire pad and end cap, but does not overhang either. Also, there is no evidence of dewetting at pad or end cap metallization.

TABLE D.5 Dimensional Criteria—Cylindrical End Cap Terminations. Dimensions in mm and [in.]

Feature	Dim.	Class 1	Class 2	Class 3
Max side overhang	A	0.25D	0.25D	0.25D
Min fillet width	B	F[a]	0.5D	0.5D
Min side fillet length[b]	L	F[a]	0.5T	0.5T
Min height	Q	F[a]	F[a]	F[a]
Max thickness	X	F[a]	F[a]	0.75 [0.03]
Min end fillet	Y	F[a]	F[a]	F[a]
Min end overlap	O	NR[a]	R[d]	R[d]

[a]Properly wetted fillet evident.
[b]Does not apply to components with end only terminations.
[c]Not required.
[d]Required.

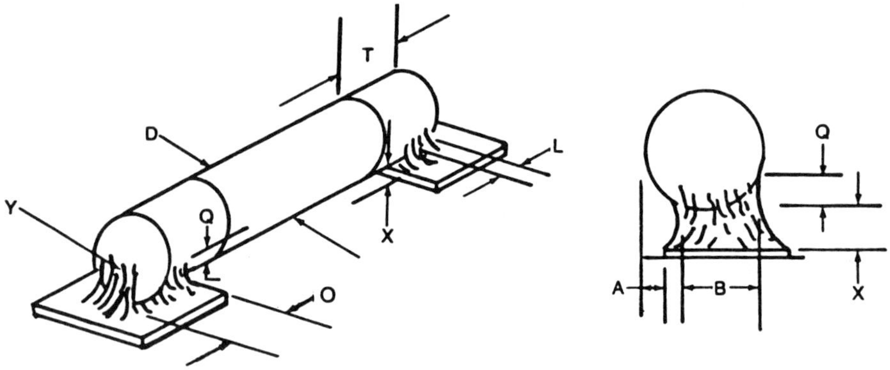

FIGURE D.9 Cylindrical end cap terminations—solder joint geometrical description.

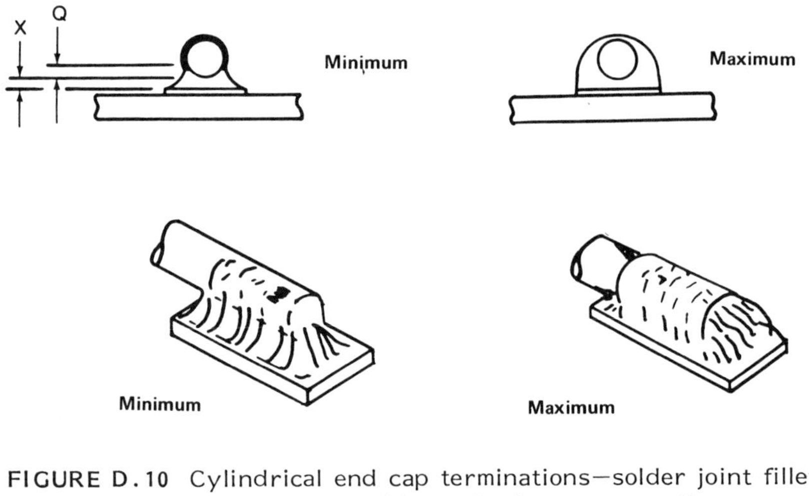

FIGURE D.10 Cylindrical end cap terminations—solder joint fillet appearance. Minimum: Solder fillet 25% of component diameter. Maximum: Solder covers cap and entire pad, but does not overhang. Toe of end cap still discernible. No evidence of dewetting.

TABLE D.6 Dimensional Criteria—Bottom-Only Terminations. Dimensions in mm and [in.] (Minimum Values Apply)

Feature	Dim.	Class 1	Class 2	Class 3
Min overlap and joint width	O	0.5 (W or P)	0.5 (W or P)	0.75 (W or P)
Max overhang	E	U[a]	U[a]	0
Max thickness	X	U[a]	U[a]	0.5 [0.02]

[a]Unspecified parameter.

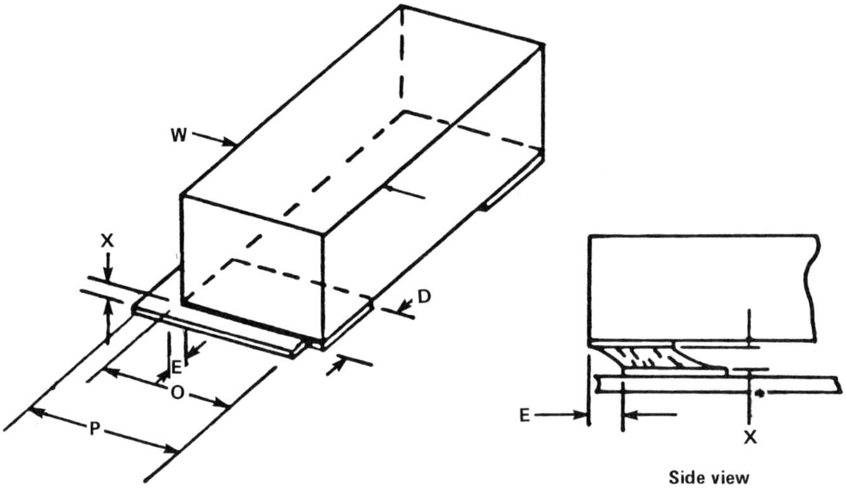

FIGURE D.11 Bottom-only terminations—solder joint description for chip components.

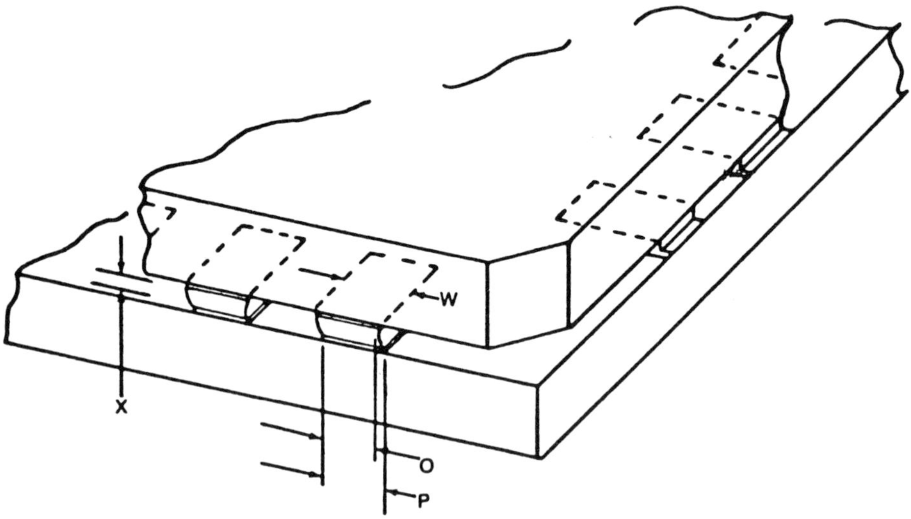

FIGURE D.12 Bottom-only terminations—solder joint description for leadless chip carriers.

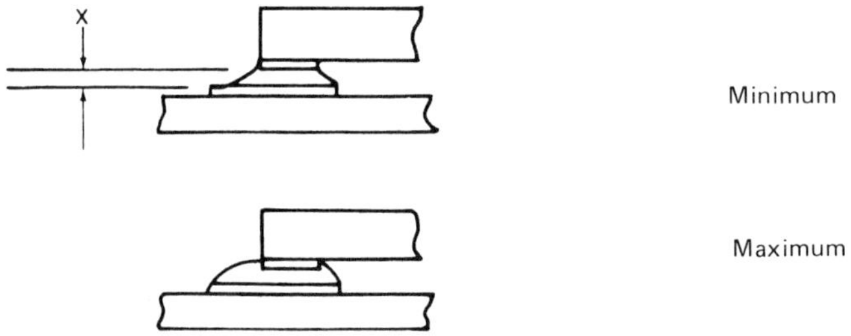

FIGURE D.13 Bottom-only terminations—solder joint fillet appearance. Minimum: Solder fillet flows to top of lead. Maximum: Solder covers entire pad, but does not overhang. Solder does not bulge at component body. No evidence of dewetting at pad.

TABLE D.7 Dimensional Criteria—Leadless Chip Carriers with Castellated Terminations. Dimensions in mm and [in.] (Minimum Values Apply)

Feature	Dim.	Class 1	Class 2	Class 3
Max side overhang	A	0.5W	0.5W	0.5W
Min fillet exten.	E	F[a]	0.5 (M + X) or P	0.5 (M + X) or P
Min fillet hgt.	M	F[a]	0.25 (H)	0.25 (H)
Max thickness	X	F[a]	F[a]	0.75 [0.03]

[a]Properly wetted fillet evident.

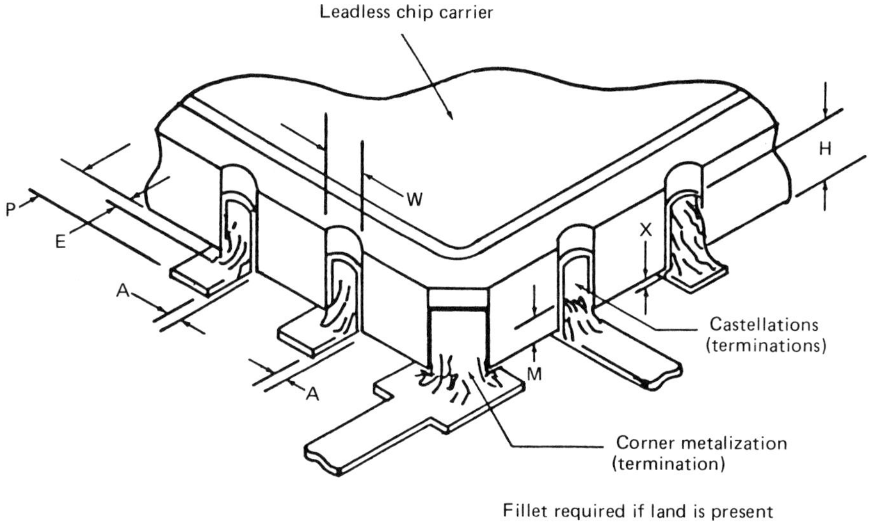

FIGURE D.14 Leadless chip carriers with castellated terminations—solder joint geometrical description.

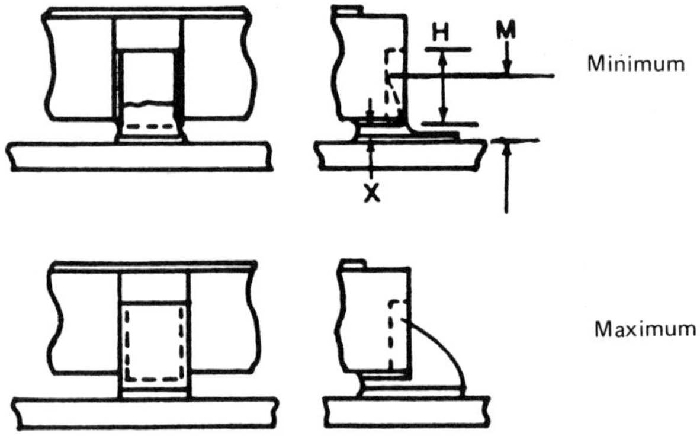

FIGURE D.15 Leadless chip carriers with castellated terminations—solder joint fillet appearance. Minimum: Solder fillet flows up termination more than 25%. Maximum: Solder covers entire pad and termination, but does not overhang front and one side of pad. Overhang permissible one side when LCC is shifted. No evidence of dewetting.

D.9 BUTT JOINTS

Solder joints formed to leads positioned perpendicular to the board land in a "butt" configuration shall meet the dimensional and solder fillet requirements of Table D.8 and Figure D.16 for each product classification. For Class 1 and Class 2 product, leads not having wettable sides by design, such as leads stamped or sheared from preplated stock, are not required to have side fillets. However, the design should permit easy inspection of wetting to wettable surfaces. Class 3 product requires side fillets.

TABLE D.8 Dimensional Criteria—Butt Joints. Dimensions in mm [in.]

Feature	Dim.	Class 1	Class 2	Class 3
Min height	H	0.25 [0.01]	0.5 [0.02]	0.5 [0.02]
Min width	B	0.5W	0.75W	W

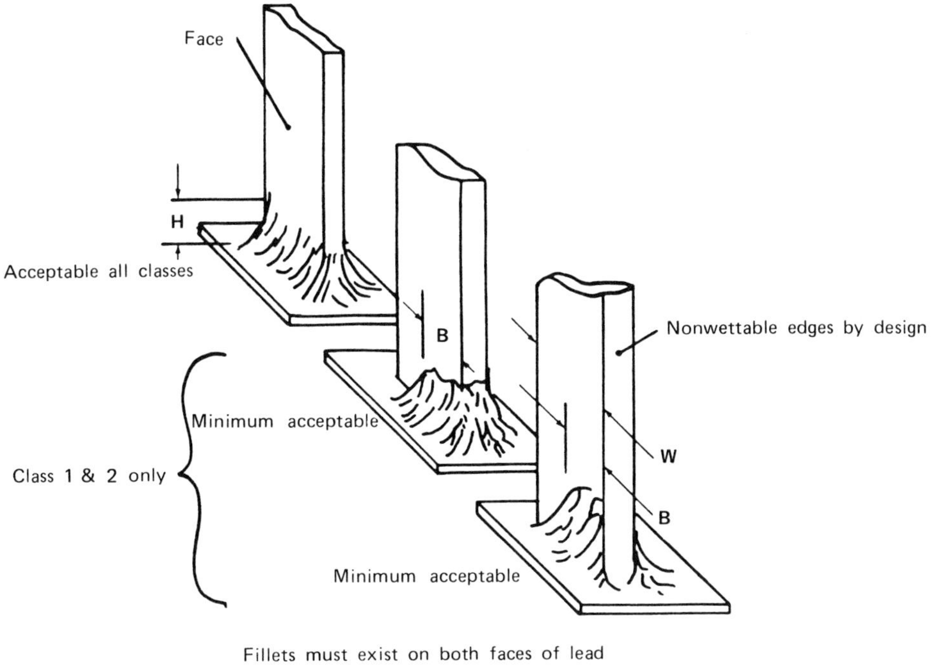

FIGURE D.16 Butt solder joint geometrical description and fillet appearance.

Index